Applications of Empirical Process Theory

CAMBRIDGE SERIES IN STATISTICAL AND PROBABILISTIC MATHEMATICS

This series of high-quality upper-division textbooks and expository monographs covers all aspects of stochastic applicable mathematics. The topics range from pure and applied statistics to probability theory, operations research, optimization, and mathematical programming. The books contain clear presentations of new developments in the field and also of the state of the art in classical methods. While emphasizing rigorous treatment of theoretical methods, the books also contain applications and discussions of new techniques made possible by advances in computational practice.

Already published

1. *Bootstrap Methods and Their Application*, A.C. Davison and D.V. Hinkley
2. *Markov Chains*, J. Norris
3. *Asymptotic Statistics*, A.W. van der Vaart

Applications of Empirical Process Theory

Sara A. van de Geer

CAMBRIDGE
UNIVERSITY PRESS

CAMBRIDGE UNIVERSITY PRESS
Cambridge, New York, Melbourne, Madrid, Cape Town, Singapore,
São Paulo, Delhi, Dubai, Tokyo

Cambridge University Press
The Edinburgh Building, Cambridge CB2 8RU, UK

Published in the United States of America by Cambridge University Press, New York

www.cambridge.org
Information on this title: www.cambridge.org/9780521123259

First published 2000
Reprinted 2006
This digitally printed version 2009

A catalogue record for this publication is available from the British Library

Library of Congress Cataloguing in Publication data
Geer, S. A. van de (Sara A.)
Applications of empirical process theory. Sara A. van de Geer.
p. cm.
Includes bibliographical references and indexes.
ISBN 0-521-65002-X
1. Nonparametric statistics. 2. Estimation theory.
3. Limit theorems (Probability theory). I. Title.
QA278.8.G44 1999
519.5'4 dc21 98-50544 CIP

ISBN 978-0-521-65002-1 Hardback
ISBN 978-0-521-12325-9 Paperback

Contents

Preface

This book is an extended version of a set of lecture notes, written for the *AIO* course 'Applications of Empirical Process Theory', which was given in the spring of 1996 in Utrecht, The Netherlands. The abbreviation *AIO* stands for *Assistent in Opleiding*, which is the Dutch equivalent of PhD student. The course was intended for students with a master of sciences in mathematics or statistics.

Nonparametric (infinite-dimensional) models provide a good alternative to the more classical parametric models, because fewer, and often more natural assumptions are imposed. In practice, nonparametric methods can be computationally complex, but this is nowadays a minor drawback. This book investigates the theoretical (asymptotic) properties of nonparametric M-estimators, with the emphasis on maximum likelihood estimators and least squares estimators. It treats the different models and estimation procedures in a unifying way, by invoking the theory on empirical processes. The general theory is illustrated with numerous examples. We hope that the methods provided will show that nonparametric models are in fact as basic as the more classical parametric models.

Empirical process theory has turned out to be a very valuable tool in asymptotic statistics. Applications include the bootstrap, the delta-method, and goodness-of-fit testing. In this book, we consider its applications to M-estimation, with special focus on maximum likelihood and least squares. We treat the latter two methods in great detail, including penalties, sieves, and semiparametrics. The description of M-estimators in general is in fact deferred to the very last chapter, where we show that, basically, the methodology allows direct generalization.

We do not assume any a priori knowledge on empirical process theory, and treat the subject, not in an exhaustive way, but with the applications always in mind. Moreover, we do not state any measurability conditions, because the formulation of these would require too many digressions. For most of the results, a full proof (modulo measurability) is presented, except when it concerns the calculation of entropy. Here, only some illustrations are given, which are meant to provide the reader with some understanding of what entropy actually is.

I am most grateful to my husband Toon, for the many fruitful discussions we had on the subject of this book, and for his very valuable suggestions on how to distinguish signal from noise.

Guide to the Reader

Chapters 2, 3, 5, 6 and 8 present empirical processes theory, and the other chapters concern applications. Some of the material in this book is added only for completeness and can certainly be skipped at first reading.

Chapter 1 provides an overview of the type of problems the book will address. Chapters 2 and 3 are essential for the rest of the book. Most of the results in Chapter 3 are along the lines of Pollard (1984, Chapter II).

Chapter 4 treats consistency of estimators, and can be seen as a preparation for the more complicated theory on rates of convergence.

Chapter 5 is one of the more technical chapters. The main result there is Theorem 5.11, which says that the increments of an empirical process behave as the integral of the (square root) entropy (with bracketing). Once this message is clear, most of what follows can be understood without knowing too many precise details.

Chapter 6 contains some of the fundamental results on empirical processes. This chapter could of course not be excluded from the book, but it plays only a small role in the subsequent chapters.

Chapter 7 derives rates of convergence for maximum likelihood estimators. Some of its examples are rather technical, whereas others are quite elegant. A reader might find some of the examples artificial, because certain constants that depend on the unknown parameter, are assumed to be known a priori. But one has to keep in mind that these models serve as a preparation for the more realistic models of Chapter 10.

Section 8.1 of Chapter 8 is needed to make the step to independent, but not identically distributed variables. The section supplies the remainder of the technical tools necessary for the statistical applications. The rest of Chapter 8 is about dependent variables, and can be skipped without being penalized for that later on in the book.

In Chapter 9, we consider rates of convergence for least squares estimators. Here, the results appear in their neatest form. A reader may choose to skip Chapters 6 and 7, and go immediately to Chapter 9 after consulting Section 8.1.

The rest of the book contains some selected topics which can be, more or less, read separately. Chapter 10 considers penalties and sieves. Again, the results for the regression estimators are more transparent than those for maximum likelihood estimators.

Chapter 11 consists of three independent parts. There are some links with the previous chapters here, because the results on rates of convergence play a prominent role in the derivation of asymptotic normality of certain functions of the curve estimators.

In Chapter 12, we first summarize the general recipe we used so far only for least squares and maximum likelihood estimators. The idea there is that the methods carry over immediately to general M-estimators. It is possible to read Section 12.1 at any stage. This might help the reader to understand what the common underlying structure actually is. The last section of the book completes the circle: we started with a parametric model, and also end with it.

Each chapter concludes with a problems section, where we sometimes also complement the theory with some auxiliary results. The level of the problems varies considerably.

1

Introduction

...where a motivation to study uniform laws of large numbers and central limit theorems is given. The examples concern parametric and nonparametric models, and contain indications on how we plan to derive consistency, rates of convergence and asymptotic normality of estimators.

Let $X_1, X_2, \ldots$ be independent copies of a random variable X with distribution P on $(\mathcal{X}, \mathcal{A})$. If $(\mathcal{X}, \mathcal{A})$ is the real line, equipped with a Borel σ-algebra, then by the strong law of large numbers, the sample mean $\bar{X} = (1/n)\sum_{i=1}^n X_i$ of the first n observations converges almost surely to the population mean $\mu = EX$, as $n \to \infty$. If moreover X has finite variance σ^2, the central limit theorem states that for n large, $\bar{X}$ is approximately normally distributed with mean μ and variance σ^2/n. For a general measurable space $(\mathcal{X}, \mathcal{A})$, such results hold for the sample mean $(1/n)\sum_{i=1}^n g(X_i)$, where $g : \mathcal{X} \to \mathbf{R}$ is some (measurable) real-valued transformation. This observation will be our starting point.

Strong law of large numbers *If $Eg(X)$ exists, then*

$$\frac{1}{n}\sum_{i=1}^n g(X_i) \to Eg(X), \text{ a.s..} \tag{1.1}$$

Central limit theorem *If $\sigma_g^2 = \mathrm{var}(g(X))$ exists, then*

$$\frac{1}{\sqrt{n}}\sum_{i=1}^n (g(X_i) - Eg(X)) \to^{\mathcal{L}} \mathcal{N}(0, \sigma_g^2). \tag{1.2}$$

For example, if we take g as the indicator function 1_A of some set $A \in \mathcal{A}$, (1.1) states the convergence of the proportion of observations falling into the set A, to the probability of the set A. Result (1.2) is then the classical central limit theorem for Bernoulli random variables.

The convergence in (1.1) and (1.2) holds for a *fixed* function g, but can be extended to hold for several g simultaneously. Consider a class $\mathcal{G} = \{g_\theta : \theta \in \Theta\}$ of functions on $\mathcal{X}$, indexed by a parameter θ in a metric space Θ. Assume that $Eg_\theta(X)$ and $\text{var}(g_\theta(X))$ exist for all $\theta \in \Theta$. We shall investigate to what extent (1.1) and (1.2) hold *uniformly* in $\theta \in \Theta$. Although the study of uniform convergence is of interest in itself, our motivation is given by the numerous applications in asymptotic statistics. Let $\{\hat{\theta}_n\}$ be some (possibly) random sequence in Θ, converging to $\theta_0 \in \Theta$ (a.s. or in probability). We have in mind the situation where $\hat{\theta}_n$ is an estimator of θ_0, based on the first n observations. Uniform results would lead to something like a law of large numbers and a central limit theorem for $g_{\hat{\theta}_n}$. In the next section, we present three examples to illustrate the importance of this. The first example sets out with a simple parametric model, where one sees that extensions of (1.1) and (1.2) can be used to prove asymptotic normality of the maximum likelihood estimator. Next, we note that a parametric model is not always appropriate. To indicate how to proceed in nonparametric models, we place ourselves in a general context in Example 1.2. We indicate there how extensions of (1.1) play a role in the development of a general theory on consistency of maximum likelihood estimators. The last example in this chapter shows that also in nonparametric models, one can exploit extensions of (1.2) to arrive at asymptotic normality.

1.1. Some examples from statistics

Example 1.1. A binary choice model Let $X_i = (Y_i, Z_i)$, with Z_i the education of individual i, $Y_i = 1$ if individual i has a job, and $Y_i = 0$ otherwise. Suppose that we asked n individuals for their education and employment status. We want to model how the probability of having a job depends on education, and estimate the parameters in the model.

Case (i) The logit model is

$$P(Y = 1 \mid Z = z) = F_0(\alpha_0 + \theta_0 z),$$

with $F_0(\xi) = e^\xi/(1+e^\xi)$ the distribution function of the logistic distribution. To avoid notational digressions, we assume that only the parameter $\theta_0 \in \mathbf{R}$ is unknown, and that $\alpha_0 = 0$.

Denote the maximum likelihood estimator of θ_0 by $\hat{\theta}_n$, i.e. $\hat{\theta}_n$ is the value of θ that maximizes the (conditional) log-likelihood

$$\sum_{i=1}^{n} \log p_\theta(Y_i \mid Z_i),$$

where

$$p_\theta(y \mid z) = F_0^y(\theta z)(1 - F_0(\theta z))^{1-y}.$$

To investigate the asymptotic behaviour of this estimator, we introduce the score function

$$l_\theta(y, z) = \frac{d}{d\theta} \log p_\theta(y \mid z) = z(y - F_0(\theta z)),$$

and let

$$g_\theta(z) = \begin{cases} -\dfrac{l_\theta(y,z) - l_{\theta_0}(y,z)}{\theta - \theta_0} = z\dfrac{F_0(\theta z) - F_0(\theta_0 z)}{\theta - \theta_0}, & \text{if } \theta \neq \theta_0, \\ z^2 F_0(\theta_0 z)(1 - F_0(\theta_0 z)), & \text{if } \theta = \theta_0 . \end{cases}$$

Note that as $\theta \to \theta_0$, $g_\theta(z) \to g_{\theta_0}(z)$ for all z. Lemma 1.1 below assumes that we have some type of law of large numbers for $g_{\hat{\theta}_n}$.

Lemma 1.1 *Suppose that*

$$\frac{1}{n}\sum_{i=1}^{n} g_{\hat{\theta}_n}(Z_i) \to_{\mathbf{P}} E g_{\theta_0}(Z) := I_{\theta_0}. \tag{1.3}$$

where $I_{\theta_0} > 0$.

Then $\sqrt{n}(\hat{\theta}_n - \theta_0)$ *is asymptotically* $\mathcal{N}(0, 1/I_{\theta_0})$*-distributed.*

Proof Since $\hat{\theta}_n$ maximizes the likelihood, the derivative of the log-likelihood at $\hat{\theta}_n$ is zero:

$$\sum_{i=1}^{n} l_{\hat{\theta}_n}(Y_i, Z_i) = 0. \tag{1.4}$$

Clearly,

$$\sum_{i=1}^{n} l_{\hat{\theta}_n}(Y_i, Z_i) = \sum_{i=1}^{n} l_{\theta_0}(Y_i, Z_i) - (\hat{\theta}_n - \theta_0)\sum_{i=1}^{n} g_{\hat{\theta}_n}(Z_i). \tag{1.5}$$

Combine (1.4) and (1.5) to give

$$\sqrt{n}(\hat{\theta}_n - \theta_0) = \frac{\frac{1}{\sqrt{n}}\sum_{i=1}^{n} l_{\theta_0}(X_i)}{\frac{1}{n}\sum_{i=1}^{n} g_{\hat{\theta}_n}(Z_i)};$$

recall that $X_i = (Y_i, Z_i)$. Note that

$$El_{\theta_0}(X) = 0, \quad \mathrm{var}\big(l_{\theta_0}(X)\big) = I_{\theta_0}. \tag{1.6}$$

By the central limit theorem, the numerator is asymptotically $\mathcal{N}(0, I_{\theta_0})$-distributed, so the result follows immediately from assumption (1.3). □

Assumption (1.3) corresponds to an extension of the law of large numbers. Another approach towards proving asymptotic normality is based on an extension of the central limit theorem. To formulate this idea, we introduce the quantities

$$m_\theta = El_\theta(X),$$

and

$$\sigma_\theta^2 = \mathrm{var}(l_\theta(X)).$$

For each θ, $\sum_{i=1}^n (l_\theta(X_i) - m_\theta)/\sqrt{n}$ converges weakly to a $\mathcal{N}(0, \sigma_\theta^2)$-distribution. If we consider $\sum_{i=1}^n (l_\theta(X_i) - m_\theta)/\sqrt{n}$ as a stochastic process indexed by θ, one can think of condition (1.7) below as asymptotic continuity of this process at θ_0.

We shall make use of the stochastic order symbol $o_{\mathbf{P}}(1)$. For example $\hat{\theta}_n = \theta_0 + o_{\mathbf{P}}(1)$ means that $\hat{\theta}_n$ converges to θ_0 in probability. For a formal definition, see Section 2.1. The general formulation of Lemma 1.2, for M-estimators of a finite-dimensional parameter, can be found in Section 12.3, Lemma 12.7.

Lemma 1.2 *Suppose that* $\hat{\theta}_n = \theta_0 + o_{\mathbf{P}}(1)$*, and that*

$$\frac{1}{\sqrt{n}} \sum_{i=1}^n \left(l_{\hat{\theta}_n}(X_i) - m_{\hat{\theta}_n} \right) = \frac{1}{\sqrt{n}} \sum_{i=1}^n (l_{\theta_0}(X_i) - m_{\theta_0}) + o_{\mathbf{P}}(1). \tag{1.7}$$

Then $\sqrt{n}(\hat{\theta}_n - \theta_0)$ *is asymptotically* $\mathcal{N}(0, 1/I_{\theta_0})$*-distributed.*

Proof Note first that

$$m_{\theta_0} = 0, \ \sigma_{\theta_0}^2 = I_{\theta_0},$$

and moreover that

$$\left.\frac{d}{d\theta} m_\theta \right|_{\theta = \theta_0} = -I_{\theta_0}. \tag{1.8}$$

Again, we rewrite the derivative of the log-likelihood at $\hat{\theta}_n$:

$$0 = \frac{1}{\sqrt{n}} \sum_{i=1}^n l_{\hat{\theta}_n}(X_i) = \frac{1}{\sqrt{n}} \sum_{i=1}^n \left(l_{\hat{\theta}_n}(X_i) - m_{\hat{\theta}_n} \right) + \sqrt{n} m_{\hat{\theta}_n}.$$

Using assumption (1.7), we see that

$$0 = \frac{1}{\sqrt{n}}\sum_{i=1}^{n}\left(l_{\theta_0}(X_i) - m_{\theta_0}\right) + o_{\mathbf{P}}(1) + \sqrt{n}m_{\hat{\theta}_n}$$

$$= \frac{1}{\sqrt{n}}\sum_{i=1}^{n} l_{\theta_0}(X_i) + o_{\mathbf{P}}(1) - \sqrt{n}(\hat{\theta}_n - \theta_0)(I_{\theta_0} + o_{\mathbf{P}}(1)),$$

where in the last step, we invoked $m_{\theta_0} = 0$ and (1.8). The result now follows from the central limit theorem for l_{θ_0}. □

Case (ii) In Case (i), we assumed a parametric model for the probability of a job, given education, but there appears to be no intrinsic reason why the probability of having a job depends on education in this specific way. All we know is for instance that the higher the education, the more likely it is to have a job. In that case the model would be

$$P(Y = 1 \mid Z = z) = F_0(z),$$

with F_0 any increasing function of z satisfying $0 \leq F_0(z) \leq 1$. The parameter space is now

$$\Lambda = \{F : \mathbf{R} \to [0,1],\ F \text{ increasing}\}.$$

The maximum likelihood estimator $\hat{F}_n$ is that value of $F \in \Lambda$ that maximizes

$$\sum_{i=1}^{n}\left(Y_i \log F(Z_i) + (1 - Y_i)\log(1 - F(Z_i))\right). \qquad (1.9)$$

It is no longer so easy to apply the argument that the derivative of the log-likelihood at $\hat{F}_n$ is zero. After all, what is a derivative in this situation? (See the first example in Section 11.2.3 for more details.) Moreover, how do we measure the distance between $\hat{F}_n$ and F_0? There are several possibilities here. In the next example, we shall indicate how one can prove consistency in the so-called Hellinger metric by employing a uniform law of large numbers. One can also think of using the $L_2(Q)$-distance, with Q the distribution of Z:

$$\|\hat{F}_n - F_0\|_Q = \left(\int \left(\hat{F}_n(z) - F_0(z)\right)^2 dQ(z)\right)^{1/2}.$$

It will be shown in Example 7.4.3 that $\|\hat{F}_n - F_0\|_Q = O_{\mathbf{P}}(n^{-1/3})$ (for an explanation of stochastic order symbols, see Section 2.1). The same rate holds true for the Hellinger metric. Compare this with Case (i), where the rate of convergence for $\hat{\theta}_n$ is $O_{\mathbf{P}}(n^{-1/2})$. So the price one has to pay for not assuming the parametric model is that the rate of convergence is much slower. We shall present a quantification of this phenomenon. The 'size' or 'richness' of parameter space will be measured by means of its *entropy* (see Section 2.3 for a definition), and this entropy will be used to calculate the rate of convergence (see for example Theorem 7.4).

Case (iii) One can think of many intermediate models between Case (i) and Case (ii). Here is an example. Suppose that the probability of having a job indeed increases with education, but the amount of increase is lower at higher education levels. This means that $P(Y = 1 \mid Z = z)$ is a concave function of z:

$$P(Y = 1 \mid Z = z) = F_0(z),$$

with

$$F_0 \in \tilde{\Lambda} = \left\{ F : \mathbf{R} \to [0,1],\ 0 \le \frac{dF(z)}{dz} \le M,\ F(z) \text{ concave} \right\}.$$

The log-likelihood is of the same form as in Case (ii), but we maximize it over a smaller parameter space $\tilde{\Lambda}$ in order to get the maximum likelihood estimator $\hat{F}_n$. It turns out that under regularity conditions on Q, the rate of convergence is $O_{\mathbf{P}}(n^{-2/5})$ (see Example 7.4.3). This is an improvement as compared to Case (ii), due to the fact that we assumed a parameter space with smaller *entropy*.

Example 1.2. Maximum likelihood In this example, we study the maximum likelihood problem in general terms. Let X have density $p_{\theta_0}(x)$, $\theta_0 \in \Theta$, with respect to a σ-finite measure μ. Here Θ may be finite-dimensional (as was the case in Example 1.1, Case (i)) or infinite-dimensional (as in Example 1.1, Case (ii) and (iii)). The maximum likelihood estimator $\hat{\theta}_n$ maximizes the log-likelihood $\sum_{i=1}^n \log p_\theta(X_i)$ over all $\theta \in \Theta$. The idea here is that the log-likelihood will be close to its expectation for large sample sizes. Because the expected log-likelihood is maximized by θ_0, $\hat{\theta}_n$ is indeed a sensible estimator. What we need to turn this idea into a rigorous argument is an extension of the law of large numbers (1.1).

The estimator $\hat{\theta}_n$ maximizes the likelihood over $\theta \in \Theta$. Because $\theta_0 \in \Theta$, we therefore have

$$\sum_{i=1}^n \log \frac{p_{\theta_0}(X_i)}{p_{\hat{\theta}_n}(X_i)} \le 0. \tag{1.10}$$

On the other hand, for all θ,

$$E \log \frac{p_{\theta_0}(X)}{p_\theta(X)} \ge 0, \tag{1.11}$$

(see Problem 1.3), with equality if $\theta = \theta_0$. The quantity given in (1.11) is the Kullback–Leibler information

$$K(p_\theta, p_{\theta_0}) = \int \left(\log \frac{p_{\theta_0}}{p_\theta}\right) p_{\theta_0}\, d\mu.$$

Let

$$g_\theta = \log \frac{p_{\theta_0}}{p_\theta},$$

so that

$$K(p_\theta, p_{\theta_0}) = E g_\theta(X).$$

Then by (1.10),

$$0 \geq \frac{1}{n}\sum_{i=1}^n g_{\hat\theta_n}(X_i) = \frac{1}{n}\sum_{i=1}^n g_{\hat\theta_n}(X_i) - K(p_{\hat\theta_n}, p_{\theta_0}) + K(p_{\hat\theta_n}, p_{\theta_0}),$$

or,

$$K(p_{\hat\theta_n}, p_{\theta_0}) \leq \left|\frac{1}{n}\sum_{i=1}^n g_{\hat\theta_n}(X_i) - K(p_{\hat\theta_n}, p_{\theta_0})\right|. \tag{1.12}$$

By the law of large numbers, for each θ,

$$\left|\frac{1}{n}\sum_{i=1}^n g_\theta(X_i) - K(p_\theta, p_{\theta_0})\right| \to 0, \quad \text{a.s.}$$

Suppose this is also true for the sequence $\{\hat\theta_n\}$:

$$\left|\frac{1}{n}\sum_{i=1}^n g_{\hat\theta_n}(X_i) - K(p_{\hat\theta_n}, p_{\theta_0})\right| \to 0, \quad \text{a.s.} \tag{1.13}$$

Then it follows from (1.12) that the Kullback–Leibler information converges to zero. The Kullback–Leibler information is not a distance function, but its convergence often implies consistency of $\hat\theta_n$ in some metric of interest. A convenient metric is the Hellinger metric, defined as

$$h(p_\theta, p_{\theta_0}) = \left(\frac{1}{2}\int (p_\theta^{1/2} - p_{\theta_0}^{1/2})^2\, d\mu\right)^{1/2}.$$

The next lemma shows that convergence of the Kullback–Leibler information always yields consistency in the Hellinger metric.

Lemma 1.3 *We have*

$$h^2(p_\theta, p_{\theta_0}) \le \frac{1}{2} K(p_\theta, p_{\theta_0}).$$

Proof Use the fact that $(1/2)\log v \le v^{1/2} - 1$ for all $v > 0$:

$$\frac{1}{2} \log \frac{p_\theta(x)}{p_{\theta_0}(x)} \le \frac{p_\theta^{1/2}(x)}{p_{\theta_0}^{1/2}(x)} - 1,$$

so

$$\frac{1}{2} K(p_\theta, p_{\theta_0}) \ge 1 - E\left(\frac{p_\theta^{1/2}(X)}{p_{\theta_0}^{1/2}(X)} \right).$$

Observe that

$$1 - E\left(\frac{p_\theta^{1/2}(X)}{p_{\theta_0}^{1/2}(X)} \right) = 1 - \int p_\theta^{1/2} p_{\theta_0}^{1/2} \, d\mu,$$

and since a density integrates to one,

$$\begin{aligned} 1 - \int p_\theta^{1/2} p_{\theta_0}^{1/2} \, d\mu &= \frac{1}{2} \int p_\theta \, d\mu + \frac{1}{2} \int p_{\theta_0} \, d\mu - \int p_\theta^{1/2} p_{\theta_0}^{1/2} \, d\mu \\ &= \frac{1}{2} \int (p_\theta^{1/2} - p_{\theta_0}^{1/2})^2 \, d\mu = h^2(p_\theta, p_{\theta_0}). \end{aligned}$$

□

In summary, the maximum likelihood estimator $\hat{\theta}_n$ maximizes an empirical average, whereas θ_0 maximizes the expectation. If averages converge to expectations in a broad enough sense, this implies consistency of $\hat{\theta}_n$. We remark here that (1.13) is sometimes difficult to prove and perhaps not true. However, a modification of the idea presented here can demonstrate consistency in Hellinger distance, although one then might lose the convergence of the Kullback–Leibler information (see Section 4.1).

Also the rate at which the maximum likelihood estimator converges can be obtained along these lines. For each θ

$$\frac{1}{\sqrt{n}} \sum_{i=1}^{n} (g_\theta(X_i) - K(p_\theta, p_{\theta_0})) \to^{\mathcal{L}} \mathcal{N}(0, \sigma_\theta^2), \tag{1.14}$$

provided $\sigma_\theta^2 = \mathrm{var}(g_\theta(X)) < \infty$. First of all, this implies that for each such θ,

$$\left| \frac{1}{n} \sum_{i=1}^{n} g_\theta(X_i) - K(p_\theta, p_{\theta_0}) \right| = O_{\mathbf{P}}(n^{-1/2}).$$

If the same is true for the sequence $\{\hat{\theta}_n\}$:

$$\left|\frac{1}{n}\sum_{i=1}^{n} g_{\hat{\theta}_n}(X_i) - K(p_{\hat{\theta}_n}, p_{\theta_0})\right| = O_{\mathbf{P}}(n^{-1/2}),$$

then (1.12), together with Lemma 1.3, would imply that $h(p_{\hat{\theta}_n}, p_{\theta_0}) = O_{\mathbf{P}}(n^{-1/4})$. In fact, one may expect that $\sigma_\theta \to 0$ as $h(p_\theta, p_{\theta_0}) \to 0$, and that in view of (1.14), $\sum_{i=1}^{n}(g_{\hat{\theta}_n}(X_i) - K(p_{\hat{\theta}_n}, p_{\theta_0}))/n$ converges with a rate faster than $O_{\mathbf{P}}(n^{-1/2})$. This would give $h(\hat{p}_n, p_0) = o_{\mathbf{P}}(n^{-1/4})$. We shall see that the rate of convergence ranges from $O_{\mathbf{P}}(n^{-1/2})$ for regular parametric models, to $o_{\mathbf{P}}(n^{-1/4})$ for moderately complex infinite-dimensional models. If the parameter space is even richer (has very large *entropy*), the rate can be even slower, or one may not have consistency at all. However, the power 1/4 appears to be something like a critical point.

Example 1.3. Estimating the mean in the binary choice model Here is another illustration of the application of extensions of the classical central limit theorem. Let us return to the situation of Example 1.1, Case (ii). There, we have

$$P(Y = 1 \mid Z = z) = F_0(z),$$

with F_0 an unknown increasing function satisfying $0 \le F_0 \le 1$. Suppose that we want to estimate the average probability of having a job. Write this as $\theta_0 = \theta_{F_0} = EY$. A good candidate for estimating θ_0 is of course $\bar{Y} = \sum_{i=1}^{n} Y_i/n$, the observed proportion of individuals with a job. We have

$$\sqrt{n}(\bar{Y} - \theta_0) \to^{\mathcal{L}} \mathcal{N}(0, \mathrm{var}(Y)),$$

with

$$\mathrm{var}(Y) = \int F_0(z)(1 - F_0(z))\, dQ(z) + \mathrm{var}(F_0(Z)). \tag{1.15}$$

Let $\hat{F}_n$ be as before the maximum likelihood estimator of F_0. Then $\sum_{i=1}^{n} \hat{F}_n(Z_i)/n$ is also a good candidate for estimating θ_0, but it turns out that this estimator is just $\bar{Y}$. Moreover, if the distribution Q of Z is completely unknown, then it is also the maximum likelihood estimator of θ_0. We shall not prove these two statements here (see Section 11.2 for more details). Instead, let us see what we can gain if we assume Q to be known. Since

$$\theta_0 = \int F_0(z)\, dQ(z),$$

the maximum likelihood estimator of θ_0 is then

$$\hat{\theta}_n = \theta_{\hat{F}_n} = \int \hat{F}_n(z)\, dQ(z).$$

More generally, we define

$$\theta_F = \int F(z)\, dQ(z).$$

Then for each $F \in \Lambda = \{F : \mathbf{R} \to [0,1],\ F \text{ increasing}\}$,

$$\frac{1}{\sqrt{n}} \sum_{i=1}^{n} \left(F(Z_i) - \theta_F\right) \to^{\mathcal{L}} \mathcal{N}(0, \mathrm{var}(F(Z))).$$

Now, suppose that the process $\{\sum_{i=1}^{n}(F(Z_i) - \theta_F)/\sqrt{n} :\ F \in \Lambda\}$ is asymptotically continuous at F_0, in the sense that

$$\frac{1}{\sqrt{n}} \sum_{i=1}^{n} \left(\hat{F}_n(Z_i) - \hat{\theta}_n\right) = \frac{1}{\sqrt{n}} \sum_{i=1}^{n} \left(F_0(Z_i) - \theta_0\right) + o_{\mathbf{P}}(1). \tag{1.16}$$

This assumption can be seen as an extension of the central limit theorem (1.2).

Lemma 1.4 *Under (1.16),*

$$\sqrt{n}(\hat{\theta}_n - \theta_0) \to^{\mathcal{L}} \mathcal{N}\left(0, \int F_0(z)(1 - F_0(z))\, dQ(z)\right). \tag{1.17}$$

Proof Since $\frac{1}{n}\sum_{i=1}^{n} \hat{F}_n(Z_i) = \bar{Y}$, we have

$$\begin{aligned}
\sqrt{n}(\hat{\theta}_n - \theta_0) &= \sqrt{n}(\bar{Y} - \theta_0) - \frac{1}{\sqrt{n}} \sum_{i=1}^{n} \left(\hat{F}_n(Z_i) - \hat{\theta}_n\right) \\
&= \sqrt{n}(\bar{Y} - \theta_0) - \frac{1}{\sqrt{n}} \sum_{i=1}^{n} \left(F_0(Z_i) - \theta_0\right) + o_{\mathbf{P}}(1) \\
&= \frac{1}{\sqrt{n}} \sum_{i=1}^{n} \left(Y_i - F_0(Z_i)\right) + o_{\mathbf{P}}(1).
\end{aligned}$$

The result now follows from the classical central limit theorem. □

So the asymptotic distribution of $\hat{\theta}_n$ has smaller variance than $\bar{Y}$. This is what we gain from knowing the distribution of Z.

1.2. Problems and complements

1.1. Verify (1.6) and (1.8). In fact, recall that if $\{p_\theta :\ \theta \in \Theta \subset \mathbf{R}\}$ is any sufficiently regular class, then for $l_\theta = d \log p_\theta / d\theta$ and $l'_\theta = dl_\theta/d\theta$, one has $El^2_{\theta_0}(X) = -El'_{\theta_0}(X) = I_{\theta_0}$. The quantity I_{θ_0} is called the Fisher information.

1.2. Find an expression for the Hellinger distance $h(p_\theta, p_{\theta_0})$ when the densities are as in Example 1.1, Case (ii).

1.3. Let Z be a positive random variable. Using the concavity of the log-function, we have by Jensen's inequality,

$$E(\log Z) \leq \log(EZ).$$

Use this to show that

$$K(p_\theta, p_{\theta_0}) \geq 0.$$

1.4. Show that $\text{var}(Y) = E\text{var}(Y \mid Z) + \text{var}(E(Y \mid Z))$, and use this to check (1.15).

2

Notation and Definitions

Stochastic order symbols are introduced, averages and expectations are written as integrals, and a definition of entropy is given. In some examples a bound for the entropy is given.

2.1. Stochastic order symbols

Let $(\Omega, \mathscr{F}, \mathbf{P})$ be a probability space, $Z_n : \Omega \to \mathbf{R}$, $n = 1, 2, \ldots$, be a sequence of random variables, and $\{k_n\}_{n=1}^{\infty}$ be a sequence of positive numbers.

We say that

$$Z_n = O_{\mathbf{P}}(k_n),$$

if

$$\lim_{T\to\infty} \limsup_{n\to\infty} \mathbf{P}(|Z_n| > T k_n) = 0.$$

Then $Z_n/k_n = O_{\mathbf{P}}(1)$.

We say that

$$Z_n = o_{\mathbf{P}}(k_n),$$

if for all $\epsilon > 0$,

$$\lim_{n\to\infty} \mathbf{P}(|Z_n| > \epsilon k_n) = 0.$$

Then $Z_n/k_n = o_{\mathbf{P}}(1)$.

2.2. The empirical process

Let $X_1, X_2, \ldots$ be independent copies of a random variable X in $(\mathcal{X}, \mathcal{A})$ with distribution P, and let $\mathcal{G} = \{g_\theta : \theta \in \Theta\}$ be a class of functions on $\mathcal{X}$. We say that $\mathcal{G}$ satisfies the uniform law of large numbers (ULLN), if

$$\sup_{\theta \in \Theta} \left| \frac{1}{n} \sum_{i=1}^{n} g_\theta(X_i) - E g_\theta(X) \right| \to 0, \quad \text{a.s.} \tag{2.1}$$

Let us, for the moment, denote the expected value of $g_\theta(X)$ as

$$m_\theta = E g_\theta(X).$$

Clearly, if (2.1) holds, then for any sequence $\{\hat{\theta}_n\} \subset \Theta$,

$$\left| \frac{1}{n} \sum_{i=1}^{n} g_{\hat{\theta}_n}(X_i) - m_{\hat{\theta}_n} \right| \to 0, \quad \text{a.s.}$$

Observe that $\hat{\theta}_n$ may even be random in this case. But it is important to realize that $m_{\hat{\theta}_n}$ is *not* the expectation of $g_{\hat{\theta}_n}(X)$ if $\hat{\theta}_n$ is random! Instead, since

$$m_\theta = \int g_\theta \, dP,$$

we have

$$m_{\hat{\theta}_n} = \int g_{\hat{\theta}_n} \, dP.$$

This is the reason why we shall often express expectations as integrals in this book.

The sample average can be seen as an empirical expectation, which can also be written as an integral. Let P_n be the empirical distribution based on $X_1, \ldots, X_n$, i.e., for each set $A \in \mathcal{A}$,

$$P_n(A) = \frac{1}{n} \{\text{number of } X_i \in A,\ 1 \le i \le n\} = \frac{1}{n} \sum_{i=1}^{n} 1_A(X_i).$$

Thus, the P_n put mass $\frac{1}{n}$ at each of the X_i, $1 \le i \le n$:

$$P_n = \frac{1}{n} \sum_{i=1}^{n} \delta_{X_i},$$

with δ_{X_i} a point mass at X_i, $i = 1, \ldots, n$. We may now use the notation

$$\frac{1}{n} \sum_{i=1}^{n} g(X_i) = \int g \, dP_n.$$

The difference between average and expectation is

$$\frac{1}{n}\sum_{i=1}^{n} g(X_i) - Eg(X) = \int g\, d(P_n - P).$$

Let $\mathcal{G}$ be a class of functions on $\mathcal{X}$. Before, we indexed $\mathcal{G}$ by a parameter $\theta \in \Theta$. Such a parametrization usually comes up naturally in statistical applications. It is a way to describe the form of the functions in $\mathcal{G}$. We do not really need it here. In fact, our natural parameter space will be the class $\mathcal{G}$ itself.

In the new formulation, we say that $\mathcal{G}$ satisfies the uniform law of large numbers (ULLN) if

$$\sup_{g \in \mathcal{G}} \left| \int g\, d(P_n - P) \right| \to 0, \quad \text{a.s.} \tag{2.2}$$

Theorem 3.7 contains the conditions on $\mathcal{G}$ for (2.2) to hold. Observe that (2.2) implies that for any sequence $\{\hat{g}_n\} \subset \mathcal{G}$,

$$\left| \int \hat{g}_n\, d(P_n - P) \right| \to 0, \quad \text{a.s.}$$

It is not so straightforward to formulate something like a uniform central limit theorem. Let us first recall that by the classical central limit theorem, for each g with $\text{var}(g(X))$ finite,

$$\sqrt{n} \int g\, d(P_n - P) \to^{\mathcal{L}} \mathcal{N}\big(0, \text{var}(g(X))\big).$$

Here, the variance of $g(X)$ can be expressed as

$$\text{var}\big(g(X)\big) = Eg^2(X) - \big(Eg(X)\big)^2.$$

We need a handy notation for this in case we are dealing with a possibly random function $\hat{g}_n$. Clearly,

$$Eg^2(X) = \int g^2\, dP.$$

This is the squared $L_2(P)$-norm of the function g, which we shall denote by $\|\cdot\|^2$. Thus

$$\|g\|^2 = \int g^2\, dP.$$

Furthermore, let us write

$$\sigma_g^2 = \int g^2\, dP - \left(\int g\, dP\right)^2.$$

Then for g fixed, σ_g^2 is the variance of $g(X)$. But if $\hat{g}_n$ is random, we do not have this type of interpretation of $\sigma_{\hat{g}_n}^2$.

One can regard $\{\nu_n(g) = \sqrt{n}\int g\, d(P_n - P) : g \in \mathcal{G}\}$ as a stochastic process indexed by $\mathcal{G}$. It is called the empirical process. The classes of functions we have in mind may be rather large, such as the class of all increasing functions g with $0 \le g \le 1$. Convergence in distribution of this process is treated in Chapter 6 (a uniform central limit theorem). In statistical problems, the important issue is often not so much the weak convergence, but rather the so-called *asymptotic equicontinuity* of the process (which is essentially a necessary condition for weak convergence). We say that $\{\nu_n(g) : g \in \mathcal{G}\}$ is asymptotically equicontinuous at $g_0 \in \mathcal{G}$ if for each (random) sequence $\{\hat{g}_n\} \subset \mathcal{G}$ with $\|\hat{g}_n - g_0\| = o_{\mathbf{P}}(1)$, we have

$$|\nu_n(\hat{g}_n) - \nu_n(g_0)| = o_{\mathbf{P}}(1). \tag{2.3}$$

Then,

$$\nu_n(\hat{g}_n) \to^{\mathcal{L}} \mathcal{N}(0, \sigma_{g_0}^2),$$

provided $\sigma_{g_0}^2 < \infty$.

In Chapter 5, we shall formulate conditions on $\mathcal{G}$ that guarantee the asymptotic equicontinuity of the empirical process indexed by $\mathcal{G}$. In particular, we shall investigate the modulus of continuity of the empirical process, i.e., the behaviour of $|\nu_n(g) - \nu_n(g_0)|$ as a function of $\|g - g_0\|$. The conditions are in terms of *entropy*, which is a measure of the complexity of $\mathcal{G}$.

We already announced in Example 1.1, what the rates of convergence are in the cases considered there. The larger the parameter space, that is, the larger its entropy, the harder it will be to estimate the true state of nature. In empirical process theory, the effect of a large $\mathcal{G}$ is that the increments of the empirical process behave irregularly. The empirical process may even not be asymptotically continuous at all.

2.3. Entropy

One can define entropy for general metric spaces, but we shall restrict ourselves to classes of functions $\mathcal{G}$. Let Q be a measure on $(\mathcal{X}, \mathcal{A})$ and $L_p(Q) = \{g : \mathcal{X} \to \mathbf{R} : \int |g|^p\, dQ < \infty\}$, $1 \le p < \infty$. For $g \in L_p(Q)$, write

$$\|g\|_{p,Q}^p = \int |g|^p\, dQ.$$

We refer to $\|\cdot\|_{p,Q}$ as the $L_p(Q)$-norm, or $L_p(Q)$-metric, and call $\|g_1 - g_2\|_{p,Q}$ the $L_p(Q)$-distance between g_1 and g_2. Actually, these are rather pseudo-norms and pseudo-distances, but we omit the term 'pseudo', although it is not our intention to identify equivalence classes.

Definition 2.1 (Entropy for the $L_p(Q)$-metric) Consider for each $\delta > 0$, a collection of functions $g_1, \ldots, g_N$, such that for each $g \in \mathcal{G}$, there is a $j = j(g) \in \{1, \ldots, N\}$, such that

$$\|g - g_j\|_{p,Q} \leq \delta.$$

Let $N_p(\delta, \mathcal{G}, Q)$ be the smallest value of N for which such a covering by balls with radius δ and centers $g_1, \ldots, g_N$ exists. (Take $N_p(\delta, \mathcal{G}, Q) = \infty$ if no finite covering by balls with radius δ exists.) Then $N_p(\delta, \mathcal{G}, Q)$ is called the *δ-covering number* and $H_p(\delta, \mathcal{G}, Q) = \log N_p(\delta, \mathcal{G}, Q)$ is called the *δ-entropy* of $\mathcal{G}$ (for the $L_p(Q)$-metric). The class $\mathcal{G}$ is called *totally bounded (for the $L_p(Q)$-metric)* if $H_p(\delta, \mathcal{G}, Q) < \infty$ for all $\delta > 0$. (The closure of $\mathcal{G}$ is then compact.)

Thus, the δ-entropy of a set $\mathcal{G}$ is the logarithm of the minimum number of balls with radius δ, necessary to cover the set. We do not insist that the centers g_j of the balls are in $\mathcal{G}$, but it can be easily seen that from any δ-covering of a set, one can construct a 2δ-covering with centers within the set.

In the next definition, we require in addition that each element of $\mathcal{G}$ is encapsulated by two functions of the covering, an upper function g^U and a lower function g^L.

Definition 2.2 (Entropy with bracketing for the $L_p(Q)$-metric) Let $N_{p,B}(\delta, \mathcal{G}, Q)$ be the smallest value of N for which there exist pairs of functions $\{[g_j^L, g_j^U]\}_{j=1}^N$ such that $\|g_j^U - g_j^L\|_{p,Q} \leq \delta$ for all $j = 1, \ldots, N$, and such that for each $g \in \mathcal{G}$, there is a $j = j(g) \in \{1, \ldots, N\}$ such that

$$g_j^L \leq g \leq g_j^U.$$

(Take $N_{p,B}(\delta, \mathcal{G}, Q) = \infty$ if no finite set of such brackets exists.) Then $H_{p,B}(\delta, \mathcal{G}, Q) = \log N_{p,B}(\delta, \mathcal{G}, Q)$ is called the *δ-entropy with bracketing* of $\mathcal{G}$.

The definition of entropy for the supremum norm is roughly Definition 2.1, with $p = \infty$. However, we shall not take the essential supremum (which depends on the measure Q), but the supremum over *all* x. Thus, our concept of entropy for the supremum norm does not depend on any measure, but only depends on $\mathcal{X}$. Denote the supremum norm of a function g by

$$|g|_\infty = \sup_{x \in \mathcal{X}} |g(x)|.$$

Definition 2.3 (Entropy for the supremum norm) Let $N_\infty(\delta, \mathcal{G})$ be the smallest value of N such that there exists $\{g_j\}_{j=1}^N$ with

$$\sup_{g\in\mathcal{G}} \min_{j=1,\dots,N} |g - g_j|_\infty \le \delta.$$

(Take $N_\infty(\delta, \mathcal{G}) = \infty$ if no finite set with this property exists.) Then $H_\infty(\delta, \mathcal{G}) = \log N_\infty(\delta, \mathcal{G})$ is called the *δ-entropy* of $\mathcal{G}$ for the supremum norm.

Lemma 2.1 *For all* $1 \le p < \infty$

$$H_p(\delta, \mathcal{G}, Q) \le H_{p,B}(\delta, \mathcal{G}, Q), \quad \textit{for all } \delta > 0.$$

Moreover, if Q is a probability measure,

$$H_{p,B}(\delta, \mathcal{G}, Q) \le H_\infty\left(\frac{\delta}{2}, \mathcal{G}\right), \quad \textit{for all } \delta > 0.$$

The proof of this lemma is left as an exercise.

The case $p = 2$ will be the most important. Whenever $p = 2$, the subscript 2 will be omitted, i.e.,

$$\|\cdot\|_Q = \|\cdot\|_{2,Q}.$$

Moreover,

$$N(\delta, \mathcal{G}, Q) = N_2(\delta, \mathcal{G}, Q),\ N_B(\delta, \mathcal{G}, Q) = N_{2,B}(\delta, \mathcal{G}, Q),$$

and

$$H(\delta, \mathcal{G}, Q) = H_2(\delta, \mathcal{G}, Q),\ H_B(\delta, \mathcal{G}, Q) = H_{2,B}(\delta, \mathcal{G}, Q).$$

Apart from entropy for the supremum norm, we shall mainly encounter entropy with bracketing for the $L_2(P)$-metric, and entropy (without bracketing) for the $L_2(P_n)$-metric. (Note that the latter is a random quantity!) The corresponding norms are written as

$$\|g\| = \|g\|_{2,P},$$

and

$$\|g\|_n = \|g\|_{2,P_n}.$$

2.4. Examples

In most cases, the derivation of a good bound for the entropy of a class of functions is rather complicated. It is not so hard to get a rough bound, but to obtain the right order, one usually needs sophisticated arguments, which may obscure the underlying idea. Therefore, we set out with some bounds for simple cases, and present some deeper results without proof. Further entropy calculations are scattered throughout the various chapters.

Lemma 2.2 *Let $\mathcal{G}$ be a class of increasing functions $g : \mathbf{R} \to [0,1]$, and let $\mathcal{X} \subset \mathbf{R}$ be a finite set with cardinality n. Then*

$$(2.4) \qquad H_\infty(\delta, \mathcal{G}) \le \frac{1}{\delta} \log\left(n + \frac{1}{\delta}\right), \quad \textit{for all } \delta > 0.$$

Proof Let $\mathcal{X}$ consist of the points $x_1 \le \cdots \le x_n$. For each g in $\mathcal{G}$, define

$$M_i = \left\lfloor \frac{g(x_i)}{\delta} \right\rfloor, \quad i = 1, \ldots, n,$$

where $\lfloor a \rfloor$ denotes the integer part of a. Take

$$\tilde{g}(x_i) = \delta M_i, \quad i = 1, \ldots, n.$$

Clearly $|\tilde{g}(x_i) - g(x_i)| \le \delta$, $i = 1, \ldots, n$.

We have $0 \le M_1 \le \cdots \le M_n \le \lfloor 1/\delta \rfloor$ and $M_i \in \mathbf{N}$, $i = 1, \ldots, n$. Therefore, as g varies, the number of functions $\tilde{g}$ is

$$\binom{n + \lfloor 1/\delta \rfloor}{\lfloor 1/\delta \rfloor}. \qquad \Box$$

In fact, Birman and Solomjak (1967) show that for $\mathcal{G} = \{g : \mathbf{R} \to [0,1],\ g \text{ increasing}\}$, one has for some constant A, and for any probability measure Q,

$$(2.5) \qquad H_B(\delta, \mathcal{G}, Q) \le A\frac{1}{\delta}, \quad \text{for all } \delta > 0.$$

So this is a much better result, but the proof is quite complicated. Birman and Solomjak do not consider entropy with bracketing, but some minor additions in their proof do give (2.5). A self-contained proof can be found in van de Geer (1991). A similar bound holds for classes of functions with bounded variation. A function g is called of bounded variation if $g = g_1 - g_2$, where g_1 and g_2 are bounded increasing functions. The total variation of g is $TV(g) = \int |\, dg \,|$. Thus, for $\mathcal{G} = \{g : \mathbf{R} \to [0,1],\ TV(g) \le 1\}$, we also have for some constant A, and any probability measure Q,

$$(2.6) \qquad H_B(\delta, \mathcal{G}, Q) \le A\frac{1}{\delta}, \quad \text{for all } \delta > 0.$$

The paper of Birman and Solomjak (1967) contains many other results on entropy, which are useful for our purposes. Let us again first consider a simple case.

Lemma 2.3 *Let* $\mathcal{G} = \{g : [0,1] \to [0,1], |g'| \leq 1\}$. *Then for some constant* A,

$$H_\infty(\delta, \mathcal{G}) \leq A\frac{1}{\delta}, \quad \textit{for all } \delta > 0. \tag{2.7}$$

Proof Take $0 = a_0 < \cdots < a_N = 1$, $a_k = k\delta$, $k = 0, \ldots, N-1$, and let $B_1 = [a_0, a_1]$ and $B_k = (a_{k-1}, a_k]$, $k = 2, \ldots, N$. For each $g \in \mathcal{G}$ define

$$\tilde{g} = \sum_{k=1}^{N} \delta \left\lfloor \frac{g(a_k)}{\delta} \right\rfloor 1_{B_k}.$$

Thus $\tilde{g}$ is constant on the intervals B_k and can only take values in the set δM with M integer. One easily sees that $|\tilde{g} - g|_\infty \leq 2\delta$.

Now, we have to count the number of $\tilde{g}$ thus obtained. There are at most $\lfloor 1/\delta \rfloor + 1$ choices for $\tilde{g}(a_1)$. We have

$$\begin{aligned}|\tilde{g}(a_k) - \tilde{g}(a_{k-1})| &\leq |\tilde{g}(a_k) - g(a_k)| + |g(a_k) - g(a_{k-1})| + |g(a_{k-1}) - \tilde{g}(a_{k-1})| \\ &\leq 3\delta.\end{aligned}$$

Therefore once a choice is made for $\tilde{g}(a_{k-1})$ there are at most seven choices left for the next value $\tilde{g}(a_k)$, $k = 2, \ldots, N$. So

$$N_\infty(2\delta, \mathcal{G}) \leq \left(\left\lfloor \frac{1}{\delta} \right\rfloor + 1 \right) 7^{\lfloor 1/\delta \rfloor}.$$

□

The result can be extended to classes of functions having m derivatives. Let $g^{(m)}$ denote the mth derivative of the function g. Consider the Sobolev class $\mathcal{G} = \{g : [0,1] \to [0,1] : \int_0^1 |g^{(m)}(x)|^2 \, dx \leq 1\}$. This class is larger than the class of functions with mth derivative bounded by 1 in absolute value. The following theorem can be found in Birman and Solomjak (1967). They also derive the extension to higher dimensions.

Theorem 2.4 *For*

$$\mathcal{G} = \left\{ g : [0,1] \to [0,1] : \int \left(g^{(m)}(x)\right)^2 dx \leq 1 \right\}, \tag{2.8}$$

we have

$$H_\infty(\delta, \mathcal{G}) \leq A\delta^{-1/m}, \quad \textit{for all } \delta > 0. \tag{2.9}$$

Suppose now that $\mathcal{G}$ is a subset of a finite-dimensional space. In that case, the entropy of $\mathcal{G}$ is roughly speaking the dimension of $\mathcal{G}$. To see this, let us first count the number of balls with radius δ, necessary to cover a ball with radius R in Euclidean space.

Lemma 2.5 *A ball $B_d(R)$ in $\mathbf{R}^d$ with Euclidean metric can be covered by*

$$\left(\frac{4R+\delta}{\delta}\right)^d$$

balls with radius δ.

Proof Let $\{c_j\}_{j=1}^N \subset B_d(R)$ be a largest set such that two distinct points c_{j_1} and c_{j_2} are at least δ apart. Then the balls with radius δ and centers c_j, $j = 1, \ldots, N$, cover $B_d(R)$. Let B_j be the ball with radius $\delta/4$ and center c_j, $j = 1, \ldots, N$. Then $B_1, \ldots, B_N$ are disjoint, and

$$\bigcup_{j=1}^N B_j \subset B_d(R+\delta/4).$$

The volume of a ball with radius ρ is $C_d\rho^d$, where the constant C_d is not of importance here, but we give it anyway: $C_d = 2/(d\Gamma(d/2))\pi^{d/2}$. So we must have

$$NC_d(\delta/4)^d \leq C_d(R+\delta/4)^d,$$

or

$$N \leq \left(\frac{R+\delta/4}{\delta/4}\right)^d. \qquad \square$$

Corollary 2.6 *Take functions $\psi_1, \ldots, \psi_d$ in $L_2(Q)$ and*

$$\mathcal{G} = \left\{ g = \sum_{k=1}^d \theta_k \psi_k \,:\, \theta = (\theta_1, \ldots, \theta_d)^T \in \mathbf{R}^d,\ \|g\|_Q \leq R \right\}.$$

Then

$$H(\delta, \mathcal{G}, Q) \leq d \log\left(\frac{4R+\delta}{\delta}\right), \quad \textit{for all } \delta > 0. \tag{2.10}$$

Clearly, if the dimension d_Q of the linear space spanned by $\psi_1, \ldots, \psi_d$ is less than d, one may replace d by d_Q in (2.10). The dimension d_Q is defined as follows. Take $\psi(x) = (\psi_1(x), \ldots, \psi_d(x))^T$, and let Σ_Q be the $d \times d$ matrix

$$\Sigma_Q = \int \psi\psi^T \, dQ.$$

Then d_Q is the rank of Σ_Q.

If $\mathcal{G}$ is indexed by a finite-dimensional parameter $\theta \in \Theta \subset \mathbf{R}^d$, then often (but not always!), one can prove a similar bound to (2.10). A bound like (2.10) also holds for so-called Vapnik–Chervonenkis subgraph classes of index d, although in general d then also appears in the log-term (see Subsection 3.7.4).

2.5. Notes

Since the ULLN involves taking the supremum over a possibly uncountable set, there may be some measurability problems in (2.1) and (2.2). The same is true for (2.3). We shall completely ignore these problems. In the book of van der Vaart and Wellner (1996), the issue is handled very elegantly, using the notion of outer probability. We shall briefly sketch this approach in Section 6.4.

There are many interesting papers on approximation in function spaces. We mention the paper of Kolmogorov and Tikhomirov (1959). They consider function spaces endowed with the supremum norm, and compute the entropy and capacity of for example, the class of differentiable functions. The δ-capacity of a set is essentially the maximum number of balls with radius δ that it can contain (see also the proof of Lemma 2.5). Clements (1963) treats for example the collection of functions of bounded variation, equipped with the L_1-norm. Birman and Solomjak (1967) derive piecewise polynomial approximations of functions on $[0, 1]^r$, in the class W_p^α. For the case $r = 1$, $p = 2$ and $\alpha = m$, this is the class given in (2.8). The class of functions of bounded variation is roughly speaking W_1^1. The results of Birman and Solomjak are very general: they consider functions on higher-dimensional spaces, with smoothness given in general L_p-norms, they allow for non-integer α and they compute entropies with respect to general L_q-norms. In Birgé and Massart (1996), entropy calculations for the even more general Besov spaces are given.

In the literature on approximation theory, it is not so common to examine bounds for the entropy with bracketing. One can either use their entropy bounds for the supremum norm, or go into their proofs, and check whether the bounds in fact hold for the entropy with bracketing. The latter strategy is adopted in van de Geer (1991).

More entropy results can be found in Sections 3.7, 7.4, 9.3, 10.1, and 10.3. We shall not attempt to provide good or even explicit constants in our entropy bounds. For example, the one given in Lemma 2.5 can certainly be improved.

2.6. Problems and complements

2.1. Prove Lemma 2.1.

2.2. Verify that the class $\mathcal{G} = \{g : \mathbf{R} \to [0,1],\ g \text{ increasing}\}$ is not totally bounded for the supremum norm on $\mathbf{R}$.

2.3. Let $\mathcal{G}$ be the class of unimodal functions $g : \mathbf{R} \to [0,1]$. Use result (2.5) of Birman and Solomjak to show that for any probability measure Q,

$$H_B(\delta, \mathcal{G}, Q) \le A\frac{1}{\delta}, \quad \text{for all } \delta > 0.$$

2.4. Let $\mathcal{G} = \{g : [0,1] \to [0,1],\ \int_0^1 |g'(x)|^2\,dx \le 1\}$, and let μ be Lebesgue measure. As a first attempt to reproduce the result (2.9) of Birman and Solomjak (with $m = 1$), show that

$$H(\delta, \mathcal{G}, \mu) \le A\frac{1}{\delta}\log\left(\frac{1}{\delta}\right), \quad \text{for all } \delta > 0.$$

(Hint: Take $0 = a_0 < a_1 < \cdots < a_N = 1$, with $a_k = k\delta$, $k = 0, 1, \ldots, N-1$, and $N \le \lfloor 1/\delta \rfloor + 1$. For $g \in \mathcal{G}$ define

$$\bar{g}_k = \frac{\int_{a_{k-1}}^{a_k} g(x)\,dx}{a_k - a_{k-1}},$$

and

$$\tilde{g} = \sum_{k=1}^{N} \bar{g}_k 1_{B_k},$$

where $B_1 = [a_0, a_1]$ and $B_k = (a_{k-1}, a_k]$, $k = 2, \ldots, N$. Prove that for $x \in B_k$,

$$|g(x) - \bar{g}_k|^2 \le \delta^2 \frac{\int_{a_{k-1}}^{a_k} |g'(u)|^2\,du}{a_k - a_{k-1}},$$

so that

$$\int_0^1 |g(x) - \tilde{g}(x)|^2\,dx \le \delta^2.)$$

2.5. Consider for $m \in \{1, 2, \ldots\}$, the class

$$\mathcal{G}^{m,\nu}_{K,L,M} = \begin{cases} \{g : [0,L] \to [0,K],\ \int_0^L |g^{(m)}(x)|^\nu\,dx \le M^\nu\}, & 1 \le \nu < \infty, \\ \{g : [0,L] \to [0,K],\ \sup_{0\le x\le L} |g^{(m)}(x)| \le M\}, & \nu = \infty. \end{cases}$$

(Note that $\mathcal{G}^{m,\infty}_{K,L,M} \subset \mathcal{G}^{m,\nu}_{K,L,\tilde{M}}$ for all $1 \le \nu < \infty$, where $\tilde{M} = L^{\frac{1}{\nu}} M$.) From (2.5), it follows that for $m\nu = 1$, and for any probability measure Q on $[0, \infty)$,

$$H_B(\delta, \mathcal{G}^{m,\nu}_{1,\infty,1}, Q) \le A\delta^{-1/m}, \quad \text{for all } \delta > 0.$$

We shall extend this to the case $m\nu > 1$ in Problem 2.6. Here, we make the necessary preparations.

Birman and Solomjak (1967) show that for $m\nu > 1$, and for some constant A_1 depending only on m and ν,

$$H_\infty(\delta, \mathcal{G}^{m,\nu}_{1,1,1}) \le A_1\delta^{-1/m}, \quad \text{for all } \delta > 0.$$

Show that this implies for $m\nu > 1$ and $L^{m-(1/\nu)}M \ge K$,

$$H_\infty(\delta, \mathcal{G}^{m,\nu}_{K,L,M}) \le A_1 L^{1-(1/m\nu)} M^{1/m} \delta^{-1/m}, \quad \text{for all } \delta > 0.$$

(Hint: use $\mathcal{G}^{m,\nu}_{K,L,M} = L^{m-(1/\nu)} M \mathcal{G}^{m,\nu}_{\tilde{K},1,1}$, with $\tilde{K} = K/(L^{m-(1/\nu)}M)$.)

2.6. (Smooth functions with unbounded support, see also van der Vaart and Wellner (1996, Corollary 2.7.4).) Let $m\nu > 1$, and

$$\mathcal{G}^{m,\nu}_{1,\infty,1} = \begin{cases} \{g : [0,\infty) \to [0,1],\ \int_0^\infty |g^{(m)}(x)|^\nu \, dx \le 1\}, & 1 \le \nu < \infty, \\ \{g : [0,\infty) \to [0,1],\ \sup_{0 \le x < \infty} |g^{(m)}(x)| \le 1\}, & \nu = \infty. \end{cases}$$

Moreover, let Q be a probability measure on $[0, \infty)$, which satisfies the following tail-condition: there exists a constant $C_Q < \infty$, such that for any $T < \infty$, there exists a partition $B_1, \ldots, B_N$ of $(0, T]$, with $B_j = (x_{j-1}, x_j]$, $0 = x_0 < \cdots < x_N = T$, such that $L_j = x_j - x_{j-1} \ge 1$, $j = 1, \ldots, N$, and such that

$$\sum_{j=1}^N L_j^{2(m\nu-1)/\nu(2m+1)} Q^{1/(2m+1)}(B_j) \le C_Q.$$

Then, for a constant A_2, depending on m, ν, and C_Q,

$$H_B(\delta, \mathcal{G}^{m,\nu}_{1,\infty,1}, Q) \le A_2\delta^{-1/m}, \quad \text{for all } \delta > 0.$$

To see this, let $\delta > 0$ be arbitrary, and take $T < \infty$, in such a way that $Q(T, \infty) \le \delta^2$. Let $B_1, \ldots, B_N$ be a partition of $(0, T]$, such that the above tail-condition holds. Without loss of generality, we may assume $Q(B_j) > 0$ for all $j = 1, \ldots, N$. Define

$$\delta_j = \frac{\delta L_j^{(m\nu-1)/\nu(2m+1)} Q^{-m/(2m+1)}(B_j)}{C_Q^{1/2}}, \quad j = 1, \ldots, N.$$

Then

$$\sum_{j=1}^{N} \delta_j^2 Q(B_j) \leq \delta^2,$$

which implies that

$$H_B\left(\sqrt{2}\delta, \mathcal{G}_{1,\infty,1}^{m,\nu}, Q\right) \leq \sum_{j=1}^{N} H_\infty(\delta_j, \mathcal{G}_j),$$

where $\mathcal{G}_j = \{g \mathrm{l}_{B_j} : g \in \mathcal{G}_{1,\infty,1}^{m,\nu}\}$, $j = 1, \ldots, N$. Now, insert the result of Problem 2.5, to find that

$$\sum_{j=1}^{N} H_\infty(\delta_j, \mathcal{G}_j) \leq A_1 \sum_{j=1}^{N} L_j^{1-1/m\nu} \delta_j^{-1/m} \leq A_1 C_Q^{1/2m} \delta^{-1/m}.$$

Let us now have a closer look at the above tail-condition on Q. Suppose that Q has density q with respect to Lebesgue measure, and that the Riemann integral $\int_0^\infty q^{1/(2m+1)}(x)\, dx$ is finite. Then there is a C_Q such that for all T, there is a partition $\{B_j = (x_{j-1}, x_j]\}_{j=1}^N$ such that for $L_j = x_j - x_{j-1}$, $j = 1, \ldots, N$,

$$\sum_{j=1}^{N} L_j^{2m/(2m+1)} Q^{1/(2m+1)}(B_j) \leq C_Q.$$

2.7. We say that $\mathcal{G}$ is of finite metric dimension at $g_0 \in \mathcal{G}$, for the $\|\cdot\|_Q$-metric, if for some constants c and d, and for all $\delta > 0$ and $R > 0$,

$$N\left(\delta, \mathcal{G} \cap \{\, \|g - g_0\|_Q \leq R \,\}, Q\right) \leq c \left(\frac{R}{\delta}\right)^d.$$

Suppose now that

$$\mathcal{G} = \{g_\theta : \theta \in \Theta\},$$

where Θ is a rectangle in $\mathbf{R}^d$, and where g_θ is differentiable in θ, with derivative $\dot{g}_\theta$. Define

$$\underline{G}(x) = \inf_{\theta \in \Theta} |\dot{g}_\theta(x)|,$$

and

$$\overline{G}(x) = \sup_{\theta \in \Theta} |\dot{g}_\theta(x)|.$$

Show that if $0 < \int \underline{G}^2\, dQ \leq \int \overline{G}^2\, dQ < \infty$, then $\mathcal{G}$ is of finite metric dimension at all $g_0 \in \mathcal{G}$.

3

Uniform Laws of Large Numbers

If the entropy with bracketing of $\mathcal{G} \subset L_1(P)$ is finite, then it is relatively easy to prove a uniform law of large numbers for $\mathcal{G}$. It is much harder to show that one also has a uniform law of large numbers under appropriate conditions on the random entropy $H_1(\delta, \mathcal{G}, P_n)$. For this purpose, we first prove a maximal inequality for weighted sums. Then we symmetrize the empirical process, work conditionally on $X_1, \ldots, X_n$, and apply Hoeffding's inequality.

Let $X_1, X_2, \ldots$ be i.i.d. with distribution P on $(\mathcal{X}, \mathcal{A})$, and let $\mathcal{G}$ be a class of functions on $\mathcal{X}$. We say that $\mathcal{G}$ satisfies the ULLN (Uniform Law of Large Numbers) if

$$\sup_{g \in \mathcal{G}} \left| \int g \, d(P_n - P) \right| \to 0, \quad \text{a.s.}$$

3.1. Uniform laws of large numbers under finite entropy with bracketing

Lemma 3.1 *Suppose that*

$$H_{1,B}(\delta, \mathcal{G}, P) < \infty, \quad \text{for all } \delta > 0. \tag{3.1}$$

Then $\mathcal{G}$ satisfies the ULLN.

Proof Let $\delta > 0$ be arbitrary, and let $\{[g_j^L, g_j^U]\}_{j=1}^N$ be a δ-bracketing set for $\mathcal{G}$, i.e., $\|g_j^U - g_j^L\|_{1,P} \leq \delta$, $j = 1, \ldots, N$, and each $g \in \mathcal{G}$ can be squeezed between a pair $[g_j^L, g_j^U]$. Then

$$\begin{aligned}\int g\, d(P_n - P) &= \int g\, dP_n - \int g\, dP \\ &\le \int g_j^U\, dP_n - \int g\, dP = \int g_j^U\, d(P_n - P) + \int (g_j^U - g)\, dP \\ &\le \int g_j^U\, d(P_n - P) + \delta.\end{aligned}$$

Similarly,

$$\int g\, d(P_n - P) \ge \int g_j^L\, d(P_n - P) - \int (g - g_j^L)\, dP \ge \int g_j^L\, d(P_n - P) - \delta.$$

Since $\{[g_j^L, g_j^U]\}_{j=1}^N$ is finite, we have that eventually

$$\max_{j=1,\ldots,N} \left| \int g_j^U\, d(P_n - P) \right| \le \delta, \quad \text{a.s.,}$$

as well as

$$\max_{j=1,\ldots,N} \left| \int g_j^L\, d(P_n - P) \right| \le \delta, \quad \text{a.s.}$$

So eventually,

$$\sup_{g \in \mathcal{G}} \left| \int g\, d(P_n - P) \right| \le 2\delta, \quad \text{a.s.} \qquad \square$$

Definition 3.1 The function

$$G = \sup_{g \in \mathcal{G}} |g|$$

is called the *envelope* of $\mathcal{G}$.

Condition (3.1) implies $G \in L_1(P)$, which we shall call the *envelope condition*. To see this, notice first of all that (3.1) implies that $\mathcal{G}$ is totally bounded. So $\mathcal{G}$ is a bounded subset of $L_1(P)$, say

$$\sup_{g \in \mathcal{G}} \|g\|_{1,P} \le R.$$

Let $\{[g_j^L, g_j^U]\}_{j=1}^N$ be a finite δ-bracketing set for $\mathcal{G}$. Take for $g \in \mathcal{G}$, $g_j^L \le g \le g_j^U$, $\|g_j^U - g_j^L\|_{1,P} \le \delta$. Then

$$|g| \le |g_j^L| + |g_j^U - g_j^L| \le \sum_{j=1}^N \left(|g_j^L| + |g_j^U - g_j^L|\right),$$

so that

$$\|G\|_{1,P} \le \sum_{j=1}^{N} \left(\|g_j^L\|_{1,P} + \|g_j^U - g_j^L\|_{1,P}\right) \le N(R+2\delta).$$

The envelope condition $G \in L_1(P)$ is rather severe, but whenever $\mathcal{G}$ is a bounded subset of $L_1(P)$, it is unavoidable if we want to prove a ULLN for $\mathcal{G}$. Summarizing: if $\sup_{g\in\mathcal{G}} \|g\|_{1,P} < \infty$ and $\mathcal{G}$ satisfies the ULLN, then $\| \sup_{g\in\mathcal{G}} |g| \|_{1,P} < \infty$. (A simple example shows that if $\mathcal{G}$ is not a bounded subset of $L_1(P)$, then the ULLN for $\mathcal{G}$ need not imply $G \in L_1(P)$: take $\mathcal{G}$ as the class of all constant functions.)

3.2. The chaining technique

Our aim is now to derive conditions for the ULLN that are weaker than condition (3.1) on the entropy with bracketing. We shall show that $\mathcal{G}$ satisfies the ULLN if the envelope condition $G \in L_1(P)$ holds, and

$$\frac{1}{n} H_1(\delta, \mathcal{G}, P_n) \to_{\mathbf{P}} 0, \quad \text{for all } \delta > 0$$

(see Theorem 3.7). The latter condition is clearly implied by (3.1).

If $\mathcal{G}$ is a finite class, then it is of course no problem to get results uniformly in $\mathcal{G}$. Now, a totally bounded class can be approximated by a finite class. The chaining technique is to apply finer and finer approximations. This works as follows. Suppose $\mathcal{G} \subset L_2(Q)$, and

$$\sup_{g\in\mathcal{G}} \|g\|_Q \le R. \tag{3.2}$$

For notational convenience, we index the functions in $\mathcal{G}$ by a parameter $\theta \in \Theta$: $\mathcal{G} = \{g_\theta : \theta \in \Theta\}$. For $s = 0, 1, 2, \ldots$, let $\{g_j^s\}_{j=1}^{N_s}$ be a minimal $2^{-s}R$-covering set of $(\mathcal{G}, \|\cdot\|_Q)$. So $N_s = N(2^{-s}R, \mathcal{G}, Q)$, and for each θ, there exists a $g_\theta^s \in \{g_1^s, \cdots, g_{N_s}^s\}$ such that $\|g_\theta - g_\theta^s\|_Q \le 2^{-s}R$. We use the parameter θ here to indicate which function in the covering set approximates a particular g. We may choose $g_\theta^0 \equiv 0$, since $\|g_\theta\|_Q \le R$. Then for any S,

$$g_\theta = \sum_{s=1}^{S} (g_\theta^s - g_\theta^{s-1}) + (g_\theta - g_\theta^S).$$

One can think of this as telescoping from g_θ to g_θ^S, i.e. we follow a path taking smaller and smaller steps. Take S sufficiently large, in such a way that $(g - g_\theta^S)$ is small enough for the purpose one has in mind. The term $\sum_{s=1}^{S} (g_\theta^s - g_\theta^{s-1})$ can be handled by exploiting the fact that as θ varies, it involves only finitely many functions.

3.3. A maximal inequality for weighted sums

In this subsection, we make an excursion to a non-i.i.d. case. As before, we let $\mathscr{G} = \{g_\theta : \theta \in \Theta\}$ be a class of functions on $\mathscr{X}$. Moreover, $\xi_1, \ldots, \xi_n$ is a set of points in $\mathscr{X}$, and Q_n is the probability measure that puts mass $(1/n)$ on $\xi_1, \ldots, \xi_n$, i.e.,

$$Q_n = \frac{1}{n} \sum_{i=1}^{n} \delta_{\xi_i}.$$

(Later on, Q_n will be the empirical measure P_n, and we shall work conditionally on the event $(X_1, \ldots, X_n) = (\xi_1, \ldots, \xi_n)$.)

Consider a random vector $W \in \mathbf{R}^n$. We assume an exponential probability inequality (in fact, a sub-Gaussian inequality) for weighted sums of the form $\sum_{i=1}^{n} W_i \gamma_i$. In the next subsection, we shall see under what conditions such a probability inequality indeed holds. The exponential probability inequality ensures that it is the logarithm of the number of functions in a covering set that governs the behaviour of empirical processes: simply note that if $\{Z_j : 1 \le j \le N\}$ satisfies for each j:

$$\mathbf{P}(Z_j \ge a) \le \exp[-a^2],$$

then

$$\mathbf{P}\left(\max_{j=1,\ldots,N} Z_j \ge a \right) \le \exp[\log N - a^2].$$

Thus, we need to take a larger than the *square root* of $\log N$, in order to have the right-hand side small.

Lemma 3.2 below states a maximal inequality for weighted sums. The proof applies the chaining technique. Our first application will be the derivation of ULLNs (see Section 3.6). For that particular application, we certainly do not use Lemma 3.2 in its full strength. A complete exploitation of the lemma will occur later on, for example when we prove asymptotic equicontinuity of the empirical process (see Theorem 5.3), or when we derive rates of convergence for least squares estimators (see Theorem 9.1). We remark moreover that Lemma 3.2 contains the main ideas for proving maximal inequalities. For example, the proof of Theorem 8.13, which concerns a maximal inequality for martingales, primarily uses the same approach, but the technical details needed there make it less transparent.

Before presenting the lemma, let us pay some attention to the entropy integral in (3.4). Due to the chaining technique, we are confronted with 2^{-s}-covering sets, for $s = 1, \ldots, S$. The total size of these covering sets should not be too large. It turns out that we need to control the weighted sum

$$\sum_{s=1}^{S} 2^{-s} R H^{1/2}(2^{-s}R, \mathscr{G}, Q_n).$$

(Indeed, this expression involves the square root of the logarithm of a number of functions.) We can replace this sum by an integral, provided such an integral is well-defined. Therefore, we assume in what follows that $H(\delta, \mathscr{G}, Q_n)$ is a continuous function of $\delta > 0$. If this is not the case, replace $H(\delta, \mathscr{G}, Q_n)$ by its smallest continuous majorant. It is easy to see that for $R > \rho$ and $S = \min\{s \geq 1 : 2^{-s}R \leq \rho\}$,

$$\sum_{s=1}^{S} 2^{-s} R H^{1/2}(2^{-s}R, \mathscr{G}, Q_n) \leq 2 \int_{\rho/4}^{R} H^{1/2}(u, \mathscr{G}, Q_n)\, du.$$

The right-hand side of the above inequality is somewhat more elegant when stating the lemma, although it is the finite sum on the left-hand side that actually plays a role.

We use the notation $a \vee b = \max\{a, b\}$, and $a \wedge b = \min\{a, b\}$.

Lemma 3.2 *Suppose that for some constants C_1 and C_2, we have that for each $\gamma \in \mathbf{R}^n$ and all $a > 0$,*

$$\mathbf{P}\left(\left| \sum_{i=1}^{n} W_i \gamma_i \right| \geq a \right) \leq C_1 \exp\left[- \frac{a^2}{C_2^2 \sum_{i=1}^{n} \gamma_i^2} \right]. \tag{3.3}$$

Assume $\sup_{g \in \mathscr{G}} \|g\|_{Q_n} \leq R$. There exists a constant C depending only on C_1 and C_2 such that for all $0 \leq \epsilon < \delta$, and $K > 0$ satisfying

$$\sqrt{n}(\delta - \epsilon) \geq C \left(\int_{\epsilon/4K}^{R} H^{1/2}(u, \mathscr{G}, Q_n)\, du \vee R \right), \tag{3.4}$$

we have

$$\mathbf{P}\left(\sup_{g \in \mathscr{G}} \left| \frac{1}{n} \sum_{i=1}^{n} W_i g(\xi_i) \right| \geq \delta \wedge \frac{1}{n} \sum_{i=1}^{n} W_i^2 \leq K^2 \right) \leq C \exp\left[- \frac{n(\delta - \epsilon)^2}{C^2 R^2} \right]. \tag{3.5}$$

Proof For once, we shall provide explicit constants. In (3.4), take C sufficiently large, such that

$$\sqrt{n}(\delta - \epsilon) \geq 12 C_2 \sum_{s=1}^{S} 2^{-s} R H^{1/2}(2^{-s}R, \mathscr{G}, Q_n) \vee (1152 \log 2)^{1/2} C_2 R, \tag{3.6}$$

Let $\{g_j^s\}_{j=1}^{N_s}$ be a minimal $2^{-s}R$-covering set of $\mathcal{G}$, $s = 0, 1, \ldots$. So $N_s = N(2^{-s}R, \mathcal{G}, Q_n)$. On the set $\{(1/n)\sum_{i=1}^n W_i^2 \le K^2\}$, one can apply the Cauchy–Schwarz inequality to obtain that for $S = \min\{s \ge 1 : 2^{-s}R \le \epsilon/K\}$,

$$\left|\frac{1}{n}\sum_{i=1}^n W_i(g_\theta(\xi_i) - g_\theta^S(\xi_i))\right| \le K\|g_\theta - g_\theta^S\|_{Q_n} \le \epsilon.$$

This shows that it suffices to prove the exponential inequality for

$$\mathbf{P}\left(\max_{j=1,\ldots,N_S}\left|\frac{1}{n}\sum_{i=1}^n W_i g_j^S(\xi_i)\right| \ge \delta - \epsilon\right).$$

Now use chaining. Write $g_\theta^S = \sum_{s=1}^S (g_\theta^s - g_\theta^{s-1})$. Note that by the triangle inequality,

$$\|g_\theta^s - g_\theta^{s-1}\|_{Q_n} \le \|g_\theta^s - g_\theta\|_{Q_n} + \|g_\theta - g_\theta^{s-1}\|_{Q_n} \le 2^{-s}R + 2^{-s+1}R$$
$$= 3(2^{-s}R).$$

Let η_s be positive numbers satisfying $\sum_{s=1}^S \eta_s \le 1$. Then

$$\mathbf{P}\left(\sup_{\theta\in\Theta}\left|\frac{1}{n}\sum_{s=1}^S\sum_{i=1}^n W_i(g_\theta^s(\xi_i) - g_\theta^{s-1}(\xi_i))\right| \ge \delta - \epsilon\right)$$
$$\le \sum_{s=1}^S \mathbf{P}\left(\sup_{\theta\in\Theta}\left|\frac{1}{n}\sum_{i=1}^n W_i(g_\theta^s(\xi_i) - g_\theta^{s-1}(\xi_i))\right| \ge (\delta - \epsilon)\eta_s\right)$$
$$\le \sum_{s=1}^S C_1 \exp\left[2H(2^{-s}R, \mathcal{G}, Q_n) - \frac{n(\delta-\epsilon)^2\eta_s^2}{9C_2^2 2^{-2s}R^2}\right].$$

What is a good choice for η_s? We take

$$\eta_s = \frac{6C_2 2^{-s}RH^{1/2}(2^{-s}R, \mathcal{G}, Q_n)}{\sqrt{n}(\delta - \epsilon)} \vee \frac{2^{-s}\sqrt{s}}{8}.$$

Then indeed, by (3.6)

$$\sum_{s=1}^S \eta_s \le \sum_{s=1}^S \frac{6C_2 2^{-s}RH^{1/2}(2^{-s}R, \mathcal{G}, Q_n)}{\sqrt{n}(\delta - \epsilon)} + \sum_{s=1}^S \frac{2^{-s}\sqrt{s}}{8} \le \frac{1}{2} + \frac{1}{2} = 1.$$

Here, we used the bound

$$\sum_{s=1}^S 2^{-s}\sqrt{s} \le 1 + \int_1^\infty 2^{-x}\sqrt{x}\,dx$$
$$\le 1 + \int_0^\infty 2^{-x}\sqrt{x}\,dx = 1 + (\pi/\log 2)^{1/2}$$
$$\le 4.$$

Observe that

$$\eta_s \geq \frac{6C_2 2^{-s} R H^{1/2}(2^{-s}R, \mathcal{G}, Q_n)}{\sqrt{n}(\delta - \epsilon)},$$

so that

$$H(2^{-s}R, \mathcal{G}, Q_n) \leq \frac{n(\delta-\epsilon)^2 \eta_s^2}{36 C_2^2 2^{-2s} R^2}.$$

Thus,

$$\sum_{s=1}^{S} C_1 \exp\left[2H(2^{-s}R, \mathcal{G}, Q_n) - \frac{n(\delta-\epsilon)^2\eta_s^2}{9C_2^2 2^{-2s}R^2}\right] \leq \sum_{s=1}^{S} C_2 \exp\left[-\frac{n(\delta-\epsilon)^2\eta_s^2}{18C_2^2 2^{-2s}R^2}\right].$$

Next, invoke that $\eta_s \geq 2^{-s}\sqrt{s}/8$:

$$\begin{aligned}
&\sum_{s=1}^{S} C_1 \exp\left[-\frac{n(\delta-\epsilon)^2\eta_s^2}{18C_2^2 2^{-2s}R^2}\right] \\
&\quad \leq \sum_{s=1}^{S} C_1 \exp\left[-\frac{n(\delta-\epsilon)^2 s}{1152 C_2^2 R^2}\right] \leq \sum_{s=1}^{\infty} C_1 \exp\left[-\frac{n(\delta-\epsilon)^2 s}{1152 C_2^2 R^2}\right] \\
&\quad = C_1 \left(1 - \exp\left[-\frac{n(\delta-\epsilon)^2}{1152C_2^2R^2}\right]\right)^{-1} \exp\left[-\frac{n(\delta-\epsilon)^2}{1152C_2^2R^2}\right] \\
&\quad \leq 2C_1 \exp\left[-\frac{n(\delta-\epsilon)^2}{1152C_2^2R^2}\right],
\end{aligned}$$

where in the last inequality, we used the assumption, included in (3.6), that

$$\frac{n(\delta-\epsilon)^2}{1152C_2^2R^2} \geq \log 2.$$

Thus, we have shown that

$$\text{(3.7)} \qquad \mathbf{P}\left(\sup_{g\in\mathcal{G}}\left|\frac{1}{n}\sum_{i=1}^{n} W_i g(\xi_i)\right| \geq \delta \wedge \frac{1}{n}\sum_{i=1}^{n} W_i^2 \leq K^2\right) \leq 2C_1 \exp\left[-\frac{n(\delta-\epsilon)^2}{1152C_2^2R^2}\right]. \qquad \square$$

3.4. Symmetrization

To fit the maximal inequality for weighted sums into our i.i.d. framework, a symmetrization device is needed. Let $X_1', \ldots, X_n'$ be i.i.d. with distribution P, independent of $X_1, \ldots, X_n$, and let P_n' be the empirical distribution based on $X_1', \ldots, X_n'$. For each $g \in \mathcal{G}$, $|\int g\, d(P_n - P)|$ is small by the ordinary law of large numbers. Assuming for a moment that the supremum is attained, write

$$\sup_{g \in \mathcal{G}} \left| \int g\, d(P_n - P) \right| = \left| \int g^*\, d(P_n - P) \right|.$$

Roughly speaking, because g^* is independent of P_n', $|\int g^*\, d(P_n' - P)|$ is small by the ordinary law of large numbers. So if $|\int g^*\, d(P_n - P)|$ is large, then so is the difference $|\int g^*\, d(P_n - P_n')|$.

Symmetrization Lemma 3.3 *Suppose that for all* $g \in \mathcal{G}$,

$$\mathbf{P}\left(\left| \int g\, d(P_n - P) \right| > \delta/2 \right) \le \frac{1}{2}. \tag{3.8}$$

Then

$$\mathbf{P}\left(\sup_{g \in \mathcal{G}} \left| \int g\, d(P_n - P) \right| > \delta \right) \le 2\mathbf{P}\left(\sup_{g \in \mathcal{G}} \left| \int g\, d(P_n - P_n') \right| > \delta/2 \right). \tag{3.9}$$

Proof Let $\mathbf{X} = (X_1, \ldots, X_n)$, $A_g = \{\mathbf{X} : |\int g\, d(P_n - P)| > \delta\}$, and $A = \bigcup_{g \in \mathcal{G}} A_g$. Then $\mathbf{X} \in A$ implies $\mathbf{X} \in A_{g^*}$ for some random function $g^* = g^*(\mathbf{X}) \in \mathcal{G}$. Moreover,

$$\begin{aligned}
&\mathbf{P}\left(A_{g^*} \text{ and } \left| \int g^*\, d(P_n' - P) \right| \le \frac{\delta}{2} \right) \\
&\quad = \mathbf{E}_{\mathbf{X}} \mathbf{P}\left(\left| \int g^*\, d(P_n' - P) \right| \le \frac{\delta}{2} \,\middle|\, \mathbf{X} \right) 1\{A_{g^*}\} \ge \frac{1}{2} \mathbf{E}_{\mathbf{X}} 1\{A_{g^*}\} = \frac{1}{2}\mathbf{P}(A_{g^*}) \\
&\quad = \frac{1}{2}\mathbf{P}\left(\left| \int g^*\, d(P_n - P) \right| > \delta \right).
\end{aligned}$$

Therefore

$$\begin{aligned}
&\mathbf{P}\left(\sup_{g \in \mathcal{G}} \left| \int g\, d(P_n - P) \right| > \delta \right) \\
&\quad \le \mathbf{P}\left(\left| \int g^*\, d(P_n - P) \right| > \delta \right) \\
&\quad \le 2\mathbf{P}\left(\left| \int g^*\, d(P_n - P) \right| > \delta \text{ and } \left| \int g^*\, d(P_n' - P) \right| \le \delta/2 \right) \\
&\quad \le 2\mathbf{P}\left(\left| \int g^*\, d(P_n - P_n') \right| > \delta/2 \right) \\
&\quad \le 2\mathbf{P}\left(\sup_{g \in \mathcal{G}} \left| \int g\, d(P_n - P_n') \right| > \delta/2 \right). \qquad \square
\end{aligned}$$

We can now transform the empirical process into a symmetrized version. Let $W_1, \ldots, W_n$ be a Rademacher sequence, i.e.,

(3.10)

$$W_1, \ldots, W_n \text{ independent}, \quad \mathbf{P}(W_i = 1) = \mathbf{P}(W_i = -1) = \frac{1}{2}, \quad i = 1, \ldots, n.$$

Take $(W_1, \ldots, W_n)$ independent of $(X_1, \ldots, X_n, X_1', \ldots, X_n')$. Clearly, for each g and i, $g(X_i) - g(X_i')$ has the same distribution as $W_i\big(g(X_i) - g(X_i')\big)$. Similarly, one finds that

$$\left\{ \big(g(X_i) - g(X_i')\big) \; : \; i = 1, \ldots, n, \; g \in \mathscr{G} \right\}$$

has the same distribution as

$$\left\{ W_i\big(g(X_i) - g(X_i')\big) \; : \; i = 1, \ldots, n, \; g \in \mathscr{G} \right\}.$$

Moreover,

$$\mathbf{P}\left(\sup_{g \in \mathscr{G}} \left| \frac{1}{n} \sum_{i=1}^{n} W_i(g(X_i) - g(X_i')) \right| > \delta/2 \right) \leq 2\mathbf{P}\left(\sup_{g \in \mathscr{G}} \left| \frac{1}{n} \sum_{i=1}^{n} W_i g(X_i) \right| > \delta/4 \right).$$

So we arrive at the following corollary.

Corollary 3.4 *Suppose* $\sup_{g \in \mathscr{G}} \|g\| \leq R$. *Then for* $n \geq 8R^2/\delta^2$,

$$\text{(3.11)} \quad \mathbf{P}\left(\sup_{g \in \mathscr{G}} \left| \int g \, d(P_n - P) \right| > \delta \right) \leq 4\mathbf{P}\left(\sup_{g \in \mathscr{G}} \left| \frac{1}{n} \sum_{i=1}^{n} W_i g(X_i) \right| > \delta/4 \right),$$

where $(W_1, \ldots, W_n)$ *is a Rademacher sequence (see* (3.10)*), independent of* $(X_1, \ldots, X_n)$.

3.5. Hoeffding's inequality

There remains the investigation of the exponential probability inequality (3.3).

Lemma 3.5 *Let* $Z_1, \ldots, Z_n$ *be independent random variables with expectation zero. Suppose that* $b_i \leq Z_i \leq c_i$ *for some* $b_i < c_i$, $i = 1, \ldots, n$. *Then for all* $a > 0$,

$$\mathbf{P}\left(\sum_{i=1}^{n} Z_i \geq a \right) \leq \exp\left[-2 \frac{a^2}{\sum_{i=1}^{n} (c_i - b_i)^2} \right].$$

The proof can be found in Hoeffding (1963). We apply the inequality to $Z_i = W_i \gamma_i$, $i = 1, \ldots, n$, where $W_1, \ldots, W_n$ is a Rademacher sequence and $\gamma_1, \ldots, \gamma_n$ are constants. Then we may choose $c_i = -b_i = \gamma_i$, $i = 1, \ldots, n$.

3.6. Uniform laws of large numbers under random entropy conditions

We now have the necessary equipment for a uniform law of large numbers under conditions on the empirical entropy. We start out with a preliminary lemma, where the entropy is with respect to the $L_2(P_n)$-norm, and where one assumes that the functions in $\mathcal{G}$ are uniformly bounded.

Lemma 3.6 *Suppose* $\sup_{g\in\mathcal{G}} |g|_\infty \le R$, *and that*

$$\frac{1}{n}H(\delta,\mathcal{G},P_n) \to_{\mathbf{P}} 0, \quad \textit{for all } \delta > 0. \tag{3.12}$$

Then $\mathcal{G}$ *satisfies the ULLN.*

Proof Martingale arguments show that $\sup_{g\in\mathcal{G}} |\int g\, d(P_n - P)|$ converges a.s. to some constant (see Pollard (1984)). So we only have to prove that $\sup_{g\in\mathcal{G}} |\int g\, d(P_n - P)|$ converges to zero in probability.

Let $\delta > 0$ be arbitrary. By Corollary 3.3, for $n \ge 8R^2/\delta^2$,

$$\mathbf{P}\left(\sup_{g\in\mathcal{G}}\left|\int g\, d(P_n - P)\right| > \delta\right) \le 4\mathbf{P}\left(\sup_{g\in\mathcal{G}}\left|\frac{1}{n}\sum_{i=1}^n W_i g(X_i)\right| > \delta/4\right),$$

where $W_1,\ldots,W_n$ are i.i.d., independent of $X_1,\ldots,X_n$, with $\mathbf{P}(W_i = 1) = \mathbf{P}(W_i = -1) = 1/2$, $i = 1,\ldots,n$.

In view of Hoeffding's inequality, for each $\gamma \in \mathbf{R}^n$, and all $a > 0$,

$$\mathbf{P}\left(\left|\sum_{i=1}^n W_i\gamma_i\right| \ge a\right) \le 2\exp\left[-\frac{a^2}{2\sum_{i=1}^n \gamma_i^2}\right].$$

In Lemma 3.2, take $K = 1$, replace δ by $\delta/4$ and take $\epsilon = \delta/8$. Clearly,

$$\int_{\frac{\delta}{32}}^R H^{1/2}(x,\mathcal{G},P_n)\, dx \le RH^{1/2}\left(\frac{\delta}{32},\mathcal{G},P_n\right).$$

Apply Lemma 3.2 conditionally on $X_1,\ldots,X_n$, on the set

$$A_n = \left\{\sqrt{n}\frac{\delta}{8} \ge C\left(RH^{1/2}\left(\frac{\delta}{32},\mathcal{G},P_n\right) \vee R\right)\right\}.$$

Then we find

$$\begin{aligned}&\mathbf{P}\left(\sup_{g\in\mathcal{G}}\left|\frac{1}{n}\sum_{i=1}^n W_i g(X_i)\right| > \delta/4\right)\\ &\le C\exp\left[-\frac{n\delta^2}{64C^2R^2}\right] + \mathbf{P}(A_n^c) \to 0, \quad \text{as } n \to \infty. \qquad \square\end{aligned}$$

It is now easy to extend the result to a class of functions with envelope $G \in L_1(P)$. Recall that this envelope is defined as

$$G = \sup_{g \in \mathcal{G}} |g|.$$

Moreover, in condition (3.12) on the entropy, we replace the $L_2(P_n)$-norm by the $L_1(P_n)$-norm.

Theorem 3.7 *Assume the envelope condition $\int G\,dP < \infty$ and suppose that*

$$\frac{1}{n} H_1(\delta, \mathcal{G}, P_n) \to_{\mathbf{P}} 0, \quad \text{for all } \delta > 0. \tag{3.13}$$

Then $\mathcal{G}$ satisfies the ULLN.

Proof For each $R > 0$ let

$$\mathcal{G}_R = \{g\mathrm{l}\{G \le R\} : g \in \mathcal{G}\},$$

be the class of truncated functions. We have for g_1, g_2 in $\mathcal{G}$,

$$\int_{G \le R} (g_1 - g_2)^2\,dP_n \le 2R \int |g_1 - g_2|\,dP_n.$$

So (3.13) implies that

$$\frac{1}{n} H(\delta, \mathcal{G}_R, P_n) \to_{\mathbf{P}} 0, \quad \text{for all } \delta > 0,\ R > 0.$$

Therefore, by Lemma 3.6, the ULLN holds for each $\mathcal{G}_R$. Now, take $\delta > 0$ arbitrary. Take R sufficiently large, such that

$$\int_{G > R} G\,dP \le \delta.$$

Next, take n sufficiently large, such that

$$\int_{G > R} G\,dP_n \le 2\delta, \ \text{a.s.},$$

and such that

$$\sup_{g \in \mathcal{G}} \left| \int_{G \le R} g\,d(P_n - P) \right| \le \delta, \quad \text{a.s.}$$

Then

$$\begin{aligned} \sup_{g \in \mathcal{G}} \left| \int g\,d(P_n - P) \right| &\le \sup_{g \in \mathcal{G}} \left| \int_{G \le R} g\,d(P_n - P) \right| + \int_{G > R} G\,dP_n + \int_{G > R} G\,dP \\ &\le 4\delta, \quad \text{a.s.} \end{aligned}$$

□

Necessity of the conditions If $\mathcal{G}$ is a bounded subset of $L_1(P)$, then the envelope condition $G \in L_1(P)$ and the entropy condition (3.13) are also necessary for the ULLN to hold. This observation is especially important when one already knows that a particular $\mathcal{G}$ satisfies the ULLN, and one wants to check the ULLN for a class of transformations of functions in $\mathcal{G}$ (see Problem 3.1 for an example).

3.7. Examples

Theorem 3.7 gives us a tool for verifying whether a given class satisfies the uniform law of large numbers. Its main condition is an entropy condition with respect to the empirical metric

$$\|\cdot\|_{1,n} = \int |\cdot|\, dP_n.$$

Lemma 3.1 assumes that the entropy with bracketing is finite, which is a stronger condition then the one of Theorem 3.7. In the first two examples, we primarily use a slight extension of the entropy results of Section 2.4. We shall see that in the first three examples, the stronger condition of Lemma 3.1 is met. The fourth example is based on empirical entropy instead of entropy with bracketing. In the fifth example, we consider ULLNs for convex hulls.

Example 3.7.1. Monotone functions Let

$$\mathcal{G} = \{\tilde{g}G : \tilde{g} \in \tilde{\mathcal{G}}\}, \tag{3.14}$$

where $\tilde{\mathcal{G}}$ is a class of increasing functions on (a subinterval of) $\mathbf{R}$ satisfying

$$\sup_{\tilde{g}\in\tilde{\mathcal{G}}} |g|_\infty \le 1,$$

and where $G \ge 0$ is a fixed function.

Lemma 3.8 *For the class* $\mathcal{G}$ *defined in* (3.14), *we have for any* $1 \le p < \infty$, *and any probability measure* Q,

$$H_{p,B}(\delta, \mathcal{G}, Q) \le A\frac{\|G\|_{p,Q}}{\delta}, \quad \textit{for all } \delta > 0.$$

Hence, if $G \in L_1(P)$, $\mathcal{G}$ *satisfies the ULLN.*

Proof Result (2.5) tells us that for any probability measure $\tilde{Q}$,

$$H_{p,B}(\delta, \tilde{\mathcal{G}}, \tilde{Q}) \le A\frac{1}{\delta}, \quad \text{for all } \delta > 0.$$

Now,

$$\int |\tilde{g}^U G - \tilde{g}^L G|^p \, dQ = \|G\|_{p,Q}^p \int |\tilde{g}^U - \tilde{g}^L|^p \, d\tilde{Q},$$

where $d\tilde{Q} = G^p dQ / \|G\|_{p,Q}^p$. So

$$H_{p,B}(\delta \|G\|_{p,Q}, \mathcal{G}, Q) \leq A \frac{1}{\delta}, \quad \text{for all } \delta > 0. \qquad \square$$

Example 3.7.2. Smooth functions Let

$$\mathcal{G} = \left\{ g : [0,1] \to \mathbf{R} : \int \left(g^{(m)}(x) \right)^2 dx \leq 1, \|g\|_Q \leq R \right\}, \tag{3.15}$$

where $m \in \mathbf{N}$ is given. Moreover, let

$$\psi_k(x) = x^{k-1}, \; k = 1, \ldots, m; \qquad \psi = (\psi_1, \ldots, \psi_m)^T,$$

and

$$\Sigma_Q = \int \psi \psi^T \, dQ.$$

Lemma 3.9 *Suppose that Σ_Q is non-singular. Then for the class $\mathcal{G}$ defined in (3.15),*

$$H_\infty(\delta, \mathcal{G}) \leq A_1 \delta^{-1/m}, \quad \textit{for all } \delta > 0,$$

so that $\mathcal{G}$ satisfies the ULLN.

Proof The result follows from Theorem 2.4 if we can show that

$$\sup_{g \in \mathcal{G}} |g|_\infty \leq K_R$$

for some constant K_R. Consider the Taylor expansion of a function $g \in \mathcal{G}$:

$$\begin{aligned} g(x) &= g(0) + x g^{(1)}(0) + \cdots + \int_0^x \frac{(x-u)^{m-1}}{(m-1)!} g^{(m)}(u) \, du \\ &= g_1(x) + g_2(x), \end{aligned}$$

where $g_1(x) = \theta^T \psi(x)$ is a polynomial of degree $m - 1$, with coefficients

$$\theta_k = \frac{g^{(k-1)}(0)}{(k-1)!}, \quad k = 1, \ldots, m$$

and where

$$|g_2(x)| \leq \left(\int_0^1 (g^{(m)}(u))^2 \, du \right)^{1/2} \leq 1.$$

So,

$$\|g_1\|_Q \le \|g\|_Q + \|g_2\|_Q \le R + 1.$$

Let $\lambda^2_{\min}$ be the smallest eigenvalue of Σ_Q. It follows that

$$\theta^T\theta \le \frac{(R+1)^2}{\lambda^2_{\min}},$$

so that

$$|\theta_k| \le \frac{R+1}{\lambda_{\min}}, \quad k = 1, \dots, m.$$

But then

$$|g(x)| \le m\frac{R+1}{\lambda_{\min}} + 1 := K_R.$$

□

Example 3.7.3. Continuity in a parameter In this example, we show that the more or less classical ULLN is also covered by Lemma 3.1. Let

$$\mathcal{G} = \{g_\theta : \theta \in \Theta\},$$

where (Θ, τ) is a compact metric space, and where the map $\theta \mapsto g_\theta(x)$ is continuous for P-almost all x. Moreover, we assume that the envelope

$$G = \sup_{\theta\in\Theta} |g_\theta|$$

is in $L_1(P)$.

Lemma 3.10 *For the class $\mathcal{G} = \{g_\theta : \theta \in \Theta\}$ defined above, we have*

$$H_{1,B}(\delta, \mathcal{G}, P) < \infty, \quad \textit{for all } \delta > 0.$$

Hence, it satisfies the ULLN.

Proof Write

$$w(\theta, \rho)(x) = \sup_{\tau(\theta,\tilde\theta)<\rho} |g_\theta(x) - g_{\tilde\theta}(x)|, \quad \theta \in \Theta, \ \rho > 0.$$

Then $w(\theta, \rho)(x) \to 0$ as $\rho \to 0$, for P-almost all x. By dominated convergence,

$$\lim_{\rho\to 0} \int w(\theta, \rho)\, dP = 0.$$

Let $\delta > 0$ be arbitrary. Take, for each θ, ρ_θ in such a way that

$$\int w(\theta, \rho_\theta)\, dP \le \delta.$$

Let $B_\theta = \{\tilde{\theta} : \tau(\tilde{\theta}, \theta) < \rho_\theta\}$ and let $B_{\theta_1}, \ldots, B_{\theta_N}$ be a finite cover of Θ. Define $g_j^L = g_{\theta_j} - w(\theta_j, \rho_{\theta_j})$, and $g_j^U = g_{\theta_j} + w(\theta_j, \rho_{\theta_j})$, $j = 1, \ldots, N$. Then $0 \le \int (g_j^U - g_j^L)\, dP \le 2\delta$ and for $\theta \in B_{\theta_j}$, $g_j^L \le g_\theta \le g_j^U$. It follows that

$$H_{1,B}(2\delta, \mathscr{G}, P) \le \log N. \qquad \square$$

Example 3.7.4. Vapnik–Chervonenkis subgraph classes Of course the empirical entropy $H_1(\delta, \mathscr{G}, P_n)$ can always be bounded by the entropy for the supremum norm. Since P_n puts all its mass on the set of observations $X_1, \ldots, X_n$, it suffices to consider only an approximation of the values of a particular function g at this finite set. But it is a random set, which leads us to considering all subsets $\mathscr{X}_n = \{\xi_1, \ldots, \xi_n\}$ of $\mathscr{X}$. We start out with the case where $\mathscr{G}$ is a collection of indicator functions. For $\delta < 1$, the δ-entropy for the supremum-norm on the set $\mathscr{X}_n$, of a collection $\{1_D : D \in \mathscr{D}\}$, is simply the number of distinct sets in $\mathscr{D}$, if we identify two sets with no points in their symmetric difference. We denote this number by $\Delta^{\mathscr{D}}(\xi_1, \ldots, \xi_n)$.

Definition 3.2 Let $\mathscr{D}$ be a collection of subsets of $\mathscr{X}$. For $\xi_1, \ldots, \xi_n \in \mathscr{X}$, define

$$\Delta^{\mathscr{D}}(\xi_1, \ldots, \xi_n) = \operatorname{card}\{D \cap \{\xi_1, \ldots, \xi_n\} : D \in \mathscr{D}\},$$

that is, $\Delta^{\mathscr{D}}(\xi_1, \ldots, \xi_n)$ is the number of different subsets of the form $D \cap \{\xi_1, \ldots, \xi_n\}$, $D \in \mathscr{D}$. Define moreover

$$m^{\mathscr{D}}(n) = \sup\{\Delta^{\mathscr{D}}(\xi_1, \ldots, \xi_n) : \xi_1, \ldots, \xi_n \in \mathscr{X}\},$$

and

$$V(\mathscr{D}) = \inf\{n \ge 1 : m^{\mathscr{D}}(n) < 2^n\}.$$

We call $V(\mathscr{D})$ the index of the class $\mathscr{D}$. The collection $\mathscr{D}$ is a *Vapnik–Chervonenkis class* if $V(\mathscr{D}) < \infty$.

In other words, a Vapnik–Chervonenkis class $\mathscr{D}$ is a collection of sets such that, if the number of points n in a configuration is large enough, one cannot separate each of the possible 2^n subsets by sets in $\mathscr{D}$. For our purposes, the Vapnik–Chervonenkis property is of interest, but we shall not need the *exact* index $V(\mathscr{D})$. To check the Vapnik–Chervonenkis property, it is enough to show that $m^{\mathscr{D}}(n)$ grows only polynomially in n. In fact, it can be shown that the latter is also a necessary condition: if $m^{\mathscr{D}}(n) < 2^n$ from certain n on, then $m^{\mathscr{D}}(n)$ can grow only polynomially in n.

Example 3.7.4a *Let $\mathscr{D} = \{(-\infty, t], \ t \in \mathbf{R}\}$. Then $m^{\mathscr{D}}(n) = n + 1$, so that $\mathscr{D}$ is a Vapnik–Chervonenkis class.*

Also in higher dimensions, the class of half-intervals or half-spaces is a Vapnik–Chervonenkis class.

Example 3.7.4b Let $\mathscr{D} = \{(-\infty, t] : t \in \mathbf{R}^d\}$. Then $m^{\mathscr{D}}(n) \leq (n+1)^d$.

Example 3.7.4c Let $\mathscr{D} = \{\{x : \theta^T x > y\} : \theta \in \mathbf{R}^d,\ y \in \mathbf{R}\}$. Then $m^{\mathscr{D}}(n) \leq 2^d \binom{n}{d}$. This is a very rough bound, which we obtained as follows. Each set of d distinct points defines a hyperplane containing those points. The hyperplane divides $\mathbf{R}^d$ in two sets, say set A and set B. Because there are 2^d possible ways to agree that one of the d points on the hyperplane lies in A or B, we find $m^{\mathscr{D}}(n) \leq 2^d \binom{n}{d}$. Thus, it grows only polynomially in n. It can be shown that $V(\mathscr{D}) \leq d+2$ (Pollard (1984)).

Remark From the above examples, one can construct more complicated Vapnik–Chervonenkis classes, by taking intersections, unions and complements. For example, if $\mathscr{D}_1$ and $\mathscr{D}_2$ are Vapnik–Chervonenkis classes, then so is $\mathscr{D}_1 \cap \mathscr{D}_2 = \{D_1 \cap D_2 : D_1 \in \mathscr{D}_1,\ D_2 \in \mathscr{D}_2\}$. The Vapnik–Chervonenkis property is preserved when taking finitely many set-theoretic operations.

Having thus defined the Vapnik–Chervonenkis property for sets, we now turn to classes of functions.

Definition 3.3 The subgraph of a function $g : \mathscr{X} \to \mathbf{R}$ is

$$\text{subgraph}(g) = \{(x, y) \in \mathscr{X} \times \mathbf{R} :\ g(x) > y\}.$$

For a class of functions $\mathscr{G}$, let $V(\mathscr{G})$ be the index of the collection of subgraphs $\{\text{subgraph}(g) : g \in \mathscr{G}\}$. A collection of functions $\mathscr{G}$ is called a *Vapnik–Chervonenkis subgraph class* if $V(\mathscr{G}) < \infty$.

Remark It is not hard to see that the Vapnik–Chervonenkis subgraph property is preserved under taking monotone transformations. The same is true for transformations of bounded variation.

Example 3.7.4d Let $\psi_k(x)$, $k = 1, \ldots, d$, be fixed functions, and let

$$\mathscr{G} = \{\theta_1\psi_1 + \cdots + \theta_d\psi_d :\ \theta \in \mathbf{R}^d\}.$$

Check that $\mathscr{G}$ is a Vapnik–Chervonenkis subgraph class. It follows from Pollard (1984) that $V(\mathscr{G}) \leq d+2$.

Theorem 3.11 *Suppose $\mathscr{G}$ is a Vapnik–Chervonenkis subgraph class. Let*

$$G = \sup_{g \in \mathscr{G}} |g|$$

be its envelope. Then for some universal constant c and all $p < \infty$, and probability measure Q,

$$N_p(\delta \|G\|_{p,Q}, \mathscr{G}, Q) \leq cV(\mathscr{G})(16e)^{V(\mathscr{G})}\delta^{-p(V(\mathscr{G})-1)}, \text{ for all } \delta > 0.$$

The proof is in van der Vaart and Wellner (1996, Theorem 2.6.7). We arrive at an important corollary.

Corollary 3.12 *Suppose $\mathcal{G}$ is a Vapnik–Chervonenkis subgraph class with envelope $G \in L_1(P)$. Then $\mathcal{G}$ satisfies the ULLN.*

Example 3.7.4d (continued) Consider again the class

$$\mathcal{G} = \{\theta_1\psi_1 + \cdots + \theta_d\psi_d : \theta \in \mathbf{R}^d\}.$$

We already presented the bound $V(\mathcal{G}) \le d + 2$. This, we shall use here to compare the bound of Lemma 2.5 for the entropy of a ball in $\mathbf{R}^d$, with the one obtained by applying Theorem 3.11.

This class $\mathcal{G}$ has no integrable envelope. Now let

$$\mathcal{G}_Q(R) = \{g \in \mathcal{G} : \ \|g\|_Q \le R\}.$$

Then we know from Corollary 2.6 that for all δ sufficiently small

$$H(\delta, \mathcal{G}_Q(R), Q) \le Ad \log\left(\frac{R}{\delta}\right). \tag{3.16}$$

Now, suppose that $\Sigma_Q = \int \psi\psi^T \, dQ = I$, where I is the $(d \times d)$-identity matrix. It is easy to see that

$$G(x) = \sup_{g \in \mathcal{G}_Q(R)} |g(x)| = \left(\sum_{k=1}^{d} \psi_k^2(x)\right)^{1/2} R.$$

Hence

$$\|G\|_Q = \sqrt{d}R.$$

Theorem 3.11 yields therefore that for all $\delta > 0$ sufficiently small

$$H(\delta, \mathcal{G}_Q(R), Q) \le \tilde{A} d \log\left(\frac{dR}{\delta}\right). \tag{3.17}$$

The bound (3.17) is not as good as (3.16), in the sense that the dimension d now appears in the log-term as well.

Example 3.7.5. Convex hulls The convex hull $\mathrm{conv}(\mathcal{K})$ of a class $\mathcal{K}$ of functions is defined as the collection of all finite convex combinations of functions in $\mathcal{K}$, i.e.,

$$\mathrm{conv}(\mathcal{K}) = \left\{\sum_{j=1}^{r} \theta_j k_j : \theta_j \ge 0, \ k_j \in \mathcal{K}, \ j = 1, \ldots, r, \ \sum_{j=1}^{r} \theta_j = 1, \ r \ge 1\right\}.$$

Lemma 3.13 *Suppose that $\mathscr{K} \subset L_1(P)$ satisfies the ULLN. Then also* $\text{conv}(\mathscr{K})$ *satisfies the ULLN.*

Proof Let $\delta > 0$ be arbitrary, and take n sufficiently large, such that

$$\sup_{k\in\mathscr{K}} \left| \int k \, d(P_n - P) \right| \leq \delta.$$

Then for $\theta_j \geq 0$, $k_j \in \mathscr{K}$, $j = 1, \ldots, r$, $\sum_{j=1}^r \theta_j = 1$, we obtain

$$\left| \int \sum_{j=1}^r \theta_j k_j \, d(P_n - P) \right| = \left| \sum_{j=1}^r \theta_j \int k_j \, d(P_n - P) \right|$$
$$\leq \left(\sum_{j=1}^r \theta_j \right) \max_{j=1,\ldots,r} \left| \int k_j \, d(P_n - P) \right| \leq \delta. \quad \Box$$

So we arrive at the ULLN for $\text{conv}(\mathscr{K})$ without having to calculate its entropy. However, for more refined results (such as increments of the empirical process), we do need entropy numbers. The following theorem can be deduced from Ball and Pajor (1990).

Theorem 3.14 *Let Q be a probability measure on $(\mathscr{X}, \mathscr{A})$ and let $\mathscr{K}$ be a subset of the unit ball in $L_2(Q)$. Suppose that for some constants $c > 0$ and $d > 0$,*

$$N(\delta, \mathscr{K}, Q) \leq c\delta^{-d}, \quad \textit{for all } \delta > 0. \tag{3.18}$$

Then there exists a constant A depending on c and d, such that

$$H(\delta, \text{conv}(\mathscr{K}), Q) \leq A\delta^{-2d/(2+d)}, \quad \textit{for all } \delta > 0. \tag{3.19}$$

Example 3.7.5a Take

$$\mathscr{G} = \{g : \mathbf{R} \to [0,1] : g \text{ increasing}\}.$$

Then $\mathscr{G} = \overline{\text{conv}}(\mathscr{K})$, where $\overline{\text{conv}}(\mathscr{K})$ is the (pointwise) closure of $\text{conv}(\mathscr{K})$, and where $\mathscr{K} = \{1_{[y,\infty)} : y \in \mathbf{R}\}$. It is easy to see that

$$N(\delta, \mathscr{K}, Q) \leq c\delta^{-2}, \text{ for all } \delta > 0. \tag{3.20}$$

Note that class $\mathscr{K}$ is indexed by a one-dimensional parameter. The fact that we find $d = 2$ in (3.20) is because the $L_2(Q)$-norm of an indicator function is the square root of its $L_1(Q)$-norm. Because taking the closure has no impact on entropy numbers, we obtain from Theorem 3.14,

$$H(\delta, \mathscr{G}, Q) \leq A\delta^{-1}, \quad \text{for all } \delta > 0. \tag{3.19}$$

This is the same bound as given in (2.5). Observe however that (2.5) presents the stronger result, because it concerns entropy with bracketing.

3.8. Notes

Almost all the results in this chapter can be found in Pollard (1984, Chapter II). Lemma 3.1 goes back to Blum (1955) and DeHardt (1971). Kolmogorov developed the chaining technique, which has become one of the essential tools in empirical process theory. Maximal inequalities like the one of Lemma 3.2 appear throughout the literature (see for example Ledoux and Talagrand (1991)). However, Lemma 3.2 does differ from most known maximal inequalities, in the sense that the lower integration limit in the entropy integral is not taken equal to zero. This allows us to recover the sufficient conditions of Theorem 3.7, for the ULLN, from it (conditions which turn out to be also necessary). Symmetrization is a well-known technique in the theory of empirical processes. Our description is very brief. More details can be found in Pollard (1984, 1990), van der Vaart and Wellner (1996), and in the references given there. We remark that symmetrization does give rise to measurability conditions, because we condition on the observations and later integrate out. In other words, we use Fubini's Theorem, which heavily relies on measurability assumptions. Theorem 3.7 is from Pollard (1984, Chapter II). A slight variant is in Giné and Zinn (1984), where also the necessity of the conditions is proved.

The results of Vapnik and Chervonenkis (1971) have led to a renewed interest in Glivenko–Cantelli theorems. Using the symmetrization technique, they prove the ULLN for classes of indicators of sets, under necessary and sufficient combinatorial conditions. In a later paper (Vapnik and Chervonenkis (1981)), they obtain necessary and sufficient conditions for more general classes of functions. Necessary and sufficient conditions of a more combinatorial nature, for general classes of functions, are presented in Talagrand (1987). Theorem 3.11 of van der Vaart and Wellner is an improvement of the Approximation Lemma of Pollard (1984, Chapter II). Pollard moreover uses the *graphs*, rather than the *subgraphs*. Theorem 3.14 is proved by Ball and Pajor (1990), but a proof which is perhaps more accessible for statisticians can be found in van der Vaart and Wellner (1996). A rougher result, but with a much easier proof, is presented in Pollard (1990).

3.9. Problems and complements

3.1. Let $\mathcal{G}$ be a subset of a ball in $L_1(P)$ with envelope G. Show that if $\mathcal{G}$ satisfies the ULLN, then also $\{h(g) : g \in \mathcal{G}\} \subset L_1(P)$ satisfies the ULLN whenever $h : \mathbf{R} \to \mathbf{R}$ is Lipschitz and satisfies $\int h(g)\, dP < \infty$ for some $g \in \mathcal{G}$.

(Hint: use the fact that $G \in L_1(P)$ and that $\mathcal{G}$ satisfies (3.13).)

3.2. Let $\{g_\theta : \theta \in \Lambda\} \subset L_1(P)$, where Λ is an open, convex subset of $\mathbf{R}^d$, and where the map $\theta \mapsto g_\theta(x)$ is convex for all $x \in \mathcal{X}$. Show for each compact subset $\Theta \subset \Lambda$,

$$\sup_{\theta\in\Theta}\left|\int g_\theta \, d(P_n - P)\right| \to 0, \quad \text{a.s.}$$

(see Rockafellar (1970)).

3.3. (Smooth functions with unbounded support.) Show that

$$\mathcal{G} = \left\{g : \mathbf{R} \to [0,1], \ \left|\frac{dg(x)}{dx}\right| \le 1, \quad \text{for all } x\right\}$$

satisfies the ULLN.

3.4. (Monotone functions with integrable envelope.) Show that

$$\mathcal{G} = \{g : \mathbf{R} \to \mathbf{R}, \ g \text{ increasing}, \ |g| \le G\},$$

with $G \in L_1(P)$, satisfies the ULLN.

3.5. Show that

$$\mathcal{G} = \left\{g : [0,1] \to [0,1], \ \frac{dg(x)}{dx} = g'(x) \text{ exists for all } x, \ TV(g') \le 1\right\},$$

satisfies the ULLN, by applying Lemma 3.13.
(Hint: consider the convex hull of $\mathcal{K} = \{k_y(x) = (x-y)\mathrm{l}\{x \ge y\}\}$.)

3.6. Consider a collection of functions $\mathcal{G}_n$, depending on n, with envelope

$$G_n = \sup_{g\in\mathcal{G}_n} |g|.$$

Suppose that for some sequence $b_n \ge 1$, $b_n = o(n^{1/2})$, we have

$$\frac{b_n^2}{n} H_1(\delta, \mathcal{G}_n, P_n) \to_{\mathbf{P}} 0, \quad \text{for all } \delta > 0,$$

and

$$\limsup_{n\to\infty} \int_{G_n > b_n} G_n \, dP = 0.$$

Then

$$\sup_{g\in\mathcal{G}_n}\left|\int g \, d(P_n - P)\right| \to_{\mathbf{P}} 0.$$

(See van de Geer (1988, Lemma 2.3.3).)

3.7. Let

$$\mathscr{G}_n = \left\{ g = \sum_{j=1}^{d_n} \theta_{j,n} A_{j,n},\ |\theta_{j,n}| \leq 1,\ j = 1, \ldots, d_n \right\},$$

where $A_{1,n}, \ldots, A_{n,d_n}$ forms a partition of $\mathscr{X}$. Use the result of Problem 3.6 to obtain that for $d_n = o(n)$,

$$\sup_{g \in \mathscr{G}_n} \left| \int g \, d(P_n - P) \right| \to_{\mathbf{P}} 0.$$

3.8. Let P be Lebesgue measure on $([0,1], \mathscr{A})$, with $\mathscr{A}$ the Borel σ-algebra. Verify that

$$\sup_{A \in \mathscr{A}} |P_n(A) - P(A)| = 1, \quad n \geq 1.$$

4

First Applications: Consistency

In this chapter, we consider consistency of maximum likelihood estimators and least squares estimators, under entropy conditions. Suppose that $\mathscr{P}$ is a class of densities. We show that the ULLN for certain transformations of $\mathscr{P}$ implies consistency of the maximum likelihood estimator. If $\mathscr{P}$ is convex, one may use a transformation that is uniformly bounded. This is desirable, because that means that the envelope condition, necessary for a ULLN to hold, is satisfied. In the regression model, let $\mathscr{G}$ be the class of all regression functions allowed by the model. We do not assume a ULLN for $\mathscr{G}$, but apply the entropy methods directly to study consistency of the least squares estimator.

4.1. Consistency of maximum likelihood estimators

Let $X_1, \ldots, X_n, \ldots$ be i.i.d. with distribution P on $(\mathscr{X}, \mathscr{A})$, and suppose

$$p_0 = \frac{dP}{d\mu} \in \mathscr{P},$$

where $\mathscr{P}$ is a given class of densities with respect to the σ-finite measure μ. Let $\hat{p}_n$ be the maximum likelihood estimator of p_0, i.e.

$$\hat{p}_n = \arg\max_{p \in \mathscr{P}} \int \log p \, dP_n.$$

Throughout, we assume that a maximizer $\hat{p}_n \in \mathscr{P}$ exists.

Our aim is to prove Hellinger consistency of $\hat{p}_n$, where the the Hellinger distance between two densities p_1 and p_2 is defined as

$$h(p_1, p_2) = \left(\frac{1}{2} \int (p_1^{1/2} - p_2^{1/2})^2 \, d\mu \right)^{1/2}.$$

The factor $\frac{1}{2}$ is a convention, which ensures that $h(p_1, p_2) \leq 1$ (Problem 4.1). The Hellinger distance describes the separation between two probability measures in a natural way, independent of a particular parametrization. But the main reason why we have chosen this distance function is because of its convenience when studying the maximum likelihood problem in a general setup.

In Example 1.2, it was shown that

$$h^2(\hat{p}_n, p_0) \leq \int_{p_0>0} \frac{1}{2} \log \frac{\hat{p}_n}{p_0} \, d(P_n - P). \tag{4.1}$$

Therefore, we have consistency in Hellinger distance if the ULLN holds for the class

$$\left\{ \frac{1}{2} \log \frac{p}{p_0} \mathrm{l}\{p_0 > 0\} : p \in \mathscr{P} \right\}. \tag{4.2}$$

However, the ULLN needs the envelope condition, whereas in most realistic statistical models, the envelope of the class (4.2) is not in $L_1(P)$. For instance, in many cases, the densities in $\mathscr{P}$ do not stay away from zero, so that the log-densities can become minus infinity. The latter problem can be easily overcome. Later on, we shall also handle the case where the densities do not have an integrable envelope, for the special situation where $\mathscr{P}$ is convex.

Consider the class

$$\mathscr{G} = \left\{ \frac{1}{2} \log \frac{p + p_0}{2p_0} \mathrm{l}\{p_0 > 0\} : p \in \mathscr{P} \right\}.$$

The functions in $\mathscr{G}$ are bounded from below:

$$g(x) \geq -\frac{1}{2} \log 2, \quad \text{for all } x \in \mathscr{X} \text{ and } g \in \mathscr{G}.$$

The next lemma presents a modification of (4.1). Because inequalities of this type will play a major role, a general terminology will help to highlight the conformity of various problems. We propose to call them *Basic Inequalities.* These are inequalities with on one side essentially the squared distance between the estimator and the true parameter, and on the other side an empirical process.

Lemma 4.1 (Basic Inequality) *We have*

$$h^2\left(\frac{\hat{p}_n+p_0}{2}, p_0\right) \le \int_{p_0>0} \frac{1}{2}\log\frac{\hat{p}_n+p_0}{2p_0}\, d(P_n-P). \tag{4.3}$$

Proof Use the concavity of the log-function to find that

$$\log\frac{\hat{p}_n+p_0}{2p_0} 1\{p_0>0\} \ge \frac{1}{2}\log\frac{\hat{p}_n}{p_0} 1\{p_0>0\}. \tag{4.4}$$

Thus,

$$\begin{aligned} 0 &\le \int_{p_0>0} \frac{1}{4}\log\frac{\hat{p}_n}{p_0}\, dP_n \le \int_{p_0>0} \frac{1}{2}\log\frac{\hat{p}_n+p_0}{2p_0}\, dP_n \\ &= \int_{p_0>0} \frac{1}{2}\log\frac{\hat{p}_n+p_0}{2p_0}\, d(P_n-P) + \int_{p_0>0} \frac{1}{2}\log\frac{\hat{p}_n+p_0}{2p_0}\, dP. \end{aligned}$$

But (see also Lemma 1.3),

$$\int_{p_0>0} \frac{1}{2}\log\frac{\hat{p}_n+p_0}{2p_0}\, dP \le -h^2\left(\frac{\hat{p}_n+p_0}{2}, p_0\right). \qquad \square$$

In the left-hand side of (4.3), one now finds the Hellinger distance between the convex combination $(\hat{p}_n+p_0)/2$ and p_0, instead of the Hellinger distance between $\hat{p}_n$ and p_0. But the next lemma shows that these behave in the same way. For the convex combination, we use the notation

$$\bar{p} = \frac{p+p_0}{2}.$$

Lemma 4.2 *We have*

$$h^2(\bar{p}_1, \bar{p}_2) \le \frac{1}{2}h^2(p_1, p_2). \tag{4.5}$$

Moreover

$$h^2(p, p_0) \le 16 h^2(\bar{p}, p_0). \tag{4.6}$$

Proof Observe that

$$\frac{p_1^{1/2}+p_2^{1/2}}{\bar{p}_1^{1/2}+\bar{p}_2^{1/2}} \le \sqrt{2}.$$

So

$$|\bar{p}_1^{1/2}-\bar{p}_2^{1/2}| = \frac{1}{2}\left(\frac{p_1^{1/2}+p_2^{1/2}}{\bar{p}_1^{1/2}+\bar{p}_2^{1/2}}\right)\left|p_1^{1/2}-p_2^{1/2}\right| \le \frac{1}{2}\sqrt{2}\left|p_1^{1/2}-p_2^{1/2}\right|. \tag{4.7}$$

This gives inequality (4.5).

The second inequality follows in the same way, using

$$\frac{\bar{p}^{1/2} + p_0^{1/2}}{p^{1/2} + p_0^{1/2}} \leq 2. \qquad \square$$

By Lemmas 4.1 and 4.2, consistency in Hellinger distance follows from a ULLN for the class

$$\mathcal{G} = \left\{ \frac{1}{2} \log \frac{p + p_0}{2p_0} 1\{p_0 > 0\} \; : \; p \in \mathcal{P} \right\}. \tag{4.8}$$

Recall that the necessary conditions for the ULLN to hold are that the empirical covering numbers of $\mathcal{G}$ do not grow exponentially fast in n, and that the envelope function

$$G = \sup_{g \in \mathcal{G}} |g| \tag{4.9}$$

is in $L_1(P)$. The following theorem summarizes the results.

Theorem 4.3 *Suppose that for the class $\mathcal{G}$ defined in* (4.8), *we have*

$$\frac{1}{n} H_1(\delta, \mathcal{G}, P_n) \to_{\mathbf{P}} 0, \quad \textit{for all } \delta > 0, \tag{4.10}$$

and

$$G \in L_1(P). \tag{4.11}$$

Then $h(\hat{p}_n, p_0) \to 0$ almost surely.

To verify conditions (4.10) and (4.11), we prove in Lemma 4.4 that it is enough to have that $H_{1,B}(\delta, \mathcal{P}, \mu_0) < \infty$ for all $\delta > 0$, where

$$d\mu_0 = 1\{p_0 > 0\} d\mu. \tag{4.12}$$

We shall show below that, when calculating an upper bound for the entropy of $\mathcal{G}$, one may replace it by a simpler class, namely

$$\mathcal{P}/p_0 = \left\{ \frac{p}{p_0} 1\{p_0 > 0\}, \; p \in \mathcal{P} \right\}.$$

Denote the envelope of the densities by

$$q(x) = \sup_{p \in \mathcal{P}} p(x), \quad x \in \mathcal{X}. \tag{4.13}$$

Lemma 4.4 *We have for all* $\delta > 0$,

$$H_1(\delta, \mathcal{G}, P_n) \le H_1(2\delta, \mathcal{P}/p_0, P_n). \tag{4.14}$$

Moreover, if $q \in L_1(\mu_0)$, *then* $G \in L_1(P)$. *Finally, if*

$$H_{1,B}(\delta, \mathcal{P}, \mu_0) < \infty \quad \text{for all } \delta > 0, \tag{4.15}$$

one has $h(\hat{p}_n, p_0) \to 0$, *a.s.*

Proof By elementary calculations (see Problem 4.2), for $\bar{p}_1 = (p_1 + p_0)/2$, and $\bar{p}_2 = (p_2 + p_0)/2$,

$$\frac{1}{2}\left|\log \frac{\bar{p}_1}{p_0} - \log \frac{\bar{p}_2}{p_0}\right| 1\{p_0 > 0\} \le \frac{1}{2}\left|\frac{p_1}{p_0} - \frac{p_2}{p_0}\right| 1\{p_0 > 0\}. \tag{4.16}$$

This yields (4.14). To prove the second assertion of the lemma, note that from (4.16)

$$G \le \frac{1}{2}\left|\frac{q}{p_0} - 1\right| 1\{p_0 > 0\},$$

so that

$$\int G \, dP \le \int q \, d\mu_0.$$

We conclude that the ULLN for $\mathcal{P}/p_0$ implies consistency in Hellinger distance (see also Problem 4.3). To verify this ULLN, we have to check the conditions of Theorem 3.7, which are implied by those of Lemma 3.1. Lemma 3.1 uses entropy with bracketing, instead of the empirical entropy. The first takes a simple form: for all $\delta > 0$,

$$H_{1,B}(\delta, \mathcal{P}/p_0, P) = H_{1,B}(\delta, \mathcal{P}, \mu_0). \tag{4.17}$$

Thus, if $H_{1,B}(\delta, \mathcal{P}, \mu_0) < \infty$ for all $\delta > 0$, then the maximum likelihood estimator $\hat{p}_n$ is consistent in Hellinger distance. □

There are still many situations where the conditions of Theorem 4.3 are not met, and where nevertheless the maximum likelihood estimator is consistent. In fact, one could expect that Theorem 4.3 is much too rough, because it only uses the following property of $\hat{p}_n$:

$$\int \log \hat{p}_n \, dP_n \ge \int \log p_0 \, dP_n. \tag{4.18}$$

This inequality yielded the Basic Inequality of Lemma 4.1. By replacing p_0 in the right hand side by some other density in $\mathcal{P}$, one can prove consistency under less strong conditions. It is clear that one has a whole range of choices here, and it depends on the model which choice gives you the best results. In the remainder of this section we shall investigate the model with $\mathcal{P}$ a convex class. Then $(\hat{p}_n + p_0)/2 \in \mathcal{P}$, which supplies us with an alternative Basic Inequality.

Lemma 4.5 (Basic Inequality) *Suppose $\mathscr{P}$ is convex. Then*

$$h^2(\hat{p}_n, p_0) \leq \int \frac{2\hat{p}_n}{\hat{p}_n + p_0}\, d(P_n - P). \tag{4.19}$$

Proof We have

$$\begin{aligned} 0 &\leq \int \log\left(\frac{2\hat{p}_n}{\hat{p}_n + p_0}\right) dP_n \leq \int \left(\frac{2\hat{p}_n}{\hat{p}_n + p_0} - 1\right) dP_n \\ &= \int \frac{2\hat{p}_n}{\hat{p}_n + p_0}\, d(P_n - P) - \int \left(\frac{p_0 - \hat{p}_n}{\hat{p}_n + p_0}\right) dP. \end{aligned}$$

But

$$\begin{aligned} \int \left(\frac{p_0 - \hat{p}_n}{\hat{p}_n + p_0}\right) dP &= \int \left(\frac{p_0 - \hat{p}_n}{\hat{p}_n + p_0}\right) p_0\, d\mu \\ &= \frac{1}{2}\int \left(\frac{p_0 - \hat{p}_n}{\hat{p}_n + p_0}\right)(p_0 + \hat{p}_n)\, d\mu + \frac{1}{2}\int \left(\frac{p_0 - \hat{p}_n}{\hat{p}_n + p_0}\right)(p_0 - \hat{p}_n)\, d\mu \\ &= \frac{1}{2}\int \frac{(\hat{p}_n - p_0)^2}{\hat{p}_n + p_0}\, d\mu \geq h^2(\hat{p}_n, p_0). \end{aligned}$$

□

From Lemma 4.5, we infer that consistency is implied by the ULLN for the class

$$\mathscr{G}^{(\mathrm{conv})} = \left\{ \frac{2p}{p + p_0} \, : \, p \in \mathscr{P} \right\}. \tag{4.20}$$

This class is uniformly bounded by 2, so its envelope is certainly in $L_1(P)$. Hence, the following theorem holds.

Theorem 4.6 *Suppose $\mathscr{P}$ is convex. Assume moreover that*

$$\frac{1}{n} H_1(\delta, \mathscr{G}^{(\mathrm{conv})}, P_n) \to_{\mathbf{P}} 0, \quad \textit{for all } \delta > 0. \tag{4.21}$$

Then $h(\hat{p}_n, p_0) \to 0$ almost surely.

4.2. Examples

Example 4.2.1. The binary choice model We return to Example 1.1. Let $Y \in \{0, 1\}$ be a binary response variable, and let $Z \in \mathbf{R}$ be the covariate. We have i.i.d. observations $X_i = (Y_i, Z_i)$, $i = 1, 2, \ldots$, of $X = (Y, Z)$. The model is

$$P(Y = 1 \mid Z = z) = F_{\theta_0}(z),$$

with $\theta_0 \in \Theta$ an unknown parameter. We assume throughout that $F_\theta(z)$ is an increasing function of z for each $\theta \in \Theta$. Now, we can take $\mu =$ (counting measure on $\{0,1\}$) $\times\, Q$ as dominating measure, where Q is the distribution of Z. In the notation of the previous section, the densities are then

$$\mathscr{P} = \{p_\theta(y,z) = yF_\theta(z) + (1-y)(1-F_\theta(z)) \,:\, \theta \in \Theta\}.$$

Because $F_\theta(z)$ is a probability, we have $0 \le F_\theta(z) \le 1$ for all $\theta \in \Theta$. Therefore, the envelope q of the densities is bounded by 1 as well:

$$q = \sup_{p \in \mathscr{P}} p \le 1.$$

Moreover, Q is a probability measure, so μ is a finite measure. Now apply (2.5) (or Lemma 3.8). The class $\mathscr{P}$ is essentially a (subset of a) class of uniformly bounded monotone functions. Therefore,

$$H_B(\delta, \mathscr{P}, \mu) \le A\frac{1}{\delta}, \quad \text{for all } \delta > 0.$$

Hence, by Lemma 4.4,

$$h(\hat{p}_n, p_0) \to 0, \quad \text{a.s.} \tag{4.22}$$

The squared Hellinger distance is in this case

$$h^2(p_\theta, p_{\theta_0}) = \frac{1}{2}\int \left(F_\theta^{1/2} - F_{\theta_0}^{1/2}\right)^2 dQ + \frac{1}{2}\int \left((1-F_\theta)^{1/2} - (1-F_{\theta_0})^{1/2}\right)^2 dQ.$$

So (4.22) is somewhat stronger than consistency of $F_{\hat{\theta}_n}$ in the $L_2(Q)$-norm. The consistency result holds no matter what the further assumptions on F_θ are. Perhaps there are no further assumptions, or we assume in addition that $F_\theta(z)$ is a concave function of z. A parametric model is also possible, for example the one of Example 1.1, Case (i).

Case (i) Suppose

$$F_\theta(z) = \frac{e^{\theta z}}{1 + e^{\theta z}},$$

with $\theta \in \mathbf{R}$ ($\Theta = \mathbf{R}$). In that case, consistency in Hellinger distance implies consistency of $\hat{\theta}_n$, because the Hellinger distance $h(p_\theta, p_{\theta_0})$ has a unique minimum in $\theta = \theta_0$ (unless we are in the degenerate case where Z concentrates on $z = 0$). Now that we have got this far, we might as well verify the assumption (1.3), which we used for proving asymptotic normality of $\hat{\theta}_n$. Because $\hat{\theta}_n$ is consistent, it stays (almost surely) in a compact set. Use Lemma 3.10, on functions that are continuous in the parameter, with the parameter varying within a compact set. The envelope condition there is satisfied if Z has finite variance. So the conclusion is now: suppose Z has finite variance, then

$$\sqrt{n}(\hat{\theta}_n - \theta_0) \to^{\mathscr{L}} \mathscr{N}(0, 1/I_{\theta_0}).$$

Example 4.2.2. Estimating a smooth function Let μ be Lebesgue measure on $[0,1]$ and let

$$\mathscr{P} = \left\{ p : [0,1] \to [0,\infty) : \int p \, d\mu = 1, \ \int_0^1 \left(p^{(m)}(x)\right)^2 dx \le M^2 \right\}. \tag{4.23}$$

Here $m \in \{1, 2, \dots\}$ and M are given. The fact that $\int p \, d\mu = 1$ implies in this case that $|p|_\infty \le K^2$ for some constant K depending on m and M (apply the same arguments as in Lemma 3.9). Application of Theorem 2.4 yields

$$H_{1,B}(\delta, \mathscr{P}, \mu) \le A\delta^{-1/m}, \quad \text{for all } \delta > 0.$$

So we find that $h(\hat{p}_n, p_0) \to 0$ almost surely. If p_0 stays away from zero, say $p_0 \ge \eta_0^2 > 0$, this in turn implies consistency in the supremum norm of all lower-order derivatives, i.e.

$$\sup_{x \in [0,1]} |\hat{p}_n^{(k)}(x) - p_0^{(k)}(x)| \to_{\text{a.s.}} 0, \quad k = 0, \dots, m-1.$$

See Lemma 10.9 for a proof.

Observe that the consistency result also holds for the maximum likelihood estimator over any subclass of $\mathscr{P}$.

In practice, one often uses a penalty, instead of restricting the density, which has the major advantage that one does not have to know a bound for $\int (p_0^{(m)}(x))^2 \, dx$. We shall study the penalized maximum likelihood estimator in Section 10.2. Another approach would be to use kernel estimators. Both penalized estimators and kernel estimators involve a tuning parameter, which has to be of the right order in order to get good asymptotic results.

Example 4.2.3. Estimating a monotone density Let μ be Lebesgue measure on $[0,1]$ and

$$\mathscr{P} = \{p \text{ is a decreasing density on } [0,1]\}.$$

Theorem 4.3 appears to be of no use here, since the densities in $\mathscr{P}$ may become arbitrarily large. However, one can exploit the convexity of $\mathscr{P}$ to arrive at consistency. To verify the entropy condition (4.21) of Theorem 4.6, note that

$$\left\{ \frac{2pp_0}{p + p_0} : p \in \mathscr{P} \right\} \tag{4.24}$$

is a class of decreasing functions. Moreover, if $|p_0|_\infty < \infty$, the class in (4.24) is also uniformly bounded. Because

$$H_{1,B}(\delta, \mathcal{G}^{(\text{conv})}, P) = H_{1,B}\left(\delta, \left\{\frac{2pp_0}{p+p_0} : p \in \mathcal{P}\right\}, \mu_0\right), \quad \text{for all } \delta > 0,$$

we obtain from (2.5) (or Lemma 3.8) that

$$H_{1,B}\left(\delta, \mathcal{G}^{(\text{conv})}, P\right) \leq A\frac{1}{\delta}, \quad \text{for all } \delta > 0.$$

Hence, under the assumption $|p_0|_\infty < \infty$, we have $h(\hat{p}_n, p_0) \to 0$, almost surely. We shall extend the result to possibly unbounded p_0 in Example 7.4.2. There, we shall also consider the case with unbounded support.

It is important to note that we used the convexity of $\mathcal{P}$ here. This means that if we replace the estimator $\hat{p}_n$ by the maximum likelihood estimator over a non-convex subset of $\mathcal{P}$, consistency is no longer guaranteed by our results.

Example 4.2.4. Mixture models Consider a random variable Y with unknown distribution F_0 on $(\mathcal{Y}, \mathcal{B})$. Suppose that $\mathcal{Y}$ is a locally compact Hausdorff space with countable base (for instance the real line). The parameter $F_0 \in \Lambda$ is completely unknown. So in this case $\Lambda = \{F \text{ is a probability measure on } (\mathcal{Y}, \mathcal{B})\}$. Instead of Y, we observe the mixture of Y with a random variable with conditional density given $Y = y$ equal to $k(x \mid y)$. We assume that $k(x \mid y)$ is known. Then

$$\mathcal{P} = \left\{p_F(x) = \int k(x \mid y)\, dF(y) : \; F \in \Lambda\right\}.$$

This is again a convex class, so that we may apply Theorem 4.6. The parameter space Λ is a subset of the compact space (Λ^*, τ), with Λ^* the class of all measures F with $F(\mathcal{Y}) \leq 1$, and with τ the metric corresponding to the vague topology. If for μ-almost all x, $k(x \mid y)$ is continuous and vanishes at infinity as a function of y, then the map $F \mapsto p_F(x)$ is τ-continuous for μ-almost all x at all $F \in \Lambda^*$. Apply Lemma 3.10 to conclude that $h(\hat{p}_n, p_0) \to 0$ almost surely. If $\hat{\theta}_n$ is identifiable (i.e., if for all $F \in \Lambda^*$, $h(p_F, p_{F_0}) = 0$ implies $\tau(F, F_0) = 0$), then the consistency in Hellinger distance implies that $\tau(\hat{F}_n, F_0) \to 0$ almost surely.

4.3. Consistency of least squares estimators

The maximum likelihood problem is in a sense less transparent than the least squares problem, but we postponed the latter because it concerns a non-i.i.d. situation. The regression model is

$$Y_i = g_0(z_i) + W_i, \quad i = 1, \ldots, n,$$

with $Y_i \in \mathbf{R}$ the observed response variable, $z_i \in \mathscr{Z}$ a covariate, and W_i measurement error, $i = 1, \ldots, n$. The covariates $z_1, \ldots, z_n$ are fixed, i.e., we consider the case of fixed design. The errors $W_1, \ldots, W_n$ are independent random variables with expectation $EW_i = 0$ and variance $\text{var}(W_i) \le \sigma_0^2 < \infty$, $i = 1, \ldots, n$. The function $g_0 : \mathscr{Z} \to \mathbf{R}$ is unknown, but we assume it to lie in a given class $\mathscr{G}$ of regression functions. It can be estimated by the least squares estimator $\hat{g}_n$, which is defined (not necessarily uniquely) by

$$\hat{g}_n = \arg\min_{g \in \mathscr{G}} \sum_{i=1}^{n} (Y_i - g(z_i))^2.$$

Throughout, we assume a minimizer $\hat{g}_n \in \mathscr{G}$ exists.

Note that we are now indeed in a non-i.i.d. situation. But this was also the case in Section 3.3, and we shall see that we can apply the results obtained there directly to the least squares estimation problem.

First, we introduce some notation. Let

$$Q_n = \frac{1}{n} \sum_{i=1}^{n} \delta_{z_i}.$$

Write, for $g : \mathscr{Z} \to \mathbf{R}$,

$$\|g\|_n^2 = \|g\|_{Q_n}^2 = \frac{1}{n} \sum_{i=1}^{n} g^2(z_i).$$

With some abuse of notation, we let

$$\|y - g\|_n^2 = \frac{1}{n} \sum_{i=1}^{n} (Y_i - g(z_i))^2.$$

Finally, let

$$(w, g)_n = \frac{1}{n} \sum_{i=1}^{n} W_i g(z_i).$$

As the Basic Inequality in this context, we propose:

Lemma 4.7. (Basic Inequality) *We have*

$$\|\hat{g}_n - g_0\|_n^2 \le 2(w, \hat{g}_n - g_0)_n. \tag{4.25}$$

Proof This is simply rewriting the inequality

$$\|y - \hat{g}_n\|_n^2 \le \|y - g_0\|_n^2.$$

□

Now, for each g, $(w, g-g_0)_n$ is the average of independent random variables with expectation zero, so it will converge to zero under some integrability conditions. What we need again is its convergence to zero uniformly over (an appropriate subclass of) $\mathcal{G}$. This would imply consistency of $\hat{g}_n$ in the $\|\cdot\|_n$-norm. However, since we are now in a non-i.i.d. situation, we cannot apply the ULLNs given in Chapter 3. Of course, one can force the situation into an i.i.d. one by assuming that $z_1, \ldots, z_n$ are realizations of i.i.d. random variables. But then one is again facing the problem that the class $\mathcal{G}$ has in general no integrable envelope.

Recall now that in order to arrive at the ULLN of Theorem 3.7, we used symmetrization and then worked conditionally on the $X_1, \ldots, X_n$. In the regression model, we can apply Lemma 3.2 directly. The condition (3.3) will in general not be true for our measurement errors W_i, but we can overcome this by truncating them. Condition (4.26) below makes this possible. It is, for example, met if $W_1, \ldots, W_n$ are the first n of an i.i.d. sequence of errors with finite variance σ^2. It turns out that in that case, and when σ^2 is sufficiently large, the entropy condition (4.27) below is also a necessary condition for consistency, provided that the errors contain no atoms (van de Geer and Wegkamp 1997).

We shall need to control the entropy, not of the whole class $\mathcal{G}$ itself, but of subclasses $\mathcal{G}_n(R)$, which are defined as

$$\mathcal{G}_n(R) = \{g \in \mathcal{G} : \|g - g_0\|_n \leq R\}.$$

Thus, $\mathcal{G}_n(R)$ is a ball with radius R around g_0, intersected with $\mathcal{G}$. Let $H(\delta, \mathcal{G}_n(R), Q_n)$ be the δ-entropy of $\mathcal{G}_n(R)$ endowed with the metric $\|\cdot\|_n$.

Theorem 4.8 *Suppose that*

$$\lim_{K\to\infty} \limsup_{n\to\infty} \frac{1}{n} \sum_{i=1}^{n} E\left(W_i^2 1\{|W_i| > K\}\right) = 0, \tag{4.26}$$

and

$$\frac{1}{n} H(\delta, \mathcal{G}_n(R), Q_n) \to 0, \quad \textit{for all } \delta > 0,\ R > 0. \tag{4.27}$$

Then $\|\hat{g}_n - g_0\|_n \to_{\mathbf{P}} 0$.

Proof Let $\eta > 0$ and $\delta > 0$ be arbitrary. First of all, we note that by (4.25)

$$\|\hat{g}_n - g_0\|_n \leq 2 \left(\frac{1}{n} \sum_{i=1}^{2} W_i^2 \right)^{1/2},$$

and for $R^2 = 4\sigma_0^2/\eta$,

$$\mathbf{P}\left(\frac{1}{n}\sum_{i=1}^n W_i^2 > \frac{R^2}{4}\right) \le \eta.$$

On the set where

$$\frac{1}{n}\sum_{i=1}^n W_i^2 \le \frac{R^2}{4},$$

we have

$$(w1\{|w| > K\}, \hat{g} - g_0)_n \le \left(\frac{1}{n}\sum_{i=1}^n W_i^2 1\{|W_i| > K\}\right)^{1/2} R.$$

For $K = K(\delta, \eta)$ sufficiently large, we find

$$\mathbf{P}\left(\frac{1}{n}\sum_{i=1}^n W_i^2 1\{|W_i| > K\} > \left(\frac{\delta^2}{4R}\right)^2\right) \le \eta.$$

This means that we can truncate the measurement error at K.

We find

$$\begin{aligned}\mathbf{P}(\|\hat{g} - g_0\|_n > \delta) &\le \mathbf{P}(\delta < \|\hat{g}_n - g_0\|_n \le R) + \eta \\ &\le \mathbf{P}\left(\sup_{g \in \mathscr{G}_n(R)} 2(w, g - g_0)_n \ge \delta^2\right) + \eta \\ &\le \mathbf{P}\left(\sup_{g \in \mathscr{G}_n(R)} (w1\{|w| \le K\}, g - g_0)_n \ge \frac{\delta^2}{4}\right) + 2\eta.\end{aligned}$$

The random variables $W_i 1\{|W_i| \le K\}$ still have expectation zero if each W_i is symmetric, which we shall assume to avoid digressions. If they are not symmetric, one can use different truncation levels to the left and right to approximately maintain zero expectation. Now apply Hoeffding's inequality (Lemma 3.5): for any function $g : \mathscr{X} \to \mathbf{R}$ and for all $\delta > 0$,

$$\mathbf{P}\left(\left|(w1\{|w| \le K\}, g)_n\right| \ge \delta\right) \le 2\exp\left[-\frac{n\delta^2}{2K^2\|g\|_n^2}\right].$$

Hence, condition (3.3) of Lemma 3.2 holds. Apply this lemma, with δ replaced by $\delta^2/4$, $\epsilon = \delta^2/8$. For $R > \delta^2/(8K)$ and

$$\sqrt{n}\frac{\delta^2}{8} \ge C\left(RH^{1/2}\left(\frac{\delta^2}{32K}, \mathscr{G}_n(R), Q_n\right) \vee R\right), \tag{4.28}$$

we find

$$\mathbf{P}\left(\sup_{g\in\mathcal{G}_n(R)}\left|(w1\{|w|\le K\}, g-g_0)_n\right| \ge \frac{\delta^2}{4}\right) \le C\exp\left[-\frac{n\delta^4}{64C^2R^2}\right].$$

Since we assumed (4.27), inequality (4.28) will be true for all n sufficiently large. □

4.4. Examples

Example 4.4.1. Linear regression Let

$$\mathcal{G} = \left\{g(z) = \theta_1\psi_1(z) + \cdots + \theta_d\psi_d(z) : \theta \in \mathbf{R}^d\right\}.$$

Then, by Corollary 2.6,

$$H(\delta, \mathcal{G}_n(R), P_n) \le d\log\left(\frac{4R+\delta}{\delta}\right), \quad \text{for all } \delta > 0.$$

So (4.27) is met. In fact, we may let the dimension d depend on n: $d = d_n$. Then we find that if $d_n/n \to 0$, one has (under condition (4.26) on the errors) $\|\hat{g}_n - g_0\|_n \to_{\mathbf{P}} 0$.

Example 4.4.2. Isotonic regression Let

$$\mathcal{G} = \{g : \mathbf{R} \to \mathbf{R} : \ g \text{ increasing}\}.$$

We assume that $g_0 \in \mathcal{G}$ is fixed, and $\sup_{z\in\mathbf{R}}|g_0(z)| \le K$ for some (unknown) constant K. Obviously, for $g \in \mathcal{G}_n(R)$,

$$\max_{i=1,\dots,n}|g(z_i) - g_0(z_i)| \le \sqrt{n}R,$$

so that

$$\max_{i=1,\dots,n}|g(z_i)| \le \sqrt{n}R + K.$$

The result (2.5) of Birman and Solomjak gives us

$$H(\delta, \mathcal{G}_n(R), Q_n) \le A\frac{\sqrt{n}R+K}{\delta}, \quad \text{for all } \delta > 0,$$

so that

$$\frac{1}{n}H(\delta, \mathcal{G}_n(R), Q_n) \to 0, \quad \text{for all } \delta > 0,\ R > 0.$$

Hence, by Theorem 4.8 (under condition (4.26) on the errors),

$$\|\hat{g}_n - g_0\|_n \to_{\mathbf{P}} 0.$$

This in turn implies uniform consistency on subintervals, provided g_0 is uniformly continuous in an appropriate sense.

Lemma 4.9 *Assume that $z_1, \ldots, z_n, \ldots$ are realizations of i.i.d. random variables $Z_1, \ldots, Z_n, \ldots$, with continuous distribution Q. Suppose that for all $\delta > 0$, there exists an $\eta_\delta > 0$, such that*

$$g_0(z) - g_0(\tilde{z}) \le \delta, \quad \text{for all } z > \tilde{z} \text{ with } Q[\tilde{z}, z] \le \eta_\delta. \tag{4.29}$$

Then for any $\eta > 0$, and any $a < b$ satisfying $Q(-\infty, a] \ge \eta$ as well as $Q[b, \infty) \ge \eta$, we have

$$\sup_{z \in [a,b]} |\hat{g}_n(z) - g_0(z)| \to_{\mathbf{P}} 0.$$

Proof Take $\eta > 0$ fixed and $a < b$ in $\mathbf{R}$ such that $Q(-\infty, a] \ge \eta$ as well as $Q[b, \infty) \ge \eta$. Let $\delta > 0$ be arbitrary, but small enough, so that $\eta_\delta \le \eta$. If for some $z_0 \in [a, b]$,

$$\hat{g}_n(z_0) - g_0(z_0) \ge 2\delta,$$

then for all $z \ge z_0$, $Q[z_0, z] \le \eta_\delta$,

$$\hat{g}_n(z) - g_0(z) \ge \hat{g}_n(z_0) - g_0(z_0) - (g_0(z) - g_0(z_0)) \ge \delta.$$

Because $z_0 \le b$, we have $Q[z_0, \infty) \ge \eta$. So there is a $z_1 > z_0$ such that $Q[z_0, z_1] = \eta_\delta$. Since eventually, $Q_n[z_0, z_1] \ge (\eta_\delta/2)$ a.s., we find that (4.29) implies

$$\|\hat{g}_n - g_0\|_n \ge \delta \left(\frac{\eta_\delta}{2}\right)^{1/2}, \quad \text{a.s.}$$

The same inequality can be derived if for some $z_0 \in [a, b]$,

$$\hat{g}_n(z_0) - g_0(z_0) \le -2\delta.$$

Because $\|\hat{g}_n - g_0\|_n \to_{\mathbf{P}} 0$, we arrive at

$$\sup_{z \in [a,b]} |\hat{g}_n(z) - g_0(z)| \to_{\mathbf{P}} 0. \qquad \square$$

4.5. Notes

The observation that the Hellinger distance is a convenient metric in maximum likelihood problems goes back to Le Cam (1970, 1973). Birgé (1983) shows that a necessary and sufficient condition for the existence of a uniform Hellinger consistent estimator is that $\mathscr{P}$ is totally bounded for the Hellinger metric. Taking the convex combination $(p + p_0)/2$ in (4.8) is an idea of Birgé and Massart (1993). An envelope condition, such as (4.11), appears in virtually all consistency proofs (see for example Huber (1967)). Theorem 4.6 gives an illustration of how to get rid of it in certain cases. The theorem is from van de Geer (1993). The consistency proof for mixture models (Example 4.2.4) is closely along the lines of Pfanzagl (1988).

Theorem 4.8 can be found in van de Geer and Wegkamp (1997), where also necessity of the conditions is examined. The case where $z_1, \ldots, z_n$ are i.i.d. realizations of a random variable Z is considered in van de Geer (1987).

We have investigated *global* consistency of estimators, in the sense that we used the *global* distance functions $h(\cdot,\cdot)$ and $\|\cdot\|_n$. They appear to be the natural distances for the problems considered. In particular cases, global consistency implies consistency of some other quantities, such as pointwise consistency.

4.6. Problems and complements

4.1. Consider two probability measures P_1 and P_2 on a measurable space $(\mathscr{X},\mathscr{A})$, dominated by a σ-finite measure μ (e.g., $\mu = P_1 + P_2$). Show that the Hellinger distance

$$\left(\frac{1}{2}\int\left(\sqrt{\frac{dP_1}{d\mu}}-\sqrt{\frac{dP_2}{d\mu}}\right)^2 d\mu\right)^{1/2}$$

does not depend on the dominating measure μ. Verify that the Hellinger distance is less than or equal to 1.

4.2. Prove inequality (4.16). (Hint: Use the fact that for $0 < x < y$, $\log y - \log x = (y-x)\frac{1}{\xi}$, with $x \le \xi \le y$.)

4.3. Let $\mathscr{G}$ and $\mathscr{G}^{(\mathrm{conv})}$ be defined in (4.8) and(4.20) respectively. Show that if $\mathscr{P}/p_0$ satisfies the ULLN, then also $\mathscr{G}$ as well as $\mathscr{G}^{(\mathrm{conv})}$ satisfy the ULLN. (Hint: use Problem 3.1.)

4.4. Let

$$\mathscr{P} = \Big\{p : [-1,1] \to [0,\infty),\ \int p\, d\mu = 1,$$
$$p(x) \text{ unimodal with a unique maximum at } x = 0\Big\}.$$

Show that the maximum likelihood estimator $\hat{p}_n$ is consistent in Hellinger distance.

4.5. Let

$$\mathscr{P} = \left\{ p : [-1,1] \to [0,\infty), \int p \, d\mu = 1, \ p \text{ unimodal} \right\}.$$

Check that the conditions of Theorem 4.3 or 4.6 are not fulfilled. This does not mean that the maximum likelihood is inconsistent though: show that if p_0 is the uniform density on $[-1,1]$, then $\hat{p}_n$ is consistent in Hellinger distance. Can you find a p_0 that is not consistently estimated by $\hat{p}_n$?

4.6. The assumption that the maximizer

$$\hat{p}_n = \arg\max_{p \in \mathscr{P}} \int \log p \, dP_n$$

exists as a member of $\mathscr{P}$ is certainly not fulfilled in all cases. A famous example is the mixture of two normals, provided by Kiefer and Wolfowitz (1956). The densities with respect to $\mu =$ Lebesgue measure on the real line are of the form

$$p(x; \mu, \sigma) = \alpha_0 \phi(x - \mu) + \frac{(1 - \alpha_0)}{\sigma} \phi\left(\frac{x - \mu}{\sigma}\right),$$

with $\alpha_0 \in (0,1)$ fixed, $\mu \in \mathbf{R}$ and $\sigma > 0$ unknown parameters, and ϕ the density of the standard normal distribution. Verify that

$$\lim_{\sigma \to 0} \int \log p(\,\cdot\,; X_1, \sigma) \, dP_n = \infty.$$

Note moreover that the conditions of Theorem 4.3 and Theorem 4.6 are not fulfilled in this case.

4.7. Let μ be a finite measure on $\mathbf{R}$ and

$$\mathscr{P} = \left\{ p : \mathbf{R} \to [0,\infty), \int p \, d\mu = 1, \ TV(p) < M \right\},$$

where $TV(p) = \int |\, dp \,|$ is the total variation of p and where M is given. Show that for $M < \infty$, $\hat{p}_n$ is Hellinger-consistent, but that for $M = \infty$ consistency fails.

4.8. The bound for $H_1(\delta, \mathscr{G}, P_n)$ given in Lemma 4.4 is not always good. Here is a simple example. Let μ be Lebesgue measure on the positive real line, and

$$\mathscr{P} = \left\{ p_\theta(x) = \frac{1}{\theta} 1\{0 < x < \theta\} : \theta > 0 \right\}.$$

Show that

$$H_1(\delta, \mathscr{G}, P_n) < \infty, \text{ a.s., for all } \delta > 0,$$

so that by Theorem 4.3, $h(\hat{p}_n, p_0) \to 0$, a.s. However, $H_1(\delta, \mathscr{P}/p_0, P_n)$ is not finite. One also easily verifies that

$$q(x) = \sup_{p \in \mathscr{P}} p(x) = \frac{1}{x}, \quad x > 0,$$

so that $q \notin L_1(\mu_0)$.

4.9. Consider the class of regression functions

$$\mathscr{G} = \left\{ g : [0,1] \to \mathbf{R} : \int_0^1 \left(g^{(m)}(z)\right)^2 dz \leq M^2 \right\}, \quad M < \infty.$$

Define $\psi_k(z) = z^{k-1}, k = 1, \ldots, m$, $\psi = (\psi_1, \ldots, \psi_{m-1})^T$ and $\Sigma_n = \int \psi\psi^T \, dQ_n$. Assuming that the eigenvalues of Σ_n stay away from zero, show that, under condition (4.26) on the errors, the least squares estimator $\hat{g}_n$ is $\|\cdot\|_n$-consistent.

4.10. Let $Y_1, \ldots, Y_n$ be independent real-valued random variables with $EY_i = \alpha_0$ for $i = 1, \ldots, \lfloor \gamma_0 n \rfloor$, and $EY_i = \beta_0$ for $i = \lfloor \gamma_0 n \rfloor + 1, \ldots, n$. We assume that α_0, β_0 and the change point γ_0 are independent of n and completely unknown. Write $g(i; \alpha, \beta, \gamma) = \alpha 1\{1 \leq i \leq \lfloor n\gamma \rfloor\} + \beta 1\{\lfloor n\gamma \rfloor + 1 \leq i \leq n\}$. Show that, under condition (4.26) on the errors $W_i = Y_i - EY_i$, the least squares estimator $\hat{g}_n = \hat{g}_n(\cdot\,; \hat{\alpha}_n, \hat{\beta}_n, \hat{\gamma}_n)$ is $\|\cdot\|_n$-consistent. Verify that this implies consistency of $(\hat{\alpha}_n, \hat{\beta}_n, \hat{\gamma}_n)$, provided that the parameters are identifiable, i.e. provided that $\alpha_0 \neq \beta_0$ and $\gamma_0 \in (0, 1)$.

4.11. Consider a regression model with $\mathscr{Z} = \mathbf{R}^d$ and

$$\mathscr{G} = \left\{ g(z) = \min\left\{\alpha^T z, \beta^T z\right\}, \begin{pmatrix} \alpha \\ \beta \end{pmatrix} \in \Theta \right\},$$

where Θ is a fixed bounded subset of $\mathbf{R}^{2d}$. Show that, under condition (4.26) on the errors, the least squares estimator is $\|\cdot\|_n$-consistent.

4.12. Consider the regression model with

$$\mathscr{G} = \{g = 1_D : D \in \mathscr{D}\},$$

where $\mathscr{D}$ is the collection of all subsets of $\mathscr{Z}$. Suppose that $g_0 \equiv 0$, and that $W_1, \ldots, W_n, \ldots$ are i.i.d. copies of a random variable W, satisfying $P(W > \frac{1}{2}) > 0$. Show that the least squares estimator is inconsistent.

5

Increments of Empirical Processes

Recall that the ULLN for $\mathcal{G}$ holds if its envelope G is in $L_1(P)$ and $H_1(\delta, \mathcal{G}, P_n)/n \to_{\mathbf{P}} 0$ for all $\delta > 0$. Now, if $G \in L_2(P)$ and $H(\delta, \mathcal{G}, P_n)/H(\delta) = O_{\mathbf{P}}(1)$, uniformly in $\delta > 0$, where $H(\delta)$ is some (non-random) function of δ satisfying $\int_0^1 H^{1/2}(u)\, du < \infty$, then the empirical process is asymptotically equicontinuous. The same conclusion holds if $\int_0^1 H_B^{1/2}(u, \mathcal{G}, P)\, du < \infty$. We also have a closer look at the increments of the empirical process, and at the ratio $\|g\|_n / \|g\|$.

Consider i.i.d. random variables $X_1, \ldots, X_n$ with distribution P on $(\mathcal{X}, \mathcal{A})$, and let $\mathcal{G} \subset L_2(P)$ be a collection of functions. The empirical process indexed by $\mathcal{G}$ is

$$\nu_n = \left\{ \nu_n(g) = \sqrt{n} \int g\, d(P_n - P) : g \in \mathcal{G} \right\}.$$

In Sections 5.1 and 5.5, we study asymptotic equicontinuity of this process, under random entropy conditions and conditions on the entropy with bracketing respectively, assuming that $X_1, \ldots, X_n$ are the first n of an infinite sequence of independent copies of a population random variable X. The other sections of this chapter include the situation of triangular arrays.

To investigate the behaviour of the empirical process near some fixed function $g_0 \in \mathcal{G}$, we consider a neighbourhood of g_0, given by

$$\mathcal{G}(\delta) = \{ g \in \mathcal{G} : \|g - g_0\| \le \delta \},$$

i.e., $\mathscr{G}(\delta)$ is a ball around g_0 intersected with $\mathscr{G}$. The radius δ is some (small) positive value.

5.1. Random entropy numbers and asymptotic equicontinuity

Let $X_1, \dots, X_n, \dots$ be i.i.d. with distribution P on $(\mathscr{X}, \mathscr{A})$. Recall Corollary 3.4, which asserts the following. Let $W_1, \dots, W_n$ be a Rademacher sequence, i.e., $W_1, \dots, W_n$ are independent random variables, with

$$\mathbf{P}(W_i = 1) = \mathbf{P}(W_i = -1) = \frac{1}{2}, \quad i = 1, \dots, n.$$

Take $(W_1, \dots, W_n)$ independent of $(X_1, \dots, X_n)$. Then for $a^2 \geq 8\delta^2$,

$$\text{(5.1)} \quad \mathbf{P}\left(\sup_{g \in \mathscr{G}(\delta)} |\nu_n(g) - \nu_n(g_0)| > a \right) \leq 4\mathbf{P}\left(\sup_{g \in \mathscr{G}(\delta)} \sqrt{n}|(w, g - g_0)_n| > a/4 \right),$$

where

$$(w, g - g_0)_n = \frac{1}{n} \sum_{i=1}^{n} W_i(g(X_i) - g_0(X_i)).$$

Lemma 5.1 below is a special case of Lemma 3.2. Due to the conditioning on $X_1, \dots, X_n$, we are confronted with the *empirical* entropy, and moreover, with the *empirical* radius

$$\text{(5.2)} \qquad \hat{\delta}_n = \sup_{g \in \mathscr{G}(\delta)} \|g - g_0\|_n.$$

Lemma 5.1 *For $\hat{\delta}_n$ given in* (5.2) *and*

$$\text{(5.3)} \qquad a \geq 8C \left(\int_{a/(32\sqrt{n})}^{\hat{\delta}_n} H^{1/2}(u, \mathscr{G}, P_n)\, du \vee \hat{\delta}_n \right),$$

we have

$$\text{(5.4)} \quad \mathbf{P}\left(\sup_{g \in \mathscr{G}(\delta)} \sqrt{n}|(w, g - g_0)_n| \geq \frac{a}{4} \mid X_1, \dots, X_n \right) \leq C \exp\left[-\frac{a^2}{64C^2\hat{\delta}_n^2} \right].$$

If the envelope function G of $\mathscr{G}$ is square integrable, and if $H(\delta, \mathscr{G}, P_n) = o_{\mathbf{P}}(n)$ for all $\delta > 0$, then it follows from the ULLN that eventually, $\hat{\delta}_n \leq 2\delta$ (Problem 5.1). Moreover, the ULLN then also gives that eventually $H(u, \mathscr{G}, P_n) \leq H(u/2, \mathscr{G}, P)$ for all $u > 0$. However, the latter does not help to check (5.3), because (5.3) involves values of u that become smaller as n

increases. Therefore, we shall impose later on the condition that for some non-random non-increasing function $H(u)$,

$$\lim_{A\to\infty} \limsup_{n\to\infty} \mathbf{P}\left(\sup_{u>0} \frac{H(u,\mathcal{G},P_n)}{H(u)} > A \right) = 0, \tag{5.5}$$

that is, $\sup_{u>0} H(u,\mathcal{G},P_n)/H(u) = O_{\mathbf{P}}(1)$. In many cases, one will in fact have that for some A fixed

$$\limsup_{n\to\infty} \mathbf{P}\left(\sup_{u>0} \frac{H(u,\mathcal{G},P_n)}{H(u)} > A \right) = 0,$$

i.e., that the second term in the right-hand side of (5.7) below vanishes. For instance, suppose a class of functions $\mathcal{G}$ satisfies Pollard's uniform entropy condition

$$\sup_{Q\in\mathcal{M}_{\text{finite}}} H(u\|G\|_Q,\mathcal{G},Q) \le \lambda(u), \quad \text{for all } u>0,$$

where $\mathcal{M}_{\text{finite}}$ is the class of all probability measures with finite support. Then $H(u,\mathcal{G},P_n) \le \lambda(u/\|G\|_n)$ and one can take $H(u) = \lambda(u/(2\|G\|))$, provided $G \in L_2(P)$. Note that Pollard's uniform entropy condition is for example met if $\mathcal{G}$ is a Vapnik–Chervonenkis subgraph class (see Theorem 3.11).

Lemma 5.2 *Suppose that $\mathcal{G}$ has envelope $G \in L_2(P)$. Moreover, assume that* (5.5) *holds. Then for each $\delta > 0$ fixed (i.e., not depending on n), and for*

$$a \ge 8CA^{1/2} \left(\int_{a/32\sqrt{n}}^{2\delta} H^{1/2}(u)\,du \vee 2\delta \right), \tag{5.6}$$

we have

$$\limsup_{n\to\infty} \mathbf{P}\left(\sup_{g\in\mathcal{G}(\delta)} |\nu_n(g)-\nu_n(g_0)| > a \right) \tag{5.7}$$
$$\le 4C \exp\left[-\frac{a^2}{256C^2\delta^2} \right] + \limsup_{n\to\infty} \mathbf{P}\left(\sup_{u>0} \frac{H(u,\mathcal{G},P_n)}{H(u)} > A \right).$$

Proof The conditions $G \in L_2(P)$ and (5.5) imply that

$$\sup_{g\in\mathcal{G}} \big| \|g-g_0\|_n - \|g-g_0\| \big| \to 0, \quad \text{a.s.}$$

So eventually, for each fixed $\delta > 0$,

$$\sup_{g\in\mathcal{G}(\delta)} \|g-g_0\|_n \le 2\delta, \quad \text{a.s.}$$

The result now follows from Lemma 5.1. □

Our next step is proving asymptotic equicontinuity of the empirical process. This means that we shall take a small in (5.6), which is possible if the entropy integral converges. Assume that

$$\int_0^1 H^{1/2}(u)\,du < \infty, \tag{5.8}$$

and define

$$J(\delta) = \left(\int_0^\delta H^{1/2}(u)\,du \vee \delta \right).$$

Roughly speaking, the increment at g_0 of the empirical process $\nu_n(g)$ behaves like $J(\delta)$ for $\|g - g_0\| \le \delta$. So, since $J(\delta) \to 0$ as $\delta \to 0$, the increments can be made arbitrarily small by taking δ small enough.

Theorem 5.3 *Suppose that $\mathcal{G}$ has envelope $G \in L_2(P)$. Assume that* (5.5) *and* (5.8) *are met. Then the empirical process ν_n is asymptotically equicontinuous at g_0, i.e., for all $\eta > 0$, there exists a $\delta > 0$ such that*

$$\limsup_{n\to\infty} \mathbf{P}\left(\sup_{g\in\mathcal{G}(\delta)} |\nu_n(g) - \nu_n(g_0)| > \eta \right) < \eta. \tag{5.9}$$

Proof Take $A \ge 1$ sufficiently large, such that

$$4C\exp[-A] < \frac{\eta}{2},$$

and

$$\limsup_{n\to\infty} \mathbf{P}\left(\sup_{u>0} \frac{H(u, \mathcal{G}, P_n)}{H(u)} > A \right) \le \frac{\eta}{2}.$$

Next, take δ sufficiently small, such that

$$8CA^{1/2}J(2\delta) < \eta.$$

Then by Lemma 5.2,

$$\limsup_{n\to\infty} \mathbf{P}\left(\sup_{g\in\mathcal{G}(\delta)} |\nu_n(g) - \nu_n(g_0)| > \eta \right) \le 4C\exp\left[-\frac{64AC^2J^2(2\delta)}{256C^2\delta^2} \right] + \frac{\eta}{2}$$
$$\le 4C\exp[-A] + \frac{\eta}{2} < \eta,$$

where we used

$$J(2\delta) \ge 2\delta.$$ □

Remark Because the conditions (5.5) and (5.8) do not depend on g_0, we have in fact shown that ν_n is asymptotically equicontinuous at *each* g_0. Moreover, the conditions then also hold with $\mathcal{G}$ replaced by $\{g_1 - g_2 : g_1, g_2 \in \mathcal{G}\}$, so that we may conclude that the process $\{\nu_n(g_1) - \nu_n(g_2) : g_1, g_2 \in \mathcal{G}\}$ is asymptotically equicontinuous at zero: for all $\eta > 0$ there exists a $\delta > 0$ such that

$$\limsup_{n\to\infty} \mathbf{P}\left(\sup_{g_1, g_2 \in \mathcal{G},\ \|g_1 - g_2\| \le \delta} |\nu_n(g_1) - \nu_n(g_2)| > \eta \right) < \eta.$$

This is related to the fact that a continuous function on a compact interval is *uniformly* continuous.

5.2. Random entropy numbers and classes depending on n

Recall that in Lemma 5.2, we assumed $\delta > 0$ fixed, so we may not apply it to sequences $\{\delta_n\}$, with $\delta_n \to 0$. However, it follows from Lemma 5.4 below that in the case $\mathcal{G}$ is uniformly bounded, we *can* apply Lemma 5.2 to sequences, provided δ_n does not converge to zero too fast (see Lemma 5.5).

We assume that for each n we have independent observations $X_1, \ldots, X_n$, with distribution P on $(\mathcal{X}, \mathcal{A})$. Here, P and the space $(\mathcal{X}, \mathcal{A})$ may also depend on n, although we do not express this in our notation. Let furthermore $\mathcal{G}$ be a class of functions on $(\mathcal{X}, \mathcal{A})$, possibly also depending on n. We do however assume that the functions are uniformly bounded by some constant independent of n, say

$$\sup_{g \in \mathcal{G}} |g - g_0|_\infty \le 1. \tag{5.10}$$

In Lemma 5.2, we used the ULLN to show that for δ fixed, eventually $\|g - g_0\|_n \le 2\delta$ whenever $\|g - g_0\| \le \delta$. We now study the case where $\|g - g_0\| \le \delta_n$, with $\delta_n \to 0$ as $n \to \infty$. The result can be found in Pollard (1984, Lemma II.6.33). We briefly present the proof, because we need some of its ingredients in Lemma 5.6, which examines the ratio $\|g\|_n / \|g\|$.

Lemma 5.4 *Let $H(u)$ be a given non-increasing function of $u > 0$. For $n\delta_n^2 \ge 2H(\delta_n)$, we have*

$$\mathbf{P}\left(\sup_{g \in \mathcal{G}(\delta_n)} \|g - g_0\|_n > 8\delta_n \right) \le 4\exp[-n\delta_n^2] + 4\mathbf{P}\left(H(u, \mathcal{G}, P_n) > H(u), \quad \text{for some } u > 0\right). \tag{5.11}$$

Proof Apply the randomization device of Pollard (1984, page 32). Let $X_{n+1}, \ldots, X_{2n}$ be an independent copy of $X_1, \ldots, X_n$, and let $\epsilon_1, \ldots, \epsilon_n$ be independent random variables, independent of $X_1,\dots,X_{2n}$, with $\mathbf{P}(\epsilon_i = 1) = \mathbf{P}(\epsilon_i = 0) = \frac{1}{2}$, $i = 1,\dots,n$. Set $\xi_i' = X_{2i-1+\epsilon_i}$, $\xi_i'' = X_{2i-\epsilon_i}$, $i = 1,\dots,n$, and $P_n' = \frac{1}{n}\sum_{i=1}^n \delta_{\xi_i'}$, $P_n'' = \frac{1}{n}\sum_{i=1}^n \delta_{\xi_i''}$, and $\bar{P}_{2n} = (P_n' + P_n'')/2$. Then

(5.12)
$$\mathbf{P}\left(\sup_{g\in\mathcal{G}(\delta_n)} \|g - g_0\|_n > 8\delta_n\right) \le 2\mathbf{P}\left(\sup_{g\in\mathcal{G}(\delta_n)} \left|\|g - g_0\|_{P_n'} - \|g - g_0\|_{P_n''}\right| > 6\delta_n\right).$$

To see this, use a symmetrization lemma (Pollard (1984, Lemma II.3.8.)). Because we assumed that $\{g - g_0 : g \in \mathcal{G}\}$ is uniformly bounded by 1, we may use the inequality in Pollard (1984, page 33), which reads

$$\text{(5.13)}\qquad \mathbf{P}\left(\sup_{g\in\mathcal{G}(\delta)} \left|\|g - g_0\|_{P_n'} - \|g - g_0\|_{P_n''}\right| > 6\delta_n \mid X_1,\dots,X_{2n}\right) \le 2\exp\left[H\left(\sqrt{2}\delta_n, \mathcal{G}, \bar{P}_{2n}\right) - 2n\delta_n^2\right].$$

Now, verify that for all $u > 0$, $H(\sqrt{2}u, \mathcal{G}, \bar{P}_n) \le H(u, \mathcal{G}, P_n') + H(u, \mathcal{G}, P_n'')$. The rest is standard. □

We are now ready to present the extension of Lemma 5.2 to the case where the radius δ_n is allowed to decrease with n. Take a non-increasing function $H(u)$ of $u > 0$, such that the integral (of the square root) converges:

$$\text{(5.14)}\qquad \int_0^1 H^{1/2}(u)\,du < \infty,$$

and define

$$\text{(5.15)}\qquad J(\delta) = \int_0^\delta H^{1/2}(u)\,du \vee \delta.$$

Lemma 5.5 *For each $A > 0$ and for $n\delta_n^2 > 2AH(\delta_n)$ and $H(8\delta_n) \to \infty$, we have*

$$\text{(5.16)}\qquad \limsup_{n\to\infty} \mathbf{P}\left(\sup_{g\in\mathcal{G}(\delta_n)} |\nu_n(g) - \nu_n(g_0)| > 4CA^{1/2}J(8\delta_n)\right) \le \limsup_{n\to\infty} 5\mathbf{P}\left(\sup_{u>0} \frac{H(u, \mathcal{G}, P_n)}{H(u)} > A\right).$$

Proof Replace in Lemma 5.4, the function $H(u)$ by $AH(u)$, where $A > 0$ is arbitrary. Then we find

$$\mathbf{P}\left(\sup_{g\in\mathcal{G}(\delta_n)} \|g - g_0\|_n > 8\delta_n\right) \le 4\exp[-n\delta_n^2] + 4\mathbf{P}\left(\sup_{u>0} \frac{H(u, \mathcal{G}, P_n)}{H(u)} > A\right).$$

Use Lemma 5.1 to arrive at

$$\mathbf{P}\left(\sup_{g \in \mathcal{G}(\delta_n)} |\nu_n(g) - \nu_n(g_0)| > 8CA^{1/2}J(8\delta_n) \right)$$
$$\leq C \exp\left[-\frac{AJ^2(8\delta_n)}{64\delta_n^2} \right] + 4\exp[-n\delta_n^2] + 5\mathbf{P}\left(\sup_{u>0} \frac{H(u, \mathcal{G}, P_n)}{H(u)} > A \right).$$

The assumptions $n\delta_n^2 \geq 2AH(\delta_n)$ and $H(8\delta_n) \to \infty$ imply that $J^2(8\delta_n)/\delta_n^2 \to \infty$ as well as $n\delta_n^2 \to \infty$. So the first two terms on the right-hand side converge to zero. □

We conclude that if the empirical entropy behaves well, in the sense that for some $A > 0$ (which without loss of generality may be taken equal to one), and for some $H(u)$ satisfying (5.14),

$$\limsup_{n\to\infty} \mathbf{P}\left(\sup_{u>0} \frac{H(u, \mathcal{G}, P_n)}{H(u)} > A \right) = 0,$$

then it follows from Lemma 5.6 that indeed, roughly speaking, the increment of the empirical process $\nu_n(g)$ at g_0 behaves like $J(\delta_n)$ for $\|g - g_0\| \leq \delta_n$. In many examples, a 'good' uniform bound

$$\sup_{Q \in \mathcal{M}} H(u, \mathcal{G}, Q) \leq H(u), \quad \text{for all } u > 0,$$

holds. Here, $\mathcal{M}$ is the collection of all probability measures on $\mathcal{X}$. By 'good' we mean that $H(u, \mathcal{G}, P_n)$ and $H(u)$ are of the same order in u. Such a 'good' bound is true for example when $\mathcal{G}$ is a uniformly bounded Vapnik–Chervonenkis subgraph class, or a class of uniformly bounded monotone functions, etc. Of course, since the empirical measure has finite support, we may weaken the uniform entropy bound somewhat by requiring only that it holds uniformly in all probability measures Q with finite support.

5.3. Empirical entropy and empirical norms

This section presents an extension of Lemma 5.4. Let $X_1, \ldots, X_n$ be i.i.d. and let $\mathcal{G}$ be a class of functions, uniformly bounded by a constant independent of n, say

$$\sup_{g \in \mathcal{G}} |g|_\infty \leq 1.$$

We shall provide a uniform bound for the ratio $\|g\|_n / \|g\|$.

In the proof of Lemma 5.6, we employ what we call the *peeling* device. Since we shall encounter this device more often, let us pay some attention to it here. Let $\tau : \mathcal{G} \to [\rho, R)$ be some function on $\mathcal{G}$, where $\rho > 0$ and where possibly $R = \infty$. Let $\{m_s\}_{s=0}^S$ be a strictly increasing sequence, with $m_0 = \rho$, $m_S = R$ (if $R = \infty$, take $S = \infty$ as well and take $\lim_{s\to\infty} m_s = \infty$). Then $\mathcal{G}$ can be peeled off into

$$\mathcal{G} = \bigcup_{s=1}^{S} \mathcal{G}_s,$$

where

$$\mathcal{G}_s = \{g \in \mathcal{G} : m_{s-1} \le \tau(g) < m_s\}, \quad s = 1, \ldots, S.$$

So, if $Z_n(g)$ is a stochastic process indexed by $\mathcal{G}$, we have for any positive a,

$$\text{(5.17)} \quad \mathbf{P}\left(\sup_{g\in\mathcal{G}} \frac{|Z_n(g)|}{\tau(g)} > a\right) \le \sum_{s=1}^{S} \mathbf{P}\left(\sup_{g\in\mathcal{G}_s} \frac{|Z_n(g)|}{\tau(g)} > a\right)$$
$$\le \sum_{s=1}^{S} \mathbf{P}\left(\sup_{g\in\mathcal{G},\ \tau(g)<m_s} |Z_n(g)| > am_{s-1}\right).$$

Thus, we obtain a probability inequality for the weighted process $Z_n(g)/\tau(g)$ from probability inequalities for the original process $Z_n(g)$.

Lemma 5.6 *Let $H(u)$ be a given non-increasing function of $u > 0$. For $n\delta_n^2 \to \infty$, $n\delta_n^2 \ge 2AH(\delta_n)$ for all n, we have*
(5.18)

$$\limsup_{n\to\infty} \mathbf{P}\left(\sup_{g\in\mathcal{G}} \frac{\|g\|_n}{\|g\| \vee \delta_n} > 14\right) \le \limsup_{n\to\infty} 4\mathbf{P}\left(\sup_{u>0} \frac{H(u,\mathcal{G},P_n)}{H(u)} > A\right).$$

Proof A symmetrization argument gives us

$$\mathbf{P}\left(\sup_{g\in\mathcal{G}} \frac{\|g\|_n}{\|g\| \vee \delta_n} > 14\right) \le 2\mathbf{P}\left(\sup_{g\in\mathcal{G}} \frac{\|g\|_{P_n'} - \|g\|_{P_n''}}{\|g\| \vee \delta_n} > 12\right).$$

Here P_n' and P_n'' are defined as in the proof of Lemma 5.4. But, using the peeling device,

$$\text{(5.19)} \quad \mathbf{P}\left(\sup_{g\in\mathcal{G}} \frac{\|g\|_{P_n'} - \|g\|_{P_n''}}{\|g\| \vee \delta_n} > 12 \mid X_1, \ldots, X_n\right)$$
$$\le \sum_{s=1}^{\infty} \mathbf{P}\left(\sup_{g\in\mathcal{G}(2^s\delta_n)} \|g\|_{P_n'} - \|g\|_{P_n''} > 6(2^s\delta_n) \mid X_1, \ldots, X_n\right)$$
$$\le \sum_{s=1}^{\infty} 2\exp\left[H(\sqrt{2}2^s\delta_n, \mathcal{G}, \bar{P}_{2n}) - 2n2^{2s}\delta_n^2\right],$$

where in the last inequality, we invoked (5.13).

If $n\delta_n^2 \geq 2AH(\delta_n)$, then also $nu^2 \geq 2AH(u)$ for all $u \geq \delta_n$, because $H(u)/u^2$ is a non-increasing function of $u > 0$. So we obtain

$$\text{(5.20)} \qquad \mathbf{P}\left(\sup_{g\in\mathcal{G}} \frac{\|g\|_{P_n'} - \|g\|_{P_n''}}{\|g\| \vee \delta_n} > 12\right)$$
$$\leq \sum_{j=1}^{\infty} 2\exp\left[-n2^{2j}\delta_n^2\right] + 2\mathbf{P}\left(\sup_{u>0} \frac{H(u,\mathcal{G},P_n)}{H(u)} > A\right).$$

Because $n\delta_n^2 \to \infty$, the first term on the right-hand side of (5.20) converges to zero. □

Example: the ratio of empirical and theoretical measure over a Vapnik–Chervonenkis class Take

$$\mathcal{G} = \{\mathrm{l}_D : D \in \mathcal{D}\},$$

where $\mathcal{D}$ is a Vapnik–Chervonenkis class. Then clearly, $\mathcal{G}$ is a Vapnik–Chervonenkis subgraph class with envelope $G \equiv 1$. Therefore,

$$H(\delta,\mathcal{G},P_n) \leq A\log\left(\frac{1}{\delta}\right),$$

for some constant A and for all $\delta > 0$ sufficiently small (see Theorem 3.11). For $g = \mathrm{l}_D$, we have

$$\|g\| = P^{1/2}(D).$$

Take

$$\delta_n^2 = A\frac{\log n}{n}.$$

Then

$$2H(\delta_n,\mathcal{G},P_n) \leq A(\log n - \log\log n - \log A) \leq A\log n = n\delta_n^2.$$

So we find from Lemma 5.6,

$$\limsup_{n\to\infty} \mathbf{P}\left(\sup_{D\in\mathcal{D}} \frac{P_n(D)}{P(D) \vee (A\log n/n)} > 196\right) = 0.$$

5.4. A uniform inequality based on entropy with bracketing

As before, $\mathcal{G}$ is a class of functions on $(\mathcal{X},\mathcal{A})$. Moreover, $X_1,\ldots,X_n$ are i.i.d. with distribution P on $(\mathcal{X},\mathcal{A})$. In this section, we derive probability inequalities for n fixed.

In Sections 5.1 and 5.2, the object of study was the behaviour of the empirical process on $\mathcal{G}(\delta) = \{g \in \mathcal{G} : \|g - g_0\| \leq \delta\}$. One arrives at the same conclusions if one first formulates the results for the entire class $\mathcal{G}$, and then applies them with $\mathcal{G}$ replaced by $\mathcal{G}(\delta)$. This is the approach we adopt here.

5.4.1. Bernstein's inequality

When using conditions on the entropy with bracketing, one no longer needs the symmetrization device. This means also that Hoeffding's inequality is not available. Instead, we shall apply Bernstein's inequality. Lemma 5.7 below presents this inequality for g fixed, and Theorem 5.11 extends it to a uniform inequality.

Lemma 5.7 (Bernstein's inequality.) *Suppose that* $\int g\, dP = 0$ *and*

$$\int |g|^m \, dP \leq \frac{m!}{2} K^{m-2} R^2, \quad m = 2, 3, \ldots . \tag{5.21}$$

Then for all $a > 0$

$$\mathbf{P}(\nu_n(g) \geq a) \leq \exp\left[-\frac{a^2}{2(aKn^{-1/2} + R^2)} \right] . \tag{5.22}$$

The proof is in Shorack and Wellner (1986). See also Lemma 8.9, where an extension to martingales is given.

The condition (5.21) is equivalent to assuming that $|g|$ has an exponential moment. Later on, it will be convenient to use the quantity

$$\rho_K^2(g) = 2K^2 \int \left(e^{|g|/K} - 1 - |g|/K \right) dP, \quad K > 0 \tag{5.23}$$

We think of $\rho_K(g)$ as an extension of the $L_2(P)$-norm, and we call $\rho_K(g_1 - g_2)$ the *Bernstein difference* between g_1 and g_2. The idea is $e^x \approx 1 + x + x^2/2$, for x small, so that

$$2K^2(e^{|g|/K} - 1 - |g|/K) \approx |g|^2$$

for K large. From the Taylor expansion $e^x - 1 - x = \sum_{m=2}^{\infty} x^m/m!$, one immediately sees that if $\rho_K(g) \leq R$, then (5.21) holds. On the other hand, (5.21) implies

$$\rho_{2K}^2(g) = 2(2K)^2 \sum_{m=2}^{\infty} \frac{\int |g|^m \, dP}{m!(2K)^m} \leq 8K^2 \sum_{m=2}^{\infty} \frac{K^{m-2} R^2}{2(2K)^m} = 2R^2.$$

For bounded g, $g(X)$ possesses an exponential moment. In many situations, we indeed shall consider (uniformly) bounded functions, so let us put this observation in a lemma.

Lemma 5.8 *Suppose that*

$$|g|_\infty \leq K, \tag{5.24}$$

and

$$\|g\| \leq R. \tag{5.25}$$

Then (5.21) *holds, so that* $\rho^2_{2K}(g) \leq 2R^2$.

Proof The conditions imply that

$$\int |g|^m \, dP \leq K^{m-2}R^2, \quad m = 2, 3, \ldots .$$

□

Of course, under (5.24), we may take $R = K$ in (5.25). But then, the lemma loses much of its power. The idea is rather that for bounded random variables, $\rho_K(g)$ behaves like $\|g\|$.

Notice that if in (5.21), $R = 1$ and $K = 1$, we have for $a \leq \sqrt{n}$,

$$\mathbf{P}(|\nu_n(g)| \geq a) \leq 2 \exp\left[-\frac{a^2}{4}\right].$$

This means that in the limit, the tails are sub-Gaussian. On the other hand, if in (5.21), $R = 1$, $K = \sqrt{n}$, we get for $a \geq 1$

$$\mathbf{P}(|\nu_n(g)| \geq a) \leq 2 \exp\left[-\frac{a}{4}\right],$$

i.e., we then have exponential tails. If $X_1, \ldots, X_n$ are the first n of an infinite sequence of i.i.d. random variables, and $g \in L_2(P)$, then roughly speaking, we indeed are in a situation where one can take $K = \sqrt{n}$. This is because

$$\begin{aligned}
&\mathbf{P}\Big(\sqrt{n}\left|\int_{|g|>\sqrt{n}} g \, dP_n\right| \geq \frac{a}{2}\Big) \\
&\leq \mathbf{P}\big(\,|g(X_i)| > \sqrt{n} \text{ for some } i \in \{1, \ldots, n\}\,\big) \leq n\mathbf{P}\big(\,|g(X_1)| > \sqrt{n}\,\big) \\
&\leq \|g\mathrm{l}\{|g| > \sqrt{n}\}\|^2 \to 0, \quad \text{as } n \to \infty.
\end{aligned}$$

We now state a corollary of Lemma 5.7, which we generalize in Theorem 5.11.

Corollary 5.9 *Suppose* $\int g\, dP = 0$, *and*

$$\rho_K(g) \leq R. \tag{5.26}$$

Then for

$$a \leq C_1 \sqrt{n} R^2 / K, \tag{5.27}$$

we have

$$\mathbf{P}\big(|\nu_n(g)| \geq a\big) \leq 2 \exp\left[-\frac{a^2}{2(C_1+1)R^2}\right]. \tag{5.28}$$

5.4.2. Generalized entropy with bracketing

We aim at extending Corollary 5.9 to hold uniformly over $\mathcal{G}$. For this purpose, we define covering numbers of $\mathcal{G}$, but we measure the closeness of two functions in $\mathcal{G}$ in terms of the Bernstein difference ρ_K. Now, ρ_K is not a distance function. Because entropies in the usual sense are based on distances, we shall use the terminology *generalized entropy*.

Definition 5.1 (Generalized entropy with bracketing) Let $\mathcal{N}_{B,K}(\delta, \mathcal{G}, P)$ be the smallest value of N for which there exist pairs of functions $\{[g_j^L, g_j^U]\}_{j=1}^N$ such that $\rho_K(g_j^U - g_j^L) \leq \delta$ for all $j = 1, \ldots, N$, and such that for each $g \in \mathcal{G}$, there is a $j = j(g) \in \{1, \ldots, N\}$ such that

$$g_j^L \leq g \leq g_j^U.$$

(Take $\mathcal{N}_{B,K}(\delta, \mathcal{G}, P) = \infty$ if no finite set of such brackets exists.) Then $\mathcal{H}_{B,K}(\delta, \mathcal{G}, P) = \log \mathcal{N}_{B,K}(\delta, \mathcal{G}, P)$ is called the *generalized entropy* with bracketing of $\mathcal{G}$.

Generalized entropy with bracketing is only of interest for classes $\mathcal{G}$ that satisfy for some K and R

$$\sup_{g \in \mathcal{G}} \rho_K(g) \leq R, \tag{5.29}$$

for instance, for uniformly bounded classes of functions. But in the latter situation, the generalized entropy with bracketing may be replaced by the (usual) entropy with bracketing, because the Bernstein difference can be bounded by the $L_2(P)$-distance (modulo some constants).

Lemma 5.10 *Suppose*

$$\sup_{g\in\mathcal{G}} |g|_\infty \le K.$$

Then

$$\mathcal{H}_{B,4K}(\sqrt{2}\delta, \mathcal{G}, P) \le H_B(\delta, \mathcal{G}, P), \quad \textit{for all } \delta > 0.$$

Proof Take a minimal δ-bracketing set $\{[g_j^L, g_j^U]\}_{j=1}^N$ for the $L_2(P)$-norm. We may choose $-K \le g_j^L \le g_j^U \le K$ for all j. Because $\|g_j^U - g_j^L\| \le \delta$, we then have (see Lemma 5.8),

$$\rho_{4K}^2(g_j^U - g_j^L) \le 2\delta^2, \quad j = 1, \ldots, N.$$ □

However, we shall see that the application to maximum likelihood uses the case where functions are *not* uniformly bounded, but *do* satisfy the uniform exponential moment condition (5.29).

5.4.3. *A uniform inequality*

Let $\mathcal{G}$ satisfy

(5.30) $$\sup_{g\in\mathcal{G}} \rho_K(g) \le R.$$

In Theorem 5.11 below, C is a (sufficiently large) universal constant, whereas a, C_0 and C_1 may be chosen, but do have to satisfy (5.31)–(5.34).

Theorem 5.11 *Take*

(5.31) $$a \le C_1 \sqrt{n} R^2 / K,$$

(5.32) $$a \le 8\sqrt{n} R,$$

(5.33) $$a \ge C_0 \left(\int_{a/(2^6\sqrt{n})}^{R} \mathcal{H}_{B,K}^{1/2}(u, \mathcal{G}, P)\, du \vee R \right),$$

and

(5.34) $$C_0^2 \ge C^2(C_1 + 1).$$

Then

(5.35) $$\mathbf{P}\left(\sup_{g\in\mathcal{G}} |\nu_n(g)| \ge a \right) \le C \exp\left[-\frac{a^2}{C^2(C_1+1)R^2} \right].$$

The theorem is a special case of Theorem 8.13.

From Corollary 5.9, it is obvious where (5.31) comes from. Condition (5.32) is only to make sure that in (5.33), the lower integration limit is smaller than the upper integration limit. If the entropy integral converges:

$$\int_0^1 \mathcal{H}_{B,K}^{1/2}(u, \mathcal{G}, P)\, du < \infty, \tag{5.36}$$

then (5.32) may be omitted and the lower integration limit in (5.33) can be taken equal to zero. Condition (5.33) is similar to the one we encountered in Section 5.1 and 5.2: the increments of the empirical process roughly behave as the integral of the (square root) entropy. Finally, condition (5.34) says that we must take our constant C_0 sufficiently large. The theorem holds for all n, and the behaviour of the empirical process is given more or less explicitly as a function of K, R and the generalized entropy $\mathcal{H}_{B,K}(\delta, \mathcal{G}, P)$, $\delta > 0$. This means that we can use the theorem for classes of functions $\mathcal{G}$ depending on n (with K and R, also depending on n, perhaps tending to zero or tending to infinity in special situations).

One may ask the question: why do we need (generalized) entropy with *bracketing* here? The reason is that we in fact only have exponential tails, and not sub-Gaussian tails. If $\{Z_j : 1 \le j \le N\}$ satisfies for each j,

$$\mathbf{P}(Z_j \ge a) \le \exp[-a],$$

then

$$\mathbf{P}\left(\max_{j=1,\dots,N} Z_j \ge a\right) \le \exp[\log N - a]$$

(compare with the loose argument in Section 3.3). In order to get a small right hand side, one should take a larger than $\log N$. For sub-Gaussian tails, a should be larger than the *square root* of $\log N$, rather than $\log N$. The idea is that if one uses entropy with bracketing, then one again has inequalities of sub-Gaussian type and one wins the square root. Bernstein's inequality is to be used in its full strength, and an adaptive truncation argument is invoked in order to be able to get good inequalities for small values of $\rho_K(\cdot)$. The adaptive truncation argument makes extensive use of (generalized) entropy with bracketing.

5.5. Entropy with bracketing and asymptotic equicontinuity

Suppose $X_1, \dots, X_n$ are the first n of an infinite i.i.d. sequence. Moreover, suppose $\mathcal{G}$ is a fixed class, i.e., not depending on n. Also assume $\mathcal{G}$ is a subset of a ball in $L_2(P)$, say

$$\sup_{g \in \mathcal{G}} \|g\| \le 1.$$

Take $g_0 \in \mathcal{G}$ fixed, and write for $\delta > 0$,

$$\mathcal{G}(\delta) = \{g \in \mathcal{G} : \|g - g_0\| \leq \delta\}.$$

We assume that

$$\int_0^1 H_B^{1/2}(u, \mathcal{G}, P)\, du < \infty. \tag{5.37}$$

Define

$$J_B(\delta, \mathcal{G}, P) = \int_0^\delta H_B^{1/2}(u, \mathcal{G}, P)\, du \vee \delta.$$

Theorem 5.12 *Suppose that* (5.37) *holds. Then ν_n is asymptotically equicontinuous at g_0, i.e., for all $\eta > 0$, there exists a $\delta > 0$ such that*

$$\limsup_{n\to\infty} \mathbf{P}\left(\sup_{g\in\mathcal{G}(\delta)} |\nu_n(g) - \nu_n(g_0)| > \eta \right) < \eta. \tag{5.38}$$

Proof Since $H_B(1, \mathcal{G}, P) < \infty$, we have that

$$\sup_{g\in\mathcal{G}} |g - g_0| \leq G_0,$$

where $\|G_0\| < \infty$. Hence also

$$\sup_{g\in\mathcal{G}} \left| g - g_0 - \int (g - g_0)\, dP \right| \leq G,$$

with $\|G\| < \infty$. Because for all $L > 0$, $\mathbf{P}(G(X_i) > L\sqrt{n},$ for some $i \in \{1, \ldots, n\}) \to 0$ as $n \to \infty$, we obtain

$$\mathbf{P}\left(\sup_{g\in\mathcal{G}} \left| \sqrt{n} \int_{G > L\sqrt{n}} (g - g_0)\, d(P_n - P) \right| > \frac{\eta}{2} \right) \to 0.$$

The class $\{g - g_0 : g \in \mathcal{G}\}$ has δ-entropy with bracketing equal to $H_B(\delta, \mathcal{G}, P)$. Recall that for

$$\mathcal{G}_K = \left\{ (g - g_0 - \int (g - g_0)\, dP) \mathrm{l}\{G \leq K/4\} : g \in \mathcal{G} \right\},$$

we have $\mathcal{H}_{B,K}(\sqrt{2}u, \mathcal{G}_K, P) \leq H_B(u, \mathcal{G}, P)$ (Lemma 5.10). We shall apply Theorem 5.11 with $C_1 = 1$, $R = \sqrt{2}\delta$, $K = 4L\sqrt{n}$ and $a = C_0\sqrt{2}J(\delta, \mathcal{G}, P)$.

Condition (5.31) is then satisfied if we take $L = \delta^2/(2a)$. Now choose C_0 sufficiently large, so that

$$C \exp\left[-\frac{C_0^2}{2C^2}\right] < \eta.$$

Due to condition (5.37), $J(\delta, \mathcal{G}, P) \to 0$ as $\delta \to 0$. Choose δ sufficiently small, so that

$$C_0\sqrt{2}J(\delta, \mathcal{G}, P) < \frac{\eta}{2}.$$

Then

$$\begin{aligned}
&\mathbf{P}\left(\sup_{g\in\mathcal{G}(\delta)} \sqrt{n}\left|\int_{G\leq L\sqrt{n}} (g-g_0)\, d(P_n - P)\right| > \frac{\eta}{2}\right)\\
&\leq \mathbf{P}\left(\sup_{g\in\mathcal{G}(\delta)} \sqrt{n}\left|\int_{G\leq L\sqrt{n}} (g-g_0)\, d(P_n - P)\right| > C_0\sqrt{2}J(\delta, \mathcal{G}, P)\right)\\
&\leq C \exp\left[-\frac{C_0^2 2J^2(\delta, \mathcal{G}, P)}{2C^2(2\delta^2)}\right] \leq C\exp\left[-\frac{C_0^2}{2C^2}\right]\\
&< \eta.
\end{aligned}$$

□

Remark One may verify that in fact the conclusion of the previous theorem can be strengthened to: for all $\eta > 0$ there exists a $\delta > 0$, such that

$$\limsup_{n\to\infty} \mathbf{P}\left(\sup_{g_1, g_2\in\mathcal{G},\ \|g_1-g_2\|\leq\delta} |\nu_n(g_1) - \nu_n(g_2)| > \eta\right) < \eta$$

(see also the remark following Theorem 5.3).

5.6. Modulus of continuity

In this section we investigate the behaviour of $|\nu_n(g) - \nu_n(g_0)|$ as a function of $\|g - g_0\|$. Consider a uniformly bounded class of functions, say

$$\sup_{g\in\mathcal{G}} |g - g_0|_\infty \leq 1. \tag{5.39}$$

Assume moreover that

$$H_B(\delta, \mathcal{G}, P) \leq A\delta^{-\alpha}, \quad \text{for all } \delta > 0, \tag{5.40}$$

for some $0 < \alpha < 2$ and some constant A. Then

$$\int_0^\delta H_B^{1/2}(u, \mathcal{G}, P) \leq \frac{A^{1/2}}{1-\alpha/2}\delta^{1-\alpha/2} = A_0\delta^{1-\alpha/2}, \quad \delta > 0.$$

Application of Theorem 5.11 to the class $\mathcal{G}(\delta) = \{g \in \mathcal{G} : \|g - g_0\| \leq \delta\}$ shows that at each g_0, the increments of the empirical process $\nu_n(g)$ for $\|g - g_0\| \leq \delta$ behave like $\delta^{1-\alpha/2}$. Because the theorem holds for each n, we can investigate what happens if δ depends on n, and converges to zero as n tends to infinity. We shall prove below that for $\delta_n = n^{-1/(2+\alpha)}$,

$$\sup_{g \in \mathcal{G}} \frac{|\nu_n(g) - \nu_n(g_0)|}{\|g - g_0\|^{1-\alpha/2} \vee \sqrt{n}\delta_n^2} = O_{\mathbf{P}}(1).$$

Thus, if $\mathcal{G}$ is small, in the sense that α is small, then the modulus of continuity of ν_n is also small. The limiting case, where $|\nu_n(g) - \nu_n(g_0)|$ behaves like $\|g - g_0\|$, uniformly in $\|g - g_0\|$ not too near to zero, is reserved for the finite-dimensional situation.

In some cases, the supremum norm decreases with the $L_2(P)$-norm. Let us capture this by assuming that for some constants $c_0 > 0$ and $0 \leq \beta \leq 1$,

(5.41) $$\sup_{g \in \mathcal{G}(\delta)} |g - g_0|_\infty \leq (c_0\delta)^\beta, \quad \text{for all } 0 < \delta \leq 1.$$

This is not really an additional assumption, because (5.39) implies (5.41) with $\beta = 0$. But for $\beta > 0$, (5.41) allows us to push $\|g - g_0\|$ a little more towards zero in the modulus of continuity result.

Lemma 5.13 *Assume* (5.39), (5.40) *and* (5.41). *For some constants c and n_0 depending on α, β, c_0 and A, we have for all $T \geq c$ and $n \geq n_0$,*

(5.42)
$$\mathbf{P}\left(\sup_{g \in \mathcal{G}\left(n^{-\frac{1}{2+\alpha-2\beta}}\right)} \left|\int (g - g_0)\, d(P_n - P)\right| \geq T n^{-\frac{2-\beta}{2+\alpha-2\beta}}\right) \leq c \exp\left[-\frac{T n^{\frac{\alpha}{2+\alpha-2\beta}}}{c^2}\right].$$

Moreover, for $T \geq c$, $n \geq n_0$,

(5.43) $$\mathbf{P}\left(\sup_{g \in \mathcal{G},\ \|g - g_0\| > n^{-\frac{1}{2+\alpha-2\beta}}} \frac{|\nu_n(g) - \nu_n(g_0)|}{\|g - g_0\|^{1-\frac{\alpha}{2}}} \geq T\right) \leq c \exp\left[-\frac{T}{c^2}\right].$$

Proof Replace $\mathcal{G}$ by $\mathcal{G}(\delta)$ in Theorem 5.11, and take $K = 4(c_0\delta)^\beta$, $R = \sqrt{2}\delta$ and $a = \frac{1}{2}C_1 c_0^{-\beta}\delta^{1-\alpha/2}$, with $C_1 = 2\sqrt{2}C_0 A_0 c_0^\beta$. Then (5.31) is satisfied for all $\delta \geq n^{-1/(2+\alpha-2\beta)}$. Condition (5.32) is satisfied if we take $n \geq n_0$, n_0 sufficiently large, and (5.33) is satisfied if we take C_0 sufficiently large. So for all $\delta \geq n^{-1/(2+\alpha-2\beta)}$, we obtain

(5.44) $$\mathbf{P}\left(\sup_{g \in \mathcal{G}(\delta)} |\nu_n(g) - \nu_n(g_0)| \geq \frac{1}{2}C_1 c_0^{-\beta}\delta^{1-\frac{\alpha}{2}}\right) \leq C \exp\left[-\frac{C_1\delta^{-\alpha}}{16C^2 c_0^{2\beta}}\right].$$

For $\delta = n^{-1/(2+\alpha-2\beta)}$, (5.44) gives (5.42).

Now, let $S = \min\{s > 1 : 2^{-s} < n^{-1/(2+\alpha-2\beta)}\}$ and apply the peeling device. Then for $T = C_1 c_0^{-\beta} 2^{-\alpha/2}$,

$$\mathbf{P}\left(\sup_{g\in\mathcal{G},\ \|g-g_0\|>n^{-\frac{1}{2+\alpha-2\beta}}} \frac{|\nu_n(g)-\nu_n(g_0)|}{\|g-g_0\|^{1-\frac{\alpha}{2}}} \geq T \right)$$

$$\leq \sum_{s=1}^{S} \mathbf{P}\left(\sup_{g\in\mathcal{G}(2^{-s+1})} |\nu_n(g)-\nu_n(g_0)| \geq \frac{1}{2} C_1 c_0^{-\beta} (2^{-s+1})^{1-\frac{\alpha}{2}} \right)$$

$$\leq \sum_{s=1}^{S} C \exp\left[-\frac{C_1 (2^{-s+1})^{-\alpha}}{16 C^2 c_0^{2\beta}} \right] \leq c \exp\left[-\frac{T}{c^2} \right]. \qquad \square$$

Lemma 5.13 can be extended to functions with uniformly bounded Bernstein difference and with generalized δ-entropy with bracketing not necessarily a polynomial in δ. We shall not pursue this here. Instead, we consider an extension in another direction.

Suppose that $\mathcal{G}$ has infinite entropy, but that

$$\mathcal{G} = \bigcup_{M\geq 1} \mathcal{G}_M,$$

where each $\mathcal{G}_M$ has finite entropy. In particular, we assume that there is a map $I : \mathcal{G} \mapsto [1,\infty)$, such that

$$\mathcal{G}_M = \{g \in \mathcal{G} : I(g) \leq M\}.$$

We think of $I(g)$ as the complexity or irregularity of the function g (for example some Sobolev or Besov norm). Now, a space of irregular functions is a rich space: we allow the entropy of $\mathcal{G}_M$ to increase with M. We shall evaluate the behaviour of the empirical process $\nu_n(g)$, not only as a function of its increments, but also as a function of $I(g)$. A slight further extension of the previous lemma is that we shall formulate the increments of the empirical process in terms of a more general distance function $d(\cdot,\cdot)$, such that

$$\|g-g_0\| \leq d(g,g_0), \quad \text{for all } g \in \mathcal{G}. \tag{5.45}$$

The following conditions will be used: there exist constants $0 < \alpha < 2$, $0 \leq \beta \leq 1$, $c_0 > 0$ and $A > 0$ such that for all $M \geq 1$,

$$\sup_{g\in\mathcal{G}_M} d(g,g_0) \leq c_0 M, \tag{5.46}$$

$$H_B(\delta, \mathcal{G}_M, P) \le A\left(\frac{M}{\delta}\right)^\alpha, \quad \text{for all } \delta > 0, \tag{5.47}$$

and

$$\sup_{g \in \mathcal{G}_M,\ d(g,g_0)\le\delta} |g - g_0|_\infty \le (c_0\delta)^\beta M^{1-\beta}, \quad \text{for all } \delta > 0. \tag{5.48}$$

Lemma 5.14 *Assume* (5.45)–(5.48). *Then, for some constants* c *and* n_0 *depending on* α, β, c_0 *and* A, *we have for all* $T \ge c$ *and* $n \ge n_0$,

$$\mathbf{P}\left(\sup_{g\in\mathcal{G},\ d(g,g_0)\le n^{-\frac{1}{2+\alpha-2\beta}} I(g)} \frac{|\int (g-g_0)\, d(P_n - P)|}{I(g)} \ge T n^{-\frac{2-\beta}{2+\alpha-2\beta}}\right) \tag{5.49}$$
$$\le c\exp\left[-\frac{T n^{\frac{\alpha}{2+\alpha-2\beta}}}{c^2}\right].$$

Moreover, for $T \ge c$, $n \ge n_0$,

$$\mathbf{P}\left(\sup_{g\in\mathcal{G},\ d(g,g_0)> n^{-\frac{1}{2+\alpha-2\beta}} I(g)} \frac{|\nu_n(g) - \nu_n(g_0)|}{d^{1-\frac{\alpha}{2}}(g,g_0) I^{\frac{\alpha}{2}}(g)} \ge T\right) \le c\exp\left[-\frac{T}{c^2}\right]. \tag{5.50}$$

Proof We only prove (5.50). Inequality (5.49) can be obtained in a similar fashion. Reasoning along the lines of Lemma 5.13, one sees that for all $M \ge 1$,

$$\mathbf{P}\left(\sup_{g\in\mathcal{G}_M,\ d(g,g_0)> n^{-\frac{1}{2+\alpha-2\beta}} \frac{M}{2}} \frac{|\nu_n(g) - \nu_n(g_0)|}{d^{1-\frac{\alpha}{2}}(g,g_0)} \ge T\left(\frac{M}{2}\right)^{\frac{\alpha}{2}}\right) \le c_1\exp\left[-\frac{T M^\alpha}{c_1^2}\right].$$

So, applying the peeling device once more,

$$\mathbf{P}\left(\sup_{g\in\mathcal{G},\ d(g,g_0)> n^{-\frac{1}{2+\alpha-2\beta}} I(g)} \frac{|\nu_n(g) - \nu_n(g_0)|}{d^{1-\frac{\alpha}{2}}(g,g_0) I^{\frac{\alpha}{2}}(g)} \ge T\right)$$
$$\le \sum_{s=0}^\infty \mathbf{P}\left(\sup_{g\in\mathcal{G},\ I(g)\le 2^{s+1},\ d(g,g_0)> n^{-\frac{1}{2+\alpha-2\beta}} 2^s} \frac{|\nu_n(g) - \nu_n(g_0)|}{d^{1-\frac{\alpha}{2}}(g,g_0)} \ge T 2^{\frac{s\alpha}{2}}\right)$$
$$\le \sum_{s=0}^\infty c_1 \exp\left[-\frac{T 2^{(s+1)\alpha}}{c_1^2}\right] \le c\exp\left[-\frac{T}{c^2}\right].$$

□

5.7. Entropy with bracketing and empirical norms

Recall that in Section 5.3, we formulated a uniform bound for the ratio $\|g\|_n/\|g\|$, based on the empirical entropy. Here, the aim is to derive such a bound from the entropy with bracketing, instead of the empirical entropy.

We again assume that $\mathcal{G}$ is uniformly bounded, say

$$\sup_{g\in\mathcal{G}} |g|_\infty \leq 1. \tag{5.51}$$

Let us start with Bernstein's inequality for a single g.

Lemma 5.15 *We have for all $a > 0$,*

$$\mathbf{P}(|\|g\|_n^2 - \|g\|^2| \geq a) \leq 2\exp\left[-\frac{na^2}{8(a+\|g\|^2)}\right]. \tag{5.52}$$

Proof This lemma can be obtained from Bernstein's inequality (see Lemma 5.7). We have

$$\int |g|^{2m}\,dP \leq \|g\|^2; \qquad \int |g^2 - \|g\|^2|^m \leq 2^{2m-2}4\|g\|^2. \qquad \square$$

Lemma 5.16 *Take $n\delta_n^2 \geq 2H_B(\delta_n, \mathcal{G}, P)$, $n \geq 1$, and $n\delta_n^2 \to \infty$. Then for each $0 < \eta < 1$, we have*

$$\limsup_{n\to\infty} \mathbf{P}\left(\sup_{g\in\mathcal{G},\ \|g\|>2^5\delta_n/\eta} \left|\frac{\|g\|_n}{\|g\|} - 1\right| \geq \eta\right) = 0. \tag{5.53}$$

Moreover,

$$\limsup_{n\to\infty} \mathbf{P}\left(\sup_{g\in\mathcal{G},\ \|g\|\leq 2^5\delta_n} |\|g\|_n - \|g\|| \geq 2\right) = 0. \tag{5.54}$$

Proof Apply the peeling device. Take $g \in \mathcal{G}$, and suppose that $(s-1)\delta_n < \|g\| \leq s\delta_n$, where $s \in \{1, 2, \dots\}$, $s \geq 2^5/\eta$. Furthermore, let $-1 \leq g^L \leq g \leq g^U \leq 1$, and $\|g^U - g^L\| \leq \delta_n$. Since

$$\|g\|_n^2 \leq \|g^U\|_n^2 + \|g^U - g^L\|_n^2,$$

the inequality

$$\frac{\|g\|_n}{\|g\|} \geq 1 + \eta,$$

implies

$$\|g^U\|_n^2 - \|g^U\|^2 + \|g^U - g^L\|_n^2 \geq (1+\eta)^2(s-1)^2\delta_n^2 - (s+1)^2\delta_n^2 \geq \eta s^2\delta_n^2.$$

Moreover,

$$\frac{s^2}{s+1} \geq \frac{s}{4}.$$

So

$$\mathbf{P}\left(\sup_{g\in\mathcal{G},\ \|g\|>2^5\delta_n/\eta} \frac{\|g\|_n}{\|g\|} \geq 1+\eta\right) \leq \sum_{s\geq\frac{2^5}{\eta}} 4\exp\left[H_B(\delta_n,\mathcal{G},P) - \frac{n^2\delta_n^2 s^2\eta^2}{2^8}\right].$$

Since $n\delta_n^2 \geq 2H_B(\delta_n,\mathcal{G},P)$, we have that $H_B(\delta_n,\mathcal{G},P) \leq (n\delta_n^2 s^2\eta^2)/2^9$, for all $s \geq 2^5/\eta$. Therefore,

$$\mathbf{P}\left(\sup_{g\in\mathcal{G},\ \|g\|>2^5\delta_n/\eta} \frac{\|g\|_n}{\|g\|} \geq 1+\eta\right) \leq \sum_{s\geq\frac{2^5}{\eta}} 4\exp\left[-\frac{n\delta_n^2 s^2\eta^2}{2^7}\right] \to 0, \quad \text{as } n\to\infty.$$

Similarly, if $(s-1)\delta_n < \|g-g_0\| \leq s\delta_n$, and $s \geq 2^5/\eta$, then the inequality

$$\frac{\|g\|_n}{\|g-g_0\|} < 1-\eta,$$

implies

$$\begin{aligned}\|g^L\|_n^2 - \|g^L\|^2 - \|g^U - g^L\|_n^2 &\leq (1-\eta)^2(s-1)^2\delta_n^2 - (s-2)^2\delta_n^2\\ &\leq -\eta\frac{s(s-2)}{2}\delta_n^2.\end{aligned}$$

Moreover,

$$\frac{s(s-2)}{2(s+1)} \geq \frac{s}{4}.$$

So the same arguments give

$$\mathbf{P}\left(\sup_{g\in\mathcal{G},\ \|g\|>2^5\delta_n/\eta} \frac{\|g\|_n}{\|g\|} < 1-\eta\right) \to 0.$$

The proof of (5.54) can be obtained in a similar fashion. □

5.8. Notes

If the functions in $\mathcal{G}$ are normalized to have expectation zero, asymptotic equicontinuity is a necessary condition for the weak convergence of the empirical process (see also Chapter 6). Therefore, the concept plays an essential role in the theory of empirical processes. We refer to Dudley (1984), and van der Vaart and Wellner (1996). An overview is given in Giné (1996). Theorem 5.3 is essentially in Pollard (1982). He imposes a stronger condition than (5.5), namely, that for each probability measure Q with finite support

$$H(u\|G\|_Q, \mathcal{G}, Q) \le \lambda(u),$$

with

$$\int_0^1 \lambda^{1/2}(u)\, du < \infty.$$

This condition is independent of the (unknown) measure P, so that its verification is feasible. The weaker condition (5.5) is in Pollard (1984). We learned the peeling device from Alexander (1985). The result on the ratio of empirical and theoretical norms over Vapnik–Chervonenkis classes can be found in Breiman, Friedman, Olshen and Stone (1984). Bernstein's inequality is proved in Bernstein (1924) and Bennett (1962). The version we present here is from Shorack and Wellner (1986). The uniform inequality of Theorem 5.11 was obtained by Birgé and Massart (1993), albeit for functions of a special form (for example, uniformly bounded functions). They also treat the case of independent, non-identically distributed variables. Theorem 5.12 is proved in Ossiander (1987). Van de Geer (1988) presents Lemma 5.15 for the case where the δ-entropy with bracketing behaves like a power of δ.

5.9. Problems and complements

In all problems, we consider a sequence $X_1, \ldots, X_n, \ldots$ of i.i.d. random variables, with distribution P.

5.1. Let $\mathcal{G}$ be a class of functions with envelope $G \in L_2(P)$, and satisfying

$$\frac{1}{n} H(\delta, \mathcal{G}, P_n) \to_{\mathbf{P}} 0, \quad \text{for all } \delta > 0.$$

Show that

$$\sup_{g \in \mathcal{G}} \big| \|g\|_n - \|g\| \big| \to 0, \quad \text{a.s.}$$

Conclude that

$$\limsup_{n \to \infty} H(\delta, \mathcal{G}, P_n) < \infty, \quad \text{a.s., for all } \delta > 0.$$

5.2. Consider a uniformly bounded Vapnik–Chervonenkis subgraph class. Prove that for some constant c, and for all $\delta \geq n^{-1/2}\sqrt{\log n}$ and $T \geq c$,

$$\mathbf{P}\left(\sup_{g\in\mathcal{G}(\delta)} \left|\nu_n(g) - \nu_n(g_0)\right| \geq T\delta\sqrt{\log\frac{1}{\delta}}\right) \leq c\exp\left[-\frac{T^2\log(1/\delta)}{c^2}\right].$$

5.3. Consider the class

$$\mathcal{G} = \{g_\theta : \mathcal{X} \to [0,1],\ \theta \in \Theta\},$$

where Θ is a rectangle in $\mathbf{R}^d$, and where the map $\theta \mapsto g_\theta$ is differentiable, with derivative $\dot{g}_\theta$. Suppose that for some constant $0 < K < \infty$,

$$\frac{1}{K} \leq \inf_{\theta\in\Theta} |\dot{g}_\theta|_\infty \leq \sup_{\theta\in\Theta} |\dot{g}_\theta|_\infty \leq K.$$

(This implies that $\mathcal{G}$ is of finite metric dimension: see Problem 2.7.) Prove that for some constant c, and for all $\delta > n^{-1/2}$ and all $T \geq c$,

$$\mathbf{P}\left(\sup_{g\in\mathcal{G}(\delta)} \left|\nu_n(g) - \nu_n(g_0)\right| \geq T\delta\right) \leq c\exp\left[-\frac{T}{c^2}\right].$$

5.4. For some $m \in \{1, 2, \ldots\}$ and $M < \infty$, let

$$\mathcal{G} = \left\{g : [0,1] \to [0,1] : \int_0^1 \left(g^{(m)}(x)\right)^2 dx \leq M^2\right\}.$$

Show that for some constant c, and for all $\delta \geq n^{-m/(2m+1)}$ and $T \geq c$,

$$\mathbf{P}\left(\sup_{g\in\mathcal{G}(\delta)} \left|\nu_n(g) - \nu_n(g_0)\right| \geq T\delta^{1-\frac{1}{2m}}\right) \leq c\exp\left[-\frac{T\delta^{-1/m}}{c^2}\right].$$

Moreover, show that for all $\eta > 0$, there exists a $C_\eta < \infty$ such that

$$\limsup_{n\to\infty} \mathbf{P}\left(\sup_{g\in\mathcal{G},\ \|g\|>C_\eta n^{-\frac{m}{2m+1}}} \left|\frac{\|g\|_n}{\|g\|} - 1\right| > \eta\right) = 0.$$

5.5. Consider a d_n-dimensional linear subspace of $L_2(P)$, say

$$\mathcal{G}_n = \left\{g(x) = \sum_{j=1}^{d_n} \theta_{j,n}\psi_{j,n}(x) : (\theta_{1,n}, \ldots, \theta_{d_n,n})^T \in \mathbf{R}^{d_n}\right\}.$$

For $g_1, g_2 \in \mathscr{G}_n$, let

$$(g_1, g_2)_n = \int g_1 g_2 \, dP_n,$$

and

$$(g_1, g_2) = \int g_1 g_2 \, dP.$$

Suppose that

$$|g|_\infty \leq A_n \|g\|, \quad \text{for all } g \in \mathscr{G}_n,$$

where $A_n^2 d_n / n \to 0$. Then for all $\eta > 0$,

$$\limsup_{n\to\infty} \left(\sup_{g_1, g_2 \in \mathscr{G}} \frac{|(g_1, g_2)_n - (g_1, g_2)|}{\|g_1\| \|g_2\|} > \eta \right) = 0.$$

In particular,

$$\limsup_{n\to\infty} \mathbf{P} \left(\sup_{g \in \mathscr{G}} \left| \frac{\|g\|_n^2}{\|g\|^2} - 1 \right| > \eta \right) = 0;$$

see Huang (1996, Lemma 8.1).

6

Central Limit Theorems

A central limit theorem that holds uniformly in $g \in \mathcal{G}$ is one of the main topics in empirical process theory. Applications to statistics can be found in goodness-of-fit testing (for distributions in higher-dimensional spaces), or in proving asymptotic normality of certain differentiable functionals (the delta-method). Such applications will not be treated in this book. Therefore, we shall only present a short discussion on weak convergence of the empirical process. Some measurability issues are also briefly addressed.

6.1. Definitions

Let $(\Omega, \mathcal{F}, \mathbf{P})$ be a probability space, $(\mathcal{X}, \mathcal{A})$ a measurable space, and for $i \geq 1$, let $X_i : \Omega \to \mathcal{X}$ be measurable maps. We assume that $X_1, X_2, \ldots$ are i.i.d. with distribution P, i.e.,

$$\mathbf{P}(X_i \in A_i,\ i \geq 1) = \prod_{i=1}^{\infty} P(A_i), \quad A_i \in \mathcal{A},\ i \geq 1,$$

where $P(A) = \mathbf{P}(X_i \in A)$, $A \in \mathcal{A}$.

Let $\mathcal{G} \subset L_2(P)$, and denote the empirical process by $\nu_n = \{\nu_n(g) : g \in \mathcal{G}\}$, with $\nu_n(g) = \sqrt{n} \int g \, d(P_n - P)$, $g \in \mathcal{G}$. We regard ν_n as an element of the space $\mathcal{K}$ of all bounded real-valued functions on $\mathcal{G}$. Equip $\mathcal{K}$ with the supremum norm, and $\mathcal{G}$ with the $L_2(P)$-norm $\|\cdot\|$. Now, ν_n is in general not Borel measurable.

Example (See Billingsley (1968).) Let $(\Omega, \mathcal{F}, \mathbf{P}) = (\mathbf{R}^\infty, \mathcal{A}^\infty, P^\infty)$ and $X_i(\omega) = \omega_i$, $i \geq 1$. Take $\mathcal{G} = \{1_{(-\infty,t]} : t \in \mathbf{R}\}$. Then ν_n is the classical empirical process $\sqrt{n}(F_n - F)$, where $F_n = \sum_{i=1}^n 1\{X_i \leq \cdot\}/n$ is the empirical distribution function, and $F = P(X_1 \leq \cdot)$ is the theoretical distribution function. We shall show that F_1 is not Borel measurable. Let B_ξ be the open ball in $\mathcal{K}$ with center $1\{\xi \leq \cdot\}$, and radius ϵ:

$$B_\xi = \left\{ k \in \mathcal{K} : \sup_{t \in \mathbf{R}} |k(1_{(-\infty,t]}) - 1\{\xi \leq t\}| < \epsilon \right\}.$$

Here, $0 < \epsilon < 1$ is fixed. Take an arbitrary subset $A \subset R$, and define $B = \bigcup_{\xi \in A} B_\xi$. Since B is open, it is Borel measurable. But we have

$$\{\omega_1 : F_1(\omega_1) \in B\} = A.$$

Because we may take A non-measurable, it follows that F_1 is not Borel measurable.

From the above example, one may get some feeling for where the problem comes from. The space $\mathcal{K}$ is in general not separable. As a result, a uncountable union of balls cannot always be written as a countable union of balls. Now, let $\bar{\mathcal{B}}$ be the σ-algebra generated by all (open or closed) balls (with centres in a separable set). Under mild regularity conditions, ν_n is measurable for $\bar{\mathcal{B}}$.

We now describe the candidate for the limiting process. Let ν be a zero-mean Gaussian process indexed by $\mathcal{G}$, with covariance structure

$$\operatorname{cov}\big(\nu(g_1), \nu(g_2)\big) = \int g_1 g_2 \, dP - \int g_1 \, dP \int g_2 \, dP, \quad g_1, g_2 \in \mathcal{G}.$$

Recall that $H(\delta, \mathcal{G}, P)$ is the δ-entropy of $(\mathcal{G}, \|\cdot\|)$. The space $(\mathcal{G}, \|\cdot\|)$ is totally bounded if $H(\delta, \mathcal{G}, P) < \infty$ for all $\delta > 0$. We assume that

$$\text{(6.1)} \qquad \int_0^1 H^{1/2}(u, \mathcal{G}, P) \, du < \infty.$$

Then we can take for ν a version with bounded and uniformly continuous sample paths (see Dudley (1967) for more refined statements). In other words, ν concentrates on $\mathcal{C}$, where $\mathcal{C}$ is the set of all bounded and uniformly continuous functions on $\mathcal{G}$. Note that $\mathcal{C}$ is a separable subset of $\mathcal{K}$.

Definition 6.1 Let ν_n be $(\mathcal{F}/\bar{\mathcal{B}})$-measurable. We say that ν_n *converges in distribution* to ν ($\nu_n \to^{\mathcal{L}} \nu$) if for every bounded, continuous, $(\bar{\mathcal{B}}/\{\text{Borel sets}\})$-measurable function $f : \mathcal{K} \to \mathbf{R}$,

$$\mathbf{E}f(\nu_n) \to \mathbf{E}f(\nu).$$

The class $\mathcal{G}$ is then called a *P-Donsker class.*

In the next two sections, we present some main results on weak convergence of the empirical process, sweeping possible measurability problems under the carpet. Section 6.4 gives a brief outline of an elegant approach to the measurability issue.

6.2. Sufficient conditions for $\mathcal{G}$ to be P-Donsker

Theorem 6.1 *Suppose that $(\mathcal{G}, \|\cdot\|)$ is totally bounded, and that for all $\eta > 0$, there exists a $\delta > 0$, such that*

$$\limsup_{n\to\infty} \mathbf{P}\left(\sup_{g_1,g_2\in\mathcal{G},\ \|g_1-g_2\|\le\delta} |\nu_n(g_1) - \nu_n(g_2)| > \eta \right) < \eta. \tag{6.2}$$

Then $\mathcal{G}$ is P-Donsker.

See Dudley (1984) for a proof of this theorem.

Theorem 6.2 *Suppose $\mathcal{G}$ has envelope $G \in L_2(P)$, and that for some (non-random) non-increasing function $H(\delta)$ with*

$$\int_0^1 H^{1/2}(u)\, du < \infty, \tag{6.3}$$

we have

$$\lim_{A\to\infty} \limsup_{n\to\infty} \mathbf{P}\left(\sup_{\delta>0} \frac{H(\delta, \mathcal{G}, P_n)}{H(\delta)} > A \right) = 0. \tag{6.4}$$

Then (6.2) *holds, and (hence) $\mathcal{G}$ is P-Donsker.*

Proof Apply Theorem 5.3 to the process $\{\nu_n(g_1 - g_2) : g_1, g_2 \in \mathcal{G}\}$. Because

$$H\left(\delta, \{g_1 - g_2 : g_1, g_2 \in \mathcal{G}\}, P_n\right) \le 2H(\delta/2, \mathcal{G}, P_n),$$

we find that this process is asymptotically equicontinuous at the origin. In other words, (6.2) holds.

Because $G \in L_2(P)$ and

$$\frac{1}{n} H(\delta, \mathcal{G}, P_n) \to_{\mathbf{P}} 0, \quad \text{for all } \delta > 0,$$

we know that

$$\sup_{g_1,g_2\in\mathcal{G}} \big|\, \|g_1 - g_2\|_n - \|g_1 - g_2\| \,\big| \to 0, \quad \text{a.s.}$$

So for each $\delta > 0$,

$$H(2\delta, \mathcal{G}, P) \leq H(\delta, \mathcal{G}, P_n)$$

(say), almost surely, for all n sufficiently large. So $H(\delta, \mathcal{G}, P)$ is finite for all $\delta > 0$, i.e., $(\mathcal{G}, \|\cdot\|)$ is totally bounded. □

Theorem 6.3 *Suppose that*

$$\int_0^1 H_B^{1/2}(u, \mathcal{G}, P)\, du < \infty. \tag{6.5}$$

Then (6.2) *holds, and (hence)* $\mathcal{G}$ *is P-Donsker.*

Proof The class $(\mathcal{G}, \|\cdot\|)$ is totally bounded, since

$$H(\delta, \mathcal{G}, P) \leq H_B(\delta, \mathcal{G}, P) < \infty,$$

for all $\delta > 0$. Now, apply Theorem 5.12 to the class $\{g_1 - g_2 : g_1, g_2 \in \mathcal{G}\}$. □

6.3. Useful theorems

Theorem 6.4 (Continuous Mapping Theorem) *Suppose that* $\mathcal{G}$ *is P-Donsker, and that* $f : \mathcal{X} \to \mathbf{R}$ *is* $(\bar{\mathcal{B}}/\{\text{Borel sets}\})$*-measurable. Assume moreover that* f *is continuous at each* $c \in \mathcal{C}$*. Then* $f(\nu_n) \to^{\mathcal{L}} f(\nu)$.

Theorem 6.5 (Almost Sure Representation Theorem) *Suppose* $\mathcal{G}$ *is P-Donsker. Then one can construct a probability space and random processes* ν_n' *and* ν' *on this space, with the same distribution as* ν_n *and* ν *respectively, such that*

$$\nu_n' \to \nu, \quad \textit{almost surely.}$$

One can find proofs of these theorems in for example Pollard (1984), and van der Vaart and Wellner (1996). The usefulness of the Continuous Mapping Theorem is obvious. For example, it implies that for a P-Donsker class $\mathcal{G}$,

$$\sup_{g \in \mathcal{G}} |\nu_n(g)| \to^{\mathcal{L}} \sup_{g \in \mathcal{G}} |\nu(g)|.$$

The Almost Sure Representation Theorem is often applied in the context of the delta-method.

6.4. Measurability

The results in this section are extracted from Hoffmann-Jørgensen (1991), and van der Vaart and Wellner (1996). They show that most of the measurability problems can be solved by taking outer probabilities whenever the maps involved are not measurable. Denote the extended real line by $\bar{\mathbf{R}} = \mathbf{R} \cup \{-\infty, \infty\}$, and let

$$Z : \Omega \to \bar{\mathbf{R}},$$

be an arbitrary map. The outer expectation of Z is defined as

$$\mathbf{E}^* Z = \inf\{\mathbf{E}\tilde{Z} \ : \ \tilde{Z} \geq Z, \ \tilde{Z} : \Omega \to \bar{\mathbf{R}} \text{ measurable, } \mathbf{E}\tilde{Z} \text{ exists}\}.$$

The outer probability of a set $A \subset \Omega$ is

$$\mathbf{P}^*(A) = \mathbf{E}^* 1_A.$$

It can be shown that if $\mathbf{E}^* Z < \infty$, then there exists a random variable Z^*, such that

$$\mathbf{E}^* Z = \mathbf{E} Z^*.$$

Now, in our considerations, it boils down to replacing the appropriate probabilities (expectations) by the outer probabilities (outer expectations). (We *do* assume however that each function $g \in \mathcal{G}$ is measurable.)

A sequence of maps $\{Z_n\}_{n=1}^{\infty}$ converges *outer almost surely* (in *outer probability*) to the map Z, if $|Z_n - Z|^*$ converges almost surely (in probability) to zero. We say that the ULLN holds for $\mathcal{G}$ if

$$\sup_{g \in \mathcal{G}} \left| \int g \, d(P_n - P) \right|$$

converges outer almost surely to zero. Lemma 3.1, which is based on entropy with bracketing, holds without measurability conditions. In Theorem 3.7 however, some mild measurability *is* needed. Recall that we used symmetrization and Fubini's Theorem, in order to arrive at this ULLN. Fubini's Theorem is no longer true if the maps involved are not measurable: it becomes an inequality. Fortunately, the inequality is in the right direction, so that the symmetrization arguments still go through.

A uniform central limit theorem can also be formulated with outer expectations. The limiting process v is Borel measurable, but the empirical process v_n may be not. We say that $v_n \to^{\mathcal{L}} v$, if for all bounded continuous functions f, $\mathbf{E}^* f(v_n) \to \mathbf{E} f(v)$. This is equivalent to $\mathbf{E}^* f(v_n) \to \mathbf{E} f(v)$ for all bounded, continuous, ($\bar{\mathcal{B}}$/Borel sets)-measurable functions f. (We remark that a continuous function $f : \mathcal{K} \to \mathbf{R}$ is always Borel measurable. The Borel σ-algebra is the smallest σ-algebra that makes all continuous functions measurable.) Theorem 6.3 does not need any measurability conditions, but Theorem 6.2 is to be understood modulo measurability, since its proof relies on Fubini's Theorem.

6.5. Notes

Dudley (1966, 1978) was the first to present generalizations of the classical Donsker Theorem: he proved, for example, a uniform central limit theorem for Vapnik–Chervonenkis classes of sets. He was also the first to realize the importance of the σ-algebra $\bar{\mathscr{B}}$. For a good overview, see also Dudley (1984). We remark here that in fact, if the functions in $\mathscr{G}$ are normalized to have expectation zero, then (6.2) is also a necessary condition for the P-Donsker property. Theorem 6.2 is essentially in Pollard (1982) (see also Pollard (1984)). Theorem 6.3 is proved in Ossiander (1987). Andersen, Giné, Ossiander and Zinn (1988) generalize Theorem 6.3. For a good and complete treatment of the subject, including many references, see the book of van der Vaart and Wellner (1996).

6.6. Problems and complements

6.1. If $\mathscr{G}$ is P-Donsker, then

$$\sup_{g\in\mathscr{G}} |\nu_n(g)| = O_{\mathbf{P}}(1). \tag{6.6}$$

Now, let $\mathscr{G}$ be a class of functions on the real line, with total variation bounded by M. Then (2.5), together with Theorem 6.3, implies that $\mathscr{G}$ is P-Donsker. In this case, result (6.6) also follows from the classical Donsker Theorem. Let $F_n = \sum_{i=1}^n 1\{Xi \le \cdot\}/n$ be the empirical distribution function, and $F = P(X_1 \le \cdot)$. Then by the classical Donsker Theorem,

$$\sup_{t\in\mathbf{R}} \sqrt{n}|F_n(t) - F(t)| = O_{\mathbf{P}}(1).$$

Use partial integration to show that this implies (6.6).

6.2. If $\mathscr{K}$ is a P-Donsker class, then so is conv($\mathscr{K}$). This can be shown by applying the Almost Sure Representation Theorem (see van der Vaart and Wellner (1996)).

6.3. For $\alpha > 0$ and $0 < L \le \infty$, let

$$\mathscr{G}_L^\alpha = \{g : [0,L) \to [0,1],\ |g(x) - g(\tilde{x})| \le |x - \tilde{x}|^\alpha,\ x, \tilde{x} \in [0,L)\}.$$

Show that

$$H_\infty(\delta, \mathscr{G}_1^\alpha) \le A\delta^{-1/\alpha}, \quad \text{for all } \delta > 0.$$

Conclude that for $\alpha > 1/2$, $\mathscr{G}_1^\alpha$ is a P-Donsker class.

Suppose that P satisfies a tail-condition as in Problem 2.6, with $\nu = \infty$: there exist constants $C < \infty$, and $\epsilon > 0$, such that for all $T < \infty$, there is a partition $B_1, \ldots, B_N$ of $(0, T]$, with $B_j = (x_{j-1}, x_j]$, $0 = x_0 < \ldots < x_N = T$, such that for $x_j - x_{j-1} \geq 1$,

$$\sum_{j=1}^{N} (x_j - x_{j-1}) p_j^{\frac{1}{2}-\epsilon} \leq C,$$

where

$$p_j = \frac{P(x_{j-1}, x_j]}{x_j - x_{j-1}}, \quad j = 1, \ldots, N.$$

Then, for $\alpha > 1/2$, $\mathcal{G}_\infty^\alpha$ is a P-Donsker class. This follows from similar entropy calculations as in Problem 2.6.

7

Rates of Convergence for Maximum Likelihood Estimators

We show that a rate of convergence in Hellinger distance for the maximum likelihood estimator can be derived from the entropy with bracketing of the class of densities $\mathcal{P}$, endowed with the Hellinger metric. If $\mathcal{P}$ is convex, these entropy conditions can be weakened, or replaced by an empirical entropy condition without bracketing.

Consider i.i.d. random variables $X_1, \ldots, X_n, \ldots$ with distribution P, and assume that

$$p_0 = \frac{dP}{d\mu} \in \mathcal{P},$$

where $\mathcal{P}$ is a given class of densities p with respect to the σ-finite dominating measure μ. The maximum likelihood estimator is

$$\hat{p}_n = \arg \max_{p \in \mathcal{P}} \sum_{i=1}^{n} \log p(X_i).$$

The Hellinger distance between two densities p_1 and p_2 is

$$h(p_1, p_2) = \left(\frac{1}{2} \int \left(p_1^{1/2} - p_2^{1/2} \right)^2 d\mu \right)^{1/2}.$$

In this chapter, we prove a rate of convergence for $h(\hat{p}_n, p_0)$, using entropy conditions on a class of transformations of the densities in $\mathcal{P}$. The transformations we shall consider are:

$$\mathscr{P}^{1/2} = \{p^{1/2} : p \in \mathscr{P}\},$$

$$\mathscr{P}^{1/2}/p_0^{1/2} = \left\{ \frac{p^{1/2}}{p_0^{1/2}} 1\{p_0 > 0\} : p \in \mathscr{P} \right\},$$

$$\bar{p} = \frac{p + p_0}{2},$$

$$\bar{\mathscr{P}} = \{\bar{p} : p \in \mathscr{P}\},$$

$$\bar{\mathscr{P}}^{1/2} = \{\bar{p}^{1/2} : \bar{p} \in \bar{P}\},$$

$$g_p = \frac{1}{2} \log \frac{\bar{p}}{p_0} 1\{p_0 > 0\},$$

and finally

$$\mathscr{G} = \{g_p : p \in \mathscr{P}\}.$$

In the entropy calculations, we only need approximations of the densities on the set where p_0 is strictly positive. We use the notation

$$d\mu_0 = 1\{p_0 > 0\} d\mu,$$

and

$$h_0(p_1, p_2) = \left(\frac{1}{2} \int (p_1^{1/2} - p_2^{1/2})^2 \, d\mu_0 \right)^{1/2}.$$

7.1. The main idea

We shall prove in Theorem 7.4 that a rate of convergence in Hellinger distance can be deduced from the entropy with bracketing of the class of densities $\mathscr{P}$, endowed with the Hellinger metric. This result follows essentially from a rather simple observation, but some technical intermediate results are needed to avoid unnecessary conditions. In this section, we take those technicalities for granted, in order to be able to highlight the main idea.

Using $\frac{1}{2} \log x \leq \sqrt{x} - 1$, $x > 0$, one may derive the Basic Inequality

$$h^2(\hat{p}_n, p_0) \leq \int_{p_0 > 0} \frac{\hat{p}_n^{1/2}}{p_0^{1/2}} \, d(P_n - P).$$

Lemma 5.13 investigates the modulus of continuity of the empirical process. We apply it to the empirical process indexed by $\mathscr{P}^{1/2}/p_0^{1/2}$, to obtain the modulus of continuity at the function $1\{p_0 > 0\}$. We moreover replace the $L_2(P)$ norm by the Hellinger distance $h(\cdot,\cdot)$. This is allowed, since

$$\left\| \left(\frac{p^{1/2}}{p_0^{1/2}} - 1 \right) 1\{p_0 > 0\} \right\|^2 = 2h_0^2(p, p_0) \leq 2h^2(p, p_0).$$

Thus, reformulating Lemma 5.13 (with $\beta = 0$), we obtain that if for some constant K,

$$\sup_{p \in \mathcal{P}} \left| \frac{p^{1/2}}{p_0^{1/2}} 1\{p_0 > 0\} \right|_\infty \leq K, \tag{7.1}$$

and if for some constants A and $0 < \alpha < 2$,

$$H_B(\delta, \mathcal{P}^{1/2}/p_0^{1/2}, P) \leq A\delta^{-\alpha}, \quad \text{for all } \delta > 0, \tag{7.2}$$

then, uniformly in $p \in \mathcal{P}$,

$$\left| \int_{p_0 > 0} \frac{p^{1/2}}{p_0^{1/2}} \, d(P_n - P) \right| = O_{\mathbf{P}}(n^{-1/2}) h^{1-\frac{\alpha}{2}}(p, p_0) \vee O_{\mathbf{P}}(n^{-\frac{2}{2+\alpha}}).$$

Therefore, under (7.1) and (7.2), we find

$$h^2(\hat{p}_n, p_0) \leq O_{\mathbf{P}}(n^{-1/2}) h^{1-\frac{\alpha}{2}}(\hat{p}_n, p_0) \vee O_{\mathbf{P}}(n^{-\frac{2}{2+\alpha}}),$$

But straightforward manipulation now yields

$$h(\hat{p}_n, p_0) = O_{\mathbf{P}}(n^{-\frac{1}{2+\alpha}}).$$

Thus, a rate of convergence follows from entropy conditions on the class $\mathcal{P}^{1/2}/p_0^{1/2}$. Note that for all $\delta > 0$,

$$H_B(\delta, \mathcal{P}^{1/2}/p_0^{1/2}, P) = H_B(\delta, \mathcal{P}^{1/2}, \mu_0) \leq H_B(\delta, \mathcal{P}^{1/2}, \mu). \tag{7.3}$$

The last expression in (7.3) is the $\delta/\sqrt{2}$-entropy of $\mathcal{P}$ endowed with the Hellinger metric.

The technical intermediate results in the next section make it possible to relax (7.1) and (7.2). Indeed, assumption (7.1) is problematic, because in many situations p_0 can become arbitrarily small. Also the assumption of a uniform bound on a class of densities is quite severe. We shall drop (7.1), and replace in (7.2) the entropy with bracketing of $(\mathcal{P}^{1/2}, \|\cdot\|_{\mu_0})$, by the smaller entropy with bracketing of $(\bar{\mathcal{P}}^{1/2}, \|\cdot\|_{\mu_0})$. However, we point out that, in the entropy with bracketing condition of Theorem 7.4, is still hidden the assumption that $\mathscr{P}$ is, in a certain sense, uniformly bounded (see also Problem 7.1).

7.2. An exponential inequality for the maximum likelihood estimator

In Lemma 4.1, we proved the Basic Inequality

$$h\left(\frac{\hat{p}_n + p_0}{p_0}, p_0\right) \leq \int g_{\hat{p}_n}\, d(P_n - P),$$

Our aim is to invoke Theorem 5.11 for $\mathscr{G} = \{g_p : p \in \mathscr{P}\}$. Theorem 5.11 presents a uniform probability inequality for the empirical process, under assumptions on the generalized entropy with bracketing. In order to be able to apply it, we have to show that, among other things, the Bernstein difference (at zero) of the functions in $\mathscr{G}$ is uniformly bounded, i.e., for some K and R,

$$\sup_{p \in \mathscr{P}} \rho_K(g_p) \leq R, \tag{7.4}$$

where

$$\rho_K^2(g) = 2K^2 \int \left(e^{|g|/K} - 1 - |g|/K\right) dP.$$

Exploiting the fact that g_p is bounded from below, it turns out that (7.4) holds with $K = 1$ and $R = 4$ (see Lemma 7.2). We first state a preliminary lemma.

Lemma 7.1 *For $x \geq -\Gamma$, $\Gamma > 0$,*

$$2(e^{|x|} - 1 - |x|) \leq c_\Gamma^2 (e^x - 1)^2,$$

where $c_\Gamma^2 = 2(e^\Gamma - 1 - \Gamma)/(e^{-\Gamma} - 1)^2$.

Proof This follows from the fact that the function

$$\frac{2(e^{|x|} - 1 - |x|)}{(e^x - 1)^2},$$

is decreasing. □

Lemma 7.2 *We have*

$$\rho_1(g_p) \leq 4h_0(\bar{p}, p_0) \quad (\leq 4h(\bar{p}, p_0)). \tag{7.5}$$

Proof First of all, the functions g_p are uniformly bounded from below: $g_p \geq -(\log 2)/2$. In view of Lemma 7.1,

$$2\int_{p_0>0}(e^{|g_p|}-1-|g_p|)\,dP \leq 8\int_{p_0>0}(e^{g_p}-1)^2\,dP = 8\int_{p_0>0}\left(\frac{\bar{p}^{1/2}}{p_0^{1/2}}-1\right)^2 dP$$
$$= 16h_0^2(\bar{p},p_0). \qquad \square$$

Thus, indeed, (7.4) is met, with $K=1$ and $R=4$, since $h(\bar{p},p_0)\leq 1$.

Next, we study the generalized entropy with bracketing (see Definition 5.1) of $\mathcal{G}$.

Lemma 7.3 *Let* $0 \leq p^L \leq p^U$. *Define* $\bar{p}^L = (p^L+p_0)/2$ *and* $\bar{p}^U = (p^U+p_0)/2$. *Also, let*

$$g^L = \frac{1}{2}\log\frac{\bar{p}^L}{p_0}1\{p_0>0\}; \quad g^U = \frac{1}{2}\log\frac{\bar{p}^U}{p_0}1\{p_0>0\}.$$

Then

$$\rho_1(g^U - g^L) \leq 2h_0(\bar{p}^L,\bar{p}^U). \tag{7.6}$$

Proof Clearly, $\bar{g}^U - \bar{g}^L \geq 0$. Apply Lemma 7.1, with $\Gamma \to 0$. Then we find

$$\rho_1^2(\bar{g}^U-\bar{g}^L) \leq \int_{p_0>0}\left(\sqrt{\frac{\bar{p}^U}{\bar{p}^L}}-1\right)^2 dP \leq 2\int_{p_0>0}(\sqrt{\bar{p}^U}-\sqrt{\bar{p}^L})^2\,d\mu$$
$$= 4h_0^2(\bar{p}^L,\bar{p}^U). \qquad \square$$

Lemma 7.3 shows that the generalized entropy with bracketing of $\mathcal{G}$ is bounded by the (usual) entropy with bracketing of $\bar{\mathcal{P}}$ endowed with the Hellinger metric:

$$\mathcal{H}_{B,1}\big(\sqrt{2}\delta,\mathcal{G},P\big) \leq H_B\big(\delta,\bar{\mathcal{P}}^{1/2},\mu_0\big) \leq H_B\big(\delta,\bar{\mathcal{P}}^{1/2},\mu\big). \tag{7.7}$$

Therefore, by Theorem 5.11, the increments of $\int g_p\,d(P_n-P)$ can be deduced from the entropy with bracketing of $\bar{\mathcal{P}}$ endowed with the Hellinger metric, or even the metric $h_0(\cdot,\cdot)$. In fact, it turns out that the *local* entropy is of main concern.

We denote a ball around p_0, intersected with $\bar{\mathcal{P}}^{1/2}$, by

$$\bar{\mathcal{P}}^{1/2}(\delta) = \left\{\bar{p}^{1/2}\in\bar{\mathcal{P}}^{1/2} : h(\bar{p},p_0)\leq\delta\right\}, \quad 0<\delta\leq 1.$$

For δ small, we refer to $H_B(u, \bar{\mathcal{P}}^{1/2}(\delta), \mu_0)$ as the *local* entropy with bracketing. The entropy integral is given by

$$J_B\big(\delta, \bar{\mathcal{P}}^{1/2}(\delta), \mu_0\big) = \int_{\delta^2/2^{13}}^{\delta} H_B^{1/2}\big(u, \bar{\mathcal{P}}^{1/2}(\delta), \mu_0\big)\, du \vee \delta, \tag{7.8}$$

It follows from Theorem 5.11 that the empirical process indexed by $\{g_p : h(\bar{p}, p_0) \le \delta\}$ behaves as $J_B(\delta, \bar{\mathcal{P}}^{1/2}(\delta), \mu_0)$. In view of the Basic Inequality of Lemma 4.1, this will supply us with the tools to establish a rate of convergence.

Theorem 7.4 *Take $\Psi(\delta) \ge J_B(\delta, \bar{\mathcal{P}}^{1/2}(\delta), \mu_0)$ in such a way that $\Psi(\delta)/\delta^2$ is a non-increasing function of δ. Then for a universal constant c, and for*

$$\sqrt{n}\delta_n^2 \ge c\Psi(\delta_n), \tag{7.9}$$

we have for all $\delta \ge \delta_n$

$$\mathbf{P}\big(h(\hat{p}_n, p_0) > \delta\big) \le c \exp\left[-\frac{n\delta^2}{c^2}\right]. \tag{7.10}$$

Proof The combination of Lemma 4.1 and Lemma 4.2 shows that it suffices to prove that

$$\mathbf{P}\left(\sup_{p\in\mathcal{P},\ h(\bar{p},p_0)>\delta/4} \nu_n(g_p) - \sqrt{n}h^2(\bar{p}, p_0) \ge 0\right) \le c \exp\left[-\frac{n\delta^2}{c^2}\right].$$

Let $S = \min\{s : 2^{s+1}\delta/4 > 1\}$. Application of the peeling device (see Section 5.3) gives

$$\begin{aligned}
&\mathbf{P}\left(\sup_{p\in\mathcal{P},\ h(\bar{p},p_0)>\delta/4} \nu_n(g_p) - \sqrt{n}h^2(\bar{p}, p_0) \ge 0\right)\\
&\le \sum_{s=0}^{S} \mathbf{P}\left(\sup_{p\in\mathcal{P},\ h(\bar{p},p_0)\le(2^{s+1})\delta/4} \nu_n(g_p) \ge \sqrt{n}2^{2s}\left(\frac{\delta}{4}\right)^2\right)\\
&= \sum_{s=0}^{S} \mathbf{P}\left(\sup_{p\in\mathcal{P},\ h(\bar{p},p_0)\le(2^{s+1})\delta/4} \nu_n(g_p) \ge \sqrt{n}2^{2s}\left(\frac{\delta}{4}\right)^2\right).
\end{aligned}$$

Next, apply Theorem 5.11, with $C_1 = 15$, $K = 1$, $R = 2^{s+1}\delta$, $a = \sqrt{n}2^{2s}\delta^2/2^4$. Indeed, if $h(\bar{p}, p_0) \le (2^{s+1})\delta/4$, then by Lemma 7.2, $\rho_1(g_p) \le 2^{s+1}\delta$. Conditions (5.31) and (5.32) are satisfied for all $s \in \{0, 1, \ldots, S\}$.

If

$$\sqrt{n}\delta_n^2 \geq \sqrt{2}2^9 C\Psi(\delta_n),$$

then for all $\delta \geq \delta_n$, and $s \geq 0$

$$\sqrt{n}2^{2s}\frac{\delta^2}{2^4} \geq 4C\left(\int_{2^{2s}\delta^2/2^{10}}^{2^{s+1}\delta} \mathcal{H}_{B,1}^{1/2}\left(u, \left\{g_p : h(\bar{p}, p_0) \leq 2^{s+1}\frac{\delta}{4}\right\}, P\right) du \vee 2^{s+1}\delta\right).$$

Here, we use the fact that

$$\mathcal{H}_{B,1}\left(u, \left\{g_p : h(\bar{p}, p_0) \leq 2^{s+1}\frac{\delta}{4}\right\}, P\right) \leq H_B\left(\frac{u}{\sqrt{2}}, \bar{\mathcal{P}}^{1/2}\left(2^{s+1}\frac{\delta}{4}\right), \mu_0\right),$$

and the assumption that $\Psi(\delta)/\delta^2$ is non-increasing. So (5.33) is satisfied for all $s \geq 0$, with $C_0 = 4C$. Because (5.34) also holds, we therefore obtain

$$\sum_{s=0}^{S} \mathbf{P}\left(\sup_{p\in\mathcal{P},\ h(\bar{p},p_0)>\delta/4} \nu_n(g_p) \geq \sqrt{n}2^{2s}\left(\frac{\delta}{4}\right)^2\right) \leq \sum_{s=0}^{\infty} C\exp\left[-\frac{2^{2s}n\delta^2}{C^2 2^{14}}\right]$$
$$\leq c\exp\left[-\frac{n\delta^2}{c^2}\right],$$

for some universal constant c. Here, we invoke the fact that (7.9) implies that $\sqrt{n}\delta_n$ stays away from zero. □

If the entropy integral converges, i.e., if

$$\int_0^\delta H_B^{1/2}\left(u, \bar{\mathcal{P}}^{1/2}(\delta), \mu_0\right) du < \infty, \quad \text{for all } 0 < \delta \leq 1,$$

then we find the rate $h(\hat{p}_n, p_0) = O_{\mathbf{P}}(\delta_n)$, with $\delta_n = o(n^{-1/4})$. We already announced this result in Example 1.2. It is a consequence of the asymptotic equicontinuity of the empirical process indexed by $\mathcal{G}$. If the entropy integral diverges, we obtain a rate slower than $o_{\mathbf{P}}(n^{-1/4})$. We shall illustrate this in Example 7.4.6.

From the proof of Theorem 7.4, one can also deduce a rate of convergence for the log-likelihood ratio. Because we need this rate in Section 11.2, we present it as a corollary.

Corollary 7.5 *Suppose that the conditions of Theorem 7.4 hold. Then for some constant* c_1 *and for all* $\delta \geq \delta_n$

$$\mathbf{P}\left(\int_{p_0>0} \log\frac{\hat{p}_n}{p_0}\, dP_n \geq \delta^2\right) \leq c_1\exp\left[-\frac{n\delta^2}{c_1^2}\right]. \tag{7.11}$$

Of course, the conditions of Theorem 7.4 are better than conditions (7.1) and (7.2) which we imposed in the previous section. However, we are still facing the rather severe envelope assumption. To be more precise, letting

$$q = \sup_{p \in \mathcal{P}} p$$

denote the envelope function of the class of densities $\mathcal{P}$, Theorem 7.4 can only be applied successfully if

$$\int q \, d\mu_0 < \infty,$$

because otherwise the entropy with bracketing of $(\bar{\mathcal{P}}^{1/2}, \|\cdot\|_{\mu_0})$ will not be finite. In fact, one can then assume without loss of generality that $\mathcal{P}$ is uniformly bounded on the support of p_0 (see Problem 7.1).

Another troublesome fact is that $\bar{\mathcal{P}}^{1/2}$ is a class of *square-root* densities. There may be some difficulties when deriving the entropy of $\bar{\mathcal{P}}^{1/2}$ from the entropy of $\mathcal{P}$, due to the non-differentiability of $\sqrt{x}$ at $x = 0$. In several cases, the following provides a method to cope with this. For each $\sigma \geq 0$, we have

$$\left|\bar{p}_1^{1/2} - \bar{p}_2^{1/2}\right| \leq 2q^{1/2} 1\{p_0 \leq \sigma\} + \left|\bar{p}_1^{1/2} - \bar{p}_2^{1/2}\right| 1\{p_0 > \sigma\}.$$

Let

$$\sigma_0(\delta) = \sup\left\{\sigma \geq 0 : \int_{p_0 \leq \sigma} q \, d\mu_0 \leq \delta^2\right\}, \tag{7.12}$$

and consider the class

$$\bar{\mathcal{P}}_\sigma^{1/2} = \left\{\bar{p}^{1/2} 1\{p_0 \geq \sigma\}\right\} : \bar{p} \in \bar{\mathcal{P}}\}.$$

Then

$$H_B\left(\sqrt{3}\delta, \bar{\mathcal{P}}^{1/2}, \mu_0\right) \leq H_B\left(\delta, \bar{\mathcal{P}}^{1/2}_{\sigma_0(\delta)}, \mu_0\right) \tag{7.13}$$

and moreover, for $dQ_\sigma = (1\{p_0 > \sigma\}/p_0) d\mu$, we find

$$H_B\left(\delta, \bar{\mathcal{P}}_\sigma^{1/2}, \mu_0\right) \leq H_B\left(\delta, \mathcal{P}, Q_\sigma\right). \tag{7.14}$$

An illustration of this argument can be found in Example 7.4.4.

Note finally that if one already proved consistency in Hellinger distance by separate means (e.g., by applying Theorem 4.3), one only needs to calculate the entropy with bracketing of $\bar{\mathcal{P}}(\delta)^{1/2}$ for δ arbitrarily small. Theorem 7.4 then produces a rate of convergence, but the exponential bound (7.10) may no longer be valid.

7.3. Convex classes of densities

If $\mathcal{P}$ is convex, one can relax the entropy condition of Theorem 7.4, and in particular circumvent the envelope condition $\int q\, d\mu_0 < \infty$. Let

$$\mathcal{G}^{(\text{conv})} = \left\{ \frac{2p}{p+p_0} : p \in \mathcal{P} \right\}.$$

This class is uniformly bounded by 2, and it is not difficult to see that the entropy of $\mathcal{G}^{(\text{conv})}$ is of smaller order than the entropy of $\bar{\mathcal{P}}^{1/2}$:

$$H_B\left(4\sqrt{2}\delta, \mathcal{G}^{(\text{conv})}, P\right) \le H_B\left(\delta, \bar{\mathcal{P}}^{1/2}, \mu_0\right), \quad \text{for all } \delta > 0.$$

Define now

$$J_B\left(\delta, \mathcal{G}^{(\text{conv})}, P\right) = \int_{\frac{\delta^2}{c}}^{\delta} H_B^{1/2}\left(u, \mathcal{G}^{(\text{conv})}, P\right) du \vee \delta. \tag{7.15}$$

Here, and later, c is again a (large) universal constant (not the same at each appearance).

Theorem 7.6 *Suppose that $\mathcal{P}$ is convex. Take $\Psi(\delta) \ge J_B(\delta, \mathcal{G}^{(\text{conv})}, P)$ in such a way that $\Psi(\delta)/\delta^2$ is a non-increasing function of δ. Then for a universal constant c and for*

$$\sqrt{n}\delta_n^2 \ge c\Psi(\delta), \tag{7.16}$$

we have for all $\delta \ge \delta_n$

$$\mathbf{P}(h(\hat{p}_n, p_0) > \delta) \le c \exp\left[-\frac{n\delta^2}{c^2}\right]. \tag{7.17}$$

Proof Use the Basic Inequality of Lemma 4.5, and the peeling device, to obtain

$$\begin{aligned}
&\mathbf{P}(h(\hat{p}_n, p_0) > \delta) \\
&\le \mathbf{P}\left(\sup_{p \in \mathcal{P}:\ h(p,p_0) > \delta} \int \left(\frac{2p}{p+p_0} - 1 \right) d(P_n - P) - h^2(p, p_0) \ge 0 \right) \\
&\le \sum_{s=0}^{S} \mathbf{P}\left(\sup_{p \in \mathcal{P},\ h(p,p_0) \le 2^{s+1}\delta} \sqrt{n} \int \left(\frac{2p}{p+p_0} - 1 \right) d(P_n - P) \ge \sqrt{n} 2^{2s} \delta^2 \right),
\end{aligned}$$

where $S = \min\{s : 2^{s+1}\delta > 1\}$.

We have

$$\left\| \frac{2p}{p+p_0} - 1 \right\|^2 \le 4h^2(p, p_0),$$

so that (since the functions involved are bounded by 1),

$$\rho_4^2 \left(\frac{2p}{p+p_0} - 1 \right) \le 8h^2(p, p_0).$$

Moreover (see Lemma 5.10),

$$\mathscr{H}_{B,4}\big(\sqrt{2}u, \mathscr{G}^{(\mathrm{conv})}, P\big) \le H_B\big(u, \mathscr{G}^{(\mathrm{conv})}, P\big).$$

So the result again follows from the application of Theorem 5.11. □

It follows from the proof of Theorem 7.6, that in (7.15), we may replace $\mathscr{G}^{(\mathrm{conv})}$ by the subset

$$\mathscr{G}^{(\mathrm{conv})}(\delta) = \left\{ \frac{2p}{p+p_0} : p \in \mathscr{P},\ h(p, p_0) \le \delta \right\},$$

i.e., we may replace (7.15) by the weaker local version.

When calculating the entropy of $\mathscr{G}^{(\mathrm{conv})}$, one again encounters the fact the p_0 appears in the denominator of the expression $2p/(p+p_0)$, and that this can cause problems when p_0 can be arbitrary small. To handle this, one may argue in the same way as in (7.13) and (7.14). Let

$$\sigma(\delta) = \sup \left\{ \sigma \ge 0 : \int_{p_0 \le \sigma} p_0 \, d\mu \le \delta^2 \right\}. \tag{7.18}$$

Consider the class

$$\mathscr{G}_\sigma^{(\mathrm{conv})} = \left\{ \frac{2p}{p+p_0} \mathrm{l}\{p_0 > \sigma\} : p \in \mathscr{P} \right\}.$$

Then,

$$H_B\big(\sqrt{3}\delta, \mathscr{G}^{(\mathrm{conv})}, P\big) \le H_B\big(\delta, \mathscr{G}_{\sigma(\delta)}^{(\mathrm{conv})}, P\big), \quad \text{for all } \delta > 0.$$

Moreover, for $dQ_\sigma = (\mathrm{l}\{p_0 > \sigma\}/p_0) d\mu$,

$$H_B\big(2\delta, \mathscr{G}_\sigma^{(\mathrm{conv})}, P\big) \le H_B\big(\delta, \mathscr{P}, Q_\sigma\big), \quad \text{for all } \delta > 0. \tag{7.19}$$

In view of (7.19), it is tempting to apply the result of Ball and Pajor (see Theorem 3.14) on convex classes to our situation. But this result only concerns entropy without bracketing. Luckily, it is indeed possible to replace our entropy with bracketing condition by an empirical entropy condition without bracketing. Let $\{\sigma_n\}$ be an appropriately chosen sequence of non-negative numbers. In order to get good rates from Theorem 7.7 below, one has to choose the sequence σ_n by a trade-off argument. If σ_n is small, condition (7.21) is easily fulfilled, but condition (7.22) becomes harder (and vice versa).

Theorem 7.7 *Suppose that $\mathcal{P}$ is convex. Let $H(\delta)$ be a non-increasing function of δ, and suppose that*

$$\sup_{\delta>0} \frac{H\big(\delta, \mathcal{G}^{(\mathrm{conv})}_{\sigma_n}, P_n\big)}{H(\delta)} = O_{\mathbf{P}}(1). \tag{7.20}$$

Take $\Psi(\delta) \geq \int_{\delta^2/c}^{\delta} H^{1/2}(u)\,du \vee \delta$ in such a way that for some $\epsilon > 0$, $\Psi(\delta)/\delta^{2-\epsilon}$ is non-increasing. Take $n\delta_n^2 \to \infty$,

$$\delta_n^2 \geq \int_{p_0 \leq \sigma_n} p_0 \, d\mu, \tag{7.21}$$

and

$$\sqrt{n}\delta_n^2 \geq \Psi(\delta_n). \tag{7.22}$$

Then $h(\hat{p}_n, p_0) = O_{\mathbf{P}}(\delta_n)$.

Proof We have

$$\int \left(\frac{2p}{p+p_0} - 1\right) d(P_n - P) = \int_{p_0 \leq \sigma_n} \left(\frac{2p}{p+p_0} - 1\right) d(P_n - P) + \int_{p_0 > \sigma_n} \left(\frac{2p}{p+p_0} - 1\right) d(P_n - P),$$

and

$$\left| \int_{p \leq \sigma_n} \left(\frac{2p}{p+p_0} - 1\right) d(P_n - P) \right| \leq \int_{p_0 \leq \sigma_n} dP_n + \int_{p_0 \leq \sigma_n} dP.$$

Now, by (7.21),

$$\mathbf{P}\left(\left| \int_{p_0 \leq \sigma_n} d(P_n - P) \right| > c_1 \delta_n^2 \right) \leq \frac{1}{n c_1^2 \delta_n^2}.$$

Hence,

$$\begin{aligned}
&\mathbf{P}(h(\hat{p}_n, p_0) > (c_1 + 2)\delta_n) \\
&\leq \frac{1}{n c_1^2 \delta_n^2} + \mathbf{P}\left(\sup_{p \in \mathcal{P}, h(p, p_0) > (c_1+2)\delta_n} \int_{p_0 > \sigma_n} \left(\frac{2p}{p+p_0} - 1\right) d(P_n - P) \right. \\
&\qquad \left. - \frac{1}{2} h^2(p, p_0) \geq 0 \right).
\end{aligned}$$

Let $\eta > 0$ be arbitrary, and take A sufficiently large, in such a way that

$$4\mathbf{P}\left(\sup_{\delta>0} \frac{H(\delta, \mathcal{G}^{(\mathrm{conv})}_{\sigma_n}, P_n)}{H(\delta)} > A \right) < \eta.$$

Because $\Psi(\delta)/\delta^{2-\epsilon}$ is non-increasing, we find from (7.22) that for some c_1 sufficiently large,

$$nc_1^2\delta_n^2 \geq 2AH(c_1\delta_n).$$

So by Lemma 5.6,

$$\mathbf{P}\left(\sup_{g\in\mathcal{G}_{\sigma_n}^{\text{conv}}} \frac{\|g - 1\{p_0 > \sigma_n\}\|_n}{\|g - 1\{p_0 > \sigma_n\}\| \vee c_1\delta_n} > 14 \right) \leq \eta.$$

Now, proceed in a similar way as in Lemma 5.5, treating the sets

$$2^s(c_1+2)\delta_n < h(p, p_0) \leq 2^{s+1}(c_1+2)\delta_n$$

separately. Thus, the proof is as in Theorem 7.6 but instead of the probability inequalities based on entropy with bracketing, use the ones based on the empirical entropy. □

Remark Again, in Theorem 7.7, one may replace the entropy condition by a local version.

We end this section with a corollary, which shows that there is a price to pay when using $\mathcal{G}^{(\text{conv})}$ instead of $\mathcal{G}$: we no longer have a rate of convergence for the log-likelihood ratio, but only for the log-ratio with p_0 replaced by $(\hat{p}_n + p_0)/2$. Fortunately, this suffices for the application we have in mind (see Section 11.2).

Corollary 7.8 *Suppose that the conditions of Theorem* 7.6 *or* 7.7 *are met. Then*

$$\int \log \frac{2\hat{p}_n}{\hat{p}_n + p_0}\, dP_n = O_{\mathbf{P}}(\delta_n^2). \tag{7.23}$$

In the case of Theorem 7.6, *one has in addition an exponential probability inequality for the expression in the right-hand side of* (7.23), *of the same type as the one occurring in Corollary* 7.5.

7.4. Examples

Let us first summarize the previous sections, and see where we shall apply what result.

In Theorem 7.4, a rate of convergence is presented under conditions on the entropy with bracketing of $\bar{\mathcal{P}}^{1/2}$. The theorem will be applied in Example 7.4.1., Case (i), and in Examples 7.4.3, 7.4.4, and 7.4.6. Example 7.4.6 will also examine the optimality of the rate thus obtained.

The major disadvantage of Theorem 7.4 is that it is only applicable in the case where $\int q\, d\mu_0 < \infty$, where q is the envelope function of $\mathscr{P}$. Theorem 7.6 replaces this condition by the condition that $\mathscr{P}$ is convex, and uses the entropy with bracketing of $\mathscr{G}^{(\text{conv})}$. Its effectiveness is illustrated in Example 7.4.1, Case (ii), and in Example 7.4.2.

Finally, Theorem 7.7 also studies the case where $\mathscr{P}$ is convex, but employs empirical entropy instead of entropy with bracketing. This is helpful, because a convex class is always the convex hull of some class, say $\mathscr{K}$. If $\mathscr{K}$ has small enough covering numbers, it allows you to insert the result of Ball and Pajor. We illustrate this in Example 7.4.5.

Example 7.4.1. Smooth densities

Case (i) Let μ be Lebesgue measure on [0, 1], and

$$\mathscr{P} = \left\{ p : [0,1] \to [0,\infty) : \int_0^1 p(x)\, dx = 1,\ \int_0^1 (p^{(m)}(x))^2\, dx \le M^2 \right\}. \tag{7.24}$$

Assume that

$$p_0 \ge \eta_0^2 > 0. \tag{7.25}$$

Then one finds from Theorem 2.4,

$$H_B(\delta, \bar{\mathscr{P}}^{1/2}, \mu) \le H_B(\delta\eta_0, \mathscr{P}, \mu) \le A\delta^{-1/m}, \quad \text{for all } \delta > 0.$$

Hence, in Theorem 7.4, one may take

$$\Psi(\delta) = A_0 \delta^{1-1/2m},$$

for some constant A_0. Inequality (7.9) then yields the rate $\delta_n = cn^{-m/(2m+1)}$: for all $T \ge c$

$$\mathbf{P}(h(\hat{p}_n, p_0) > Tn^{-\frac{m}{2m+1}}) \le c \exp\left[-\frac{T^2 n^{\frac{1}{2m+1}}}{c^2} \right]. \tag{7.26}$$

We did not use the convexity of $\mathscr{P}$ here, so the same rate holds for the maximum likelihood estimator over any subclass of $\mathscr{P}$. In Case (ii) below, we consider a related model with $m = 1$, where we can drop the assumption $p_0 \ge \eta_0^2 > 0$, but where we *do* need the convexity in order to arrive at the rate $O_{\mathbf{P}}(n^{-1/3})$.

Case (ii) Let μ be Lebesgue measure, and

$$\mathscr{P} = \left\{ p : [0,1] \to [0,\infty),\ \int_0^1 p(x)\, dx = 1,\ |p'|_\infty \le M \right\}. \tag{7.27}$$

Here, p' is the derivative of p with respect to Lebesgue measure. We do not assume that p_0 stays away from zero, so that a good entropy bound for the square-root densities does not immediately follow from an entropy bound for the densities themselves. Now, it is clear that for some constant A,

$$H_B(2\delta, \mathcal{P}, \mu) \leq H_\infty(\delta, \mathcal{P}) \leq A\delta^{-1}, \quad \text{for all } \delta > 0.$$

(This bound cannot be improved.) We show below that the bound $A\delta^{-1}$ also holds for the entropy with bracketing of $\mathcal{G}^{(\mathrm{conv})}$ (with a different constant A), even when p_0 does not stay away from zero (but *does* satisfy (7.28)), and thus that $h(\hat{p}_n, p_0) = O_{\mathbf{P}}(n^{-1/3})$. The entropy result requires a rather lengthy proof, which should not be understood as an indication of its importance. We rather want to illustrate that the maximum likelihood estimator can behave like a kernel estimator with locally adaptive bandwidth (see the arguments following Lemma 7.9).

Lemma 7.9 *Assume that the differential equation*

$$z'(y) = p_0^{1/3}(z(y)) \tag{7.28}$$

has a solution $z : [a,b] \to [0,1]$, $-\infty < a < b < \infty$, *with boundary conditions* $z(a) = 0$, $z(b) = 1$. *Then for some constant* A,

$$H_B\left(\delta, \mathcal{G}^{(\mathrm{conv})}, P\right) \leq A\delta^{-1}, \quad \text{for all } \delta > 0. \tag{7.29}$$

Proof For $p \in \mathcal{P}$,

$$\left|\left(\frac{pp_0}{p+p_0}\right)'\right|_\infty = \left|\frac{p_0^2 p' + p^2 p_0'}{(p+p_0)^2}\right|_\infty \leq 2M.$$

Hence, the class

$$\mathcal{F} = \left\{\frac{pp_0}{p+p_0} : p \in \mathcal{P}\right\} \tag{7.30}$$

is a (subclass of a) class of functions f with $|f'|_\infty \leq 2M$ and with $f \leq p_0$.

Assume without loss of generality that $2M = 1$. Define

$$B = \left\{ y : p_0(z(y)) \geq \delta^{3/2} \right\}.$$

Then,

$$B = \bigcup_{j=1}^{J} [a_j, b_j],$$

for some disjoint intervals $[a_j, b_j]$, $j = 1, \ldots, J$, and for some finite J. We shall assume $J = 1$. For general J, the proof carries through by treating each interval $[a_j, b_j]$ separately.

Let

$$y_k = a_1 + k\delta, \quad k = 1, \ldots, N-1; \qquad y_{N-1} < b_1, \; y_N = b_1.$$

Hence,

$$N \leq \left\lfloor \frac{b_1 - a_1}{\delta} \right\rfloor + 1 \leq \frac{b-a}{\delta} + 1,$$

where $\lfloor x \rfloor$ denotes the integer part of $x \geq 0$.

For $f \in \mathcal{F}$ take

$$\tilde{f}(z(y)) = \left\lfloor \frac{f(z(y_k))}{\delta z'(y_k)} \right\rfloor \delta z'(y_k), \quad y \in (y_{k-1}, y_k], \; k = 1, \ldots, N.$$

Note that

$$|z''(y)| \leq \frac{1}{3z'(y)} \leq \frac{1}{3\sqrt{\delta}}, \quad \text{for } y \in [a_1, b_1].$$

Therefore,

$$\left| \frac{z'(y)}{z'(\tilde{y})} \right| \leq 1 + \left| \frac{z'(y) - z'(\tilde{y})}{z'(\tilde{y})} \right| \leq \frac{4}{3} \leq 2,$$

for $y, \tilde{y} \in [a_1, b_1]$, $|y - \tilde{y}| \leq \delta$. So for $y \in (y_{k-1}, y_k]$,

$$\begin{aligned} |f(z(y)) - \tilde{f}(z(y))| &\leq \delta z'(y_k) + \left| f(z(y)) - f(z(y_k)) \right| \\ &\leq 3\delta z'(y_k), \quad k = 1, \ldots, N. \end{aligned}$$

Now, take

$$f^U(z(y)) = \begin{cases} p_0(z(y)), & y \notin (a_1, b_1], \\ \tilde{f}(z(y)) + 3\delta z'(y_k), & y \in (y_{k-1}, y_k], \; k = 1, \ldots, N, \end{cases}$$

and

$$f^L(z(y)) = \begin{cases} 0, & y \notin (a_1, b_1], \\ \tilde{f}(z(y)) - 3\delta z'(y_k), & y \in (y_{k-1}, y_k], \; k = 1, \ldots, N. \end{cases}$$

Clearly, using the fact that $f \leq p_0$,

$$f^L \leq f \leq f^U.$$

Moreover,

$$\begin{aligned}
&\int_{y\in(a_1,b_1]} \left(f^U(z(y)) - f^L(z(y))\right)^2 \frac{1}{p_0(z(y))}\, dz(y)\\
&= \sum_{k=1}^{N} \int_{y_{k-1}}^{y_k} \left(6\delta z'(y_k)\right)^2 \frac{1}{p_0\big(z(y)\big)} z'(y)\, dy\\
&\le 144\delta^2 \int_{a_1}^{b_1} \frac{\left(z'(y)\right)^3}{p_0\big(z(y)\big)}\, dy\\
&= 144\delta^2(b_1 - a_1),
\end{aligned}$$

and

$$\begin{aligned}
&\int_{y\notin(a_1,b_1],\ p_0(z(y))>0} \left(f^U(z(y)) - f^L(z(y))\right)^2 \frac{1}{p_0\big(z(y)\big)}\, dz(y)\\
&= \int_{y\notin(a_1,b_1]} p_0\big(z(y)\big) z'(y)\, dy \le \delta^2(b-a).
\end{aligned}$$

So,

$$\int_{p_0(x)>0} \left(f^U(x) - f^L(x)\right)^2 \frac{1}{p_0(x)}\, dx \le 144\delta^2(b_1 - a_1) + \delta^2(b-a) \le 169\delta^2(b-a).$$

It remains to calculate an upper bound for the number of brackets $[f^L, f^U]$ thus obtained. Note first of all that the number of $[f^L, f^U]$ does not exceed the number of $\tilde{f}$, where for some $f \in \mathscr{F}$,

$$\tilde{f}(z(y)) = \left\lfloor \frac{f(z(y_k))}{\delta z'(y_k)} \right\rfloor \delta z'(y_k) = M_k \delta z'(y_k), \quad y \in (y_{k-1}, y_k],$$

with $M_k \in \{0, 1, \ldots\}$, $k = 1, \ldots, N$. Because $p_0 \le K^2$ for some constant K,

$$\left\lfloor \frac{f(z(y_1))}{\delta z'(y_1)} \right\rfloor \le \left\lfloor \frac{p_0(z(y_1))}{\delta z'(y_1)} \right\rfloor = \left\lfloor \frac{p_0^{2/3}(z(y_1))}{\delta} \right\rfloor \le \left\lfloor \frac{K^{4/3}}{\delta} \right\rfloor,$$

we have at most $\lfloor K^{4/3}/\delta \rfloor + 1$ choices for M_1. Moreover, for $k = 2, \ldots, N$,

$$\begin{aligned}
\left| \left\lfloor \frac{f\big(z(y_k)\big)}{\delta z'(y_k))} \right\rfloor - \left\lfloor \frac{f\big(z(y_{k-1})\big)}{\delta z'(y_{k-1})} \right\rfloor \right| &\le 1 + \left| \frac{f\big(z(y_k)\big) - f\big(z(y_{k-1})\big)}{\delta z'(y_k)} \right|\\
&\quad + \left| \frac{f\big(z(y_{k-1})\big)\big(z'(y_k) - z'(y_{k-1})\big)}{\delta z'(y_k) z'(y_{k-1})} \right|\\
&\le 1 + 2 + \frac{4p_0\big(z(y_{k-1})\big)}{z'(y_{k-1})} \le 7.
\end{aligned}$$

So given M_{k-1}, we have at most 15 choices for M_k, $k = 2, \ldots, N$. Therefore, the total number of functions $\tilde{f}$ as $f \in \mathcal{F}$ varies is at most

$$\left(\left\lfloor \frac{K^{4/3}}{\delta} \right\rfloor + 1 \right) (15)^{N-1} \leq \left(\left\lfloor \frac{K^{4/3}}{\delta} \right\rfloor + 1 \right) (15)^{b-a/\delta}.$$

Hence,

$$H_B\big(13\sqrt{b-a}\delta, \mathcal{G}^{(\mathrm{conv})}, P\big) \leq \frac{b-a}{\delta} \log(15) + \log\left(\left\lfloor \frac{K^{4/2}}{\delta} \right\rfloor + 1 \right) \leq A\delta^{-1},$$

for some constant A. □

In view of Theorem 7.6, the conclusion is that under condition (7.28), $h(\hat{p}_n, p_0) = O_{\mathbf{P}}(n^{-1/3})$. Condition (7.28) is for example fulfilled for $p_0(x) = (m+1)x^m$, with $1 \leq m < 3$ or $m = 0$. In that case one finds

$$z(y) = \left(\frac{(m+1)^{\frac{1}{3}}(3-m)}{3} y \right)^{\frac{3}{3-m}}, \quad 0 \leq y \leq \frac{3}{(m+1)^{\frac{1}{3}}(3-m)}.$$

It does not hold for $p_0(x) = 4x^3$. The differential equation then has the solution

$$z(y) = e^{4^{\frac{1}{3}} y}, \quad -\infty < y \leq 0,$$

which has unbounded support. However, we require in (7.28) that the support $[a, b]$ is bounded. For $p_0(x) = 4x^3$, one can show in a similar way as in Lemma 7.9 that

$$H_B\big(\delta, \mathcal{G}^{(\mathrm{conv})}, P\big) \leq A \frac{1}{\delta} \log \frac{1}{\delta}, \quad \text{for all } \delta > 0.$$

Theorem 7.6 then gives the rate $h(\hat{p}_n, p_0) = O_{\mathbf{P}}\big(n^{-1/3} (\log n)^{1/3}\big)$.

Let us now briefly compare the result with a histogram estimator. Define $x_0 = 0$, $x_k = x_{k-1} + \Delta_k$ and let

$$\tilde{p}_n(x_k) = \frac{P_n(x_{k-1}, x_k]}{\Delta_k}, \quad k = 1, 2, \ldots,$$

be the histogram estimator of $p_0(x) = 2x$, $x \in (x_{k-1}, x_k]$, Δ_k being the bandwidth at x_k, $k = 1, 2, \ldots$. Because

$$\big| \operatorname{bias}(\tilde{p}_n(x_k)) \big|^2 = \Delta_k^2 (1 + o(1)),$$

and

$$\operatorname{var}(\tilde{p}_n(x_k)) = \frac{p_0(x_k)}{n\Delta_k} (1 + o(1)),$$

the optimal choice (in the sense of minimizing the mean square error) for the bandwidth is

$$\Delta_k = n^{-1/3} p_0^{1/3}(x_k).$$

So we find

$$x_k - x_{k-1} = n^{-1/3} p_0^{1/3}(x_k), \quad k = 1, 2, \ldots .$$

This is just the discrete analog of (7.28). In other words, (7.28) requires that one should be able to build a histogram with optimal bandwidth at each cell. Indeed, if one takes the suboptimal bandwidth $\Delta_k = n^{-1/3}$ for all k, then the histogram estimator converges with rate $O_{\mathbf{P}}(n^{-1/3} \log n^{1/2})$ in Hellinger distance:

$$\mathbf{E}\big(h^2(\tilde{p}_n, p_0)\big) = O\big(n^{-2/3} \log n\big).$$

Note that in general, the optimal choice of the bandwidth depends on p_0'. But this has no influence on the rate of convergence for the Hellinger distance. Moreover, given the model (7.27), p_0' cannot be estimated consistently by the maximum likelihood estimator.

Example 7.4.2. Decreasing densities Let

$$\mathscr{P} = \left\{ p : [0,\infty) \to [0,\infty),\ \int p \, d\mu = 1,\ p \text{ decreasing} \right\}. \tag{7.31}$$

Also in this case, any kernel estimator needs a locally adaptive bandwidth, because otherwise it will not have the (optimal) rate of convergence $O_{\mathbf{P}}(n^{-1/3})$ at most p_0 (see Section 10.3 for more details in a comparable situation). The maximum likelihood estimator converges with rate $O_{\mathbf{P}}(n^{-1/3})$ under mild conditions on p_0, which are for example met if p_0 is bounded and $d\mu_0 = \mathrm{l}\{p_0 > 0\} d\mu$ is a finite measure. This follows from application of Theorem 7.6. Note that

$$\left\{ \frac{p p_0}{p + p_0} : p \in \mathscr{P} \right\}$$

is a class of decreasing functions, with envelope p_0. If p_0 is bounded and μ_0 is a finite measure, it is not hard to prove that

$$H_B\big(\delta, \mathscr{G}^{(\mathrm{conv})}, P\big) \leq A\delta^{-1}, \quad \text{for all } \delta > 0. \tag{7.32}$$

To see this, first use

$$\left| \frac{p_1}{p_1 + p_0} - \frac{p_2}{p_2 + p_0} \right| \leq 2 \left| \left(\frac{p_1}{p_1 + p_0} \right)^{1/2} - \left(\frac{p_2}{p_2 + p_0} \right)^{1/2} \right| \tag{7.33}$$

and then apply Lemma 3.8. By Theorem 7.6, the rate of convergence is $O_{\mathbf{P}}(n^{-1/3})$. We now introduce two lemmas to relax the assumptions on p_0 and μ. The idea in Lemma 7.10 is to replace the power $\frac{1}{2}$ in (7.33) by the power $1/(2\alpha)$, with $0 < \alpha \leq 1$. This handles the situation where $\mu(\{p_0 > 0\})$ is infinite. Lemma 7.11 uses a similar idea to treat the case where p_0 is unbounded. Theorem 7.12 combines the two lemmas.

Lemma 7.10 *Let*

$$\mathcal{F} = \{f : [0, \infty) \to [0, \infty),\ f \text{ decreasing},\ f \le F\},$$

where $0 \le F \le 1$ *and* $\int F^{2(1-\alpha)}\, d\mu < \infty$ *for some* $0 < \alpha \le 1$. *Then for some* $\tilde{A}$,

$$H_B(\delta, \mathcal{F}, \mu) \le \tilde{A}\delta^{-1}, \quad \text{for all } \delta > 0.$$

Proof The class $\mathcal{F}^\alpha = \{f^\alpha : f \in \mathcal{F}\}$ is a collection of decreasing functions uniformly bounded by 1. Moreover, $dQ = F^{2(1-\alpha)} d\mu$ is a finite measure. Hence, by (2.5),

$$H_B(\delta, \mathcal{F}^\alpha, Q) \le A\delta^{-1}, \quad \text{for all } \delta > 0.$$

The result now follows from the fact that for $0 \le f^L \le f^U \le F$,

$$\int (f^U - f^L)^2\, d\mu \le \frac{1}{\alpha^2} \int \left((f^U)^\alpha - (f^L)^\alpha\right)^2 dQ. \qquad \square$$

Lemma 7.11 *Let*

$$\mathcal{F} = \{f : [0, \infty) \to [0, \infty),\ f \text{ decreasing},\ f \le F\},$$

with F *decreasing,* $F \ge 1$, *and* $\int F^{2(1+\alpha)}\, d\mu < \infty$ *for some* $\alpha > 0$. *Then for some* $\tilde{A}$,

$$H_B(\delta, \mathcal{F}, \mu) \le \tilde{A}\delta^{-1}, \quad \text{for all } \delta > 0.$$

Proof The class $(\mathcal{F} + F)^{-\alpha} = \{(f + F)^{-\alpha} : f \in \mathcal{F}\}$ is a collection of increasing functions uniformly bounded by 1, and $dQ = F^{2(1+\alpha)} d\mu$ is a finite measure. So by (2.5),

$$H_B(\delta, (\mathcal{F} + F)^{-\alpha}, \mu) \le A\delta^{-1}, \quad \text{for all } \delta > 0.$$

Furthermore, for $0 \le f^L \le f^U \le F$,

$$\int (f^U - f^L)^2\, d\mu \le \frac{2^{2(\alpha+1)}}{\alpha^2} \int \left((f^L + F)^{-\alpha} - (f^U + F)^{-\alpha}\right)^2 dQ. \qquad \square$$

Theorem 7.12 *Let*

$$\mathcal{P} = \left\{p : [0, \infty) \to [0, \infty),\ \int p\, d\mu = 1,\ p \text{ decreasing}\right\}.$$

Suppose that for some $0 < \alpha \le 1$,

$$\text{(7.34)} \qquad \int_{p_0 \le 1} p_0^{(1-\alpha)}\, d\mu < \infty, \quad \int_{p_0 > 1} p_0^{(1+\alpha)}\, d\mu < \infty.$$

Then for some constant c and for all $T \geq c$,

$$\text{(7.35)} \qquad \mathbf{P}(h(\hat{p}_n, p_0) > Tn^{-1/3}) \leq c \exp\left[-\frac{T^2 n^{1/3}}{c^2}\right].$$

Proof Apply Lemmas 7.10 and 7.11 to

$$\mathscr{F} = \left\{\left(\frac{pp_0}{p+p_0}\right)^{1/2} : p \in \mathscr{P}\right\},$$

with $F = p_0^{1/2}$. Theorem 7.6 then yields (7.35). □

In the special case that μ is Lebesgue measure, condition (7.34) is fulfilled if for some $m > 1$,

$$\limsup_{x\to\infty} x^m p_0(x) < \infty, \quad \limsup_{x\to 0} x^{1/m} p_0(x) < \infty.$$

So far, we did not make use of local entropies. In the following special case, however, the local entropy is much smaller, resulting in a much faster rate of convergence. Let μ be Lebesgue measure on $[0, \infty)$ and $p_0 = \mathrm{l}_{[0,1]}$. Then it is shown in van de Geer (1993, Lemma 4.7) that

$$H_B(u, \mathscr{G}^{(\mathrm{conv})}(\delta), P) \leq A\frac{\delta}{u} \log\left(\frac{1}{\delta}\right), \quad \text{for all } 0 < u \leq \delta \leq 1.$$

So the maximum likelihood estimator of the uniform density has rate

$$h(\hat{p}_n, p_0) = O_{\mathbf{P}}\left(n^{-\frac{1}{2}} \log^{\frac{1}{2}} n\right).$$

Example 7.4.3. Interval censoring case I, and the binary choice model Consider a group of n patients, which are susceptible to a certain disease. Let Y_i be the time of onset of the disease for patient i, and let Z_i be the time patient i is examined. We assume that Y_i and Z_i are independent, and that we observe Z_i and $\Delta_i = \mathrm{l}\{Y_i \leq Z_i\}$. Let Q be the (unknown) distribution of Z_i, and F_0 the unknown distribution of Y_i. In the first instance we assume nothing is known about F_0, i.e.,

$$F_0 \in \Lambda = \{\text{all distributions } F \text{ on } \mathbf{R}\}.$$

We write $F(y) = F(-\infty, y]$, $y \in \mathbf{R}$, for the corresponding distribution functions. The observations are $X_i = (\Delta_i, Z_i)$, $i = 1, \ldots, n$ (also known as current status data), which are i.i.d., with density of the form

$$\text{(7.36)} \qquad p_F(\Delta, z) = \Delta F(z) + (1 - \Delta)(1 - F(z)),$$

with respect to $\mu =$ (counting measure on $\{0,1\}) \times Q$. Because a distribution function is a monotone function with values between 0 and 1, and in view of (2.5), it is easy to see that for some constant A,

$$H_B(\delta, \mathcal{P}^{1/2}, \mu) \leq A\frac{1}{\delta}, \quad \text{for all } \delta > 0.$$

By Theorem 7.4, one therefore finds the rate

$$h(\hat{p}_n, p_0) = O_{\mathbf{P}}(n^{-1/3}).$$

In Example 11.2.3c, we shall use the rate of convergence to prove asymptotic normality of the maximum likelihood estimator of for example the mean $\int y \, dF_0(y)$.

From the fact that the densities in the binary choice model of Example 4.2.1 are of the same form, we know that this rate is also valid there. Because we did not use the convexity of $\mathcal{P}$ here, the rate remains true, and may in fact be faster, if we use a maximum likelihood estimator over a smaller class $\tilde{\Lambda} \subset \Lambda$. For instance, suppose that in the binary choice model it is known that F_0 is in the set

$$\tilde{\Lambda} = \{0 \leq F \leq 1 : 0 \leq F'(z) \leq M, \ F(z) \quad \text{is a concave function of } z\}.$$

Assume that Z has bounded support, say $Z \in [0,1]$, and that the density with respect to Lebesgue measure of Q exists and is bounded. Moreover, assume that for some $\eta_0 > 0$,

$$\eta_0^2 \leq F_0(z) \leq 1 - \eta_0^2, \quad \text{for } Q\text{-almost all } z \in [0,1].$$

Then

$$H_B\left(\delta, \tilde{\mathcal{P}}^{1/2}, \mu\right) \leq H_B\left(\delta\eta_0, \mathcal{P}, \mu\right) \leq A\delta^{-\frac{1}{2}}, \quad \text{for all } \delta > 0.$$

The latter bound can be derived from Birman and Solomjak (1967) (see also Example 9.3.4). So the rate now becomes

$$h(\hat{p}_n, p_0) = O_{\mathbf{P}}\left(n^{-2/5}\right).$$

Example 7.4.4. Interval censoring case II Consider again a group of n patients which are susceptible to a certain disease. In this example, we consider the situation where each patient has been infected, and will get the disease within a given time period. This difference with the previous example is that we now have an additional examination time per patient at our disposal. Each patient visits his/her medical doctor, who examines whether he/she has developed the disease. If not, the patient returns for a second check-up on a later occasion. Let $Y_i \in [0,2]$ be the time of onset of the disease for patient i. We assume that Y_i is a random variable with unknown distribution $F_0 \in \Lambda = \{\text{all distributions } F \text{ on } [0,2]\}$, $i = 1, \ldots, n$. We write $F(y) = F[0, y]$, $y \in [0,2]$, for the corresponding distribution functions. We observe for $i = 1, \ldots, n$,

$$Z_i = (U_i, V_i) \in [0,2]^2,$$

with $V_i > U_i$ the examination times, which are assumed to be independent of Y_i, and

$$\beta_i = 1\{Y_i \le U_i\}, \qquad \gamma_i = 1\{U_i < Y_i \le V_i\}.$$

The distribution of (U_i, V_i) is denoted by $\tilde{Q}$, and we suppose that $\tilde{Q}$ has density $\tilde{q}$ with respect to Lebesgue measure. In the notation of this chapter, our observations are $X_i = (\beta_i, \gamma_i, U_i, V_i)$, $i = 1, \ldots, n$.

The density with respect to $\mu = $ (counting measure on $\{(1,0),(0,1),(0,0)\}$) $\times\, \tilde{Q}$ is

$$p_0(x) = p_{F_0}(\beta, \gamma, u, v) = \beta F_0(u) + \gamma(F_0(v) - F_0(u)) + (1 - \beta - \gamma)(1 - F_0(v)), \quad F_0 \in \Lambda.$$

Suppose that

$$\tilde{q}(u, v) \le c_0, \quad \text{for all } (u, v) \in [0,2]^2, \text{ with } v > u \tag{7.37}$$

and

$$\frac{1}{c_1} \le f_0(y) \le c_1, \quad \text{for all } y \in [0,1], \tag{7.38}$$

where f_0 is the density of F_0 with respect to Lebesgue measure. Then,

$$\int_{p_0 \le \delta^2/(6c_0c_1)} d\mu \le \delta^2.$$

Recall the notation

$$dQ_\sigma = \frac{1}{p_0} 1\{p_0 > \sigma\} d\mu.$$

We have

$$\begin{aligned}\int dQ_\sigma &= \int_{F_0(u) > \sigma} \frac{1}{F_0(u)}\, d\tilde{Q}(u, v) + \int_{F_0(v) - F_0(u) > \sigma} \frac{1}{F_0(v) - F_0(u)}\, d\tilde{Q}(u, v) \\ &\quad + \int_{1 - F_0(v) > \sigma} \frac{1}{1 - F_0(v)}\, d\tilde{Q}(u, v) \\ &\le 3c_0c_1 \log\left(\frac{c_1}{\sigma}\right).\end{aligned}$$

The application of Lemma 3.8 now yields that for some constant A_1,

$$H_B(\delta,\mathcal{P},Q_\sigma) \le A\frac{1}{\delta}\log^{1/2}\left(\frac{1}{\sigma}\right), \quad \text{for all } \delta > 0.$$

Therefore, taking $\sigma_0(\delta) = \delta^2/(6c_0c_1)$, we have

$$H_B(\delta,\bar{\mathcal{P}}^{1/2},\mu) \le A_1\frac{1}{\delta}\log^{1/2}\left(\frac{1}{\delta}\right), \quad \text{for all } \delta > 0.$$

Inserting Theorem 7.4, we conclude that under conditions (7.37) and (7.38), the rate of convergence for the Hellinger distance is

$$h(\hat{p}_n, p_0) = O_{\mathbf{P}}\big(n^{-1/3}\log^{1/6} n\big). \tag{7.39}$$

Notice that

$$\begin{aligned} h^2(\hat{p}_n, p_0) &= \frac{1}{2}\int (\hat{F}_n^{1/2}(u) - F_0^{1/2}(u))^2\, d\tilde{Q}(u,v) \\ &\quad + \frac{1}{2}\int ((\hat{F}_n(v) - \hat{F}_n(u))^{1/2} - (F_0(v) - F_0(u))^{1/2})^2\, d\tilde{Q}(u,v) \\ &\quad + \frac{1}{2}\int ((1 - \hat{F}_n(v))^{1/2} - (1 - F_0(v))^{1/2})^2\, d\tilde{Q}(u,v). \end{aligned} \tag{7.40}$$

In other words, the Hellinger distance is actually a very strong metric, as compared to, for example, the L_2(Lebesgue measure)-norm between two distribution functions. The rate $O_{\mathbf{P}}\big(n^{-1/3}\log^{1/6} n\big)$ for the Hellinger distance suggests the rate $O_{\mathbf{P}}\big(n^{-1/3}\log^{-1/3} n\big)$ for $\big(\int_0^2(\hat{F}_n(y) - F_0(y))^2\, dy\big)^{1/2}$.

If the two examination times U_i and V_i cannot be arbitrary close to one another, i.e. if for some $\eta > 0$,

$$\mathbf{P}(V_i - U_i < \eta) = 0, \tag{7.41}$$

then under condition (7.38), $F_0(V_i) - F_0(U_i)$ stays away from zero as well. It is not hard to see that in that case one finds

$$H\big(\delta,\bar{\mathcal{P}}^{1/2},\mu\big) \le A_2\frac{1}{\delta}, \quad \text{for all } \delta > 0,$$

so that the rate is

$$h(\hat{p}_n, p_0) = O_{\mathbf{P}}(n^{-1/3}). \tag{7.42}$$

Although this is a faster rate than the one in (7.39), the result is actually worse, because under (7.41) the Hellinger metric is less strong.

As in Case I, the obtained rates can be applied to prove asymptotic normality of certain functions of $\hat{F}_n$ (see Example 11.2.3d).

Example 7.4.5. A convolution model Let Y and Z be independent random variables on $[0,1]$. Suppose that Z has a given density k with respect to Lebesgue measure. The distribution F_0 of Y is unknown. We observe independent copies $X_1, \ldots, X_n$ of $X = Y + Z$. The density of X is then of the form

$$p_F(x) = \int_0^1 k(x-y)\, dF(y), \quad 0 \le x \le 2, \tag{7.43}$$

with

$$F \in \Lambda = \{\text{all distributions on } [0,1]\}.$$

Note that $\mathscr{P} = \{p_F : F \in \Lambda\}$ is a convex class. In fact, we can write $\mathscr{P} = \overline{\text{conv}}(\mathscr{K})$, with $\mathscr{K} = \{k(\cdot - y) : y \in [0,1]\}$. We shall apply the result of Ball and Pajor (Theorem 3.14) to calculate the appropriate entropy.

Throughout, we assume that the density f_0, with respect to Lebesgue measure, of F_0 exists, and that for some constant $c_1 > 0$,

$$\frac{1}{c_1} \le |f_0(y)| \le c_1, \quad \text{for all } y \in [0,1]. \tag{7.44}$$

Case (i) Suppose that Z has the beta distribution with density

$$k(z) = \frac{\Gamma(2\beta+2)}{\Gamma^2(\beta+1)} z^\beta (1-z)^\beta, \quad 0 \le z \le 1, \tag{7.45}$$

where $\beta > 0$ is fixed. For simplicity we assume that the two parameters of the beta distribution are equal. The rate of convergence is determined by the behaviour of k near $x = 0$ and $x = 1$. Let $c_2, c_3, \ldots$ be suitable, strictly positive constants depending on c_1 and β. One easily verifies that under condition (7.44), for $0 \le x \le 1$,

$$p_0(x) \le c_2 x^{\beta+1},$$

and

$$p_0(x) \ge c_3 x^{\beta+1}.$$

Define

$$G(x) = \sup_{y \in [0,1]} \frac{k(x-y)}{p_0(x)}.$$

Then for $0 \le x \le 1$,

$$G(x) \le c_4 \frac{1}{x},$$

so that

$$\int_0^1 G^2(x)dP(x) \le c_2 c_4^2 \int_0^1 x^{\beta-1}\,dx \le c_5^2.$$

The same arguments can be applied to the case $1 \le x \le 2$, so that one finds

$$\int_0^2 G^2(x)\,dP(x) \le c_6^2. \tag{7.46}$$

Write $\tilde{\mathcal{G}} = \{k(\cdot - y)/p_0 \,:\, y \in [0,1]\}$. Because

$$|k(x-y_1) - k(x-y_2)| \le \begin{cases} c_7|y_1 - y_2|, & \beta \ge 1, \\ c_8|y_1 - y_2|^\beta, & 0 < \beta \le 1\,, \end{cases}$$

we find

$$N_\infty(\delta, \mathcal{K}) \le \begin{cases} c_9\delta^{-1}, & \beta \ge 1, \\ c_{10}\delta^{-1/\beta}, & 0 < \beta \le 1. \end{cases}$$

So, in view of Theorem 3.14, on the set $\{\int G^2\,dP_n \le (2c_6)^2\}$,

$$H\big(\delta, \overline{\text{conv}}(\tilde{\mathcal{G}}), P_n\big) \le \begin{cases} A_1\delta^{-2/3}, & \beta \ge 1, \\ A_2\delta^{-2/(2\beta+1)}, & 0 < \beta \le 1. \end{cases}$$

Since

$$H\big(2\delta, \mathcal{G}^{(\text{conv})}, P_n\big) \le H\big(\delta, \overline{\text{conv}}(\tilde{\mathcal{G}}), P_n\big),$$

(see also (7.19)), we find from Theorem 7.7 (with $\sigma_n = 0$),

$$h(\hat{p}_n, p_0) = \begin{cases} O_{\mathbf{P}}(n^{-3/8}), & \beta \ge 1, \\ O_{\mathbf{P}}\big(n^{-(2\beta+1)/(4\beta+4)}\big), & 0 < \beta \le 1\,. \end{cases}$$

However, for $0 < \beta \le \frac{1}{2}$, we can improve the rate. We have that $p_\theta(x)$ is increasing for $0 \le x \le 1$, and decreasing for $1 \le x \le 2$. So it follows from Theorem 7.4 that

$$h(\hat{p}_n, p_0) = O_{\mathbf{P}}(n^{-1/3}), \quad 0 < \beta \le \frac{1}{2}.$$

In fact, the same is true when $\beta = 0$. In that case, k is the density of the uniform distribution, and the model is of the same form as the one in Example 7.4.3.

Clearly, if k is smooth, then the entropy of $\mathcal{P}$ is small, resulting in a fast rate of convergence in Hellinger distance. On the other hand, if k is smooth, the inverse problem is hard. Therefore, a fast rate in Hellinger distance can correspond to a slow rate for, say, $\left(\int(\hat{F}_n(y) - F_0(y))^2\, dy\right)^{1/2}$.

Case (ii) In Case (i), the density k is concave. Let us now see what happens if k is convex. We examine the case

$$k(z) = 3 - 12z(1-z).$$

Let

$$G(x) = \sup_{y\in[0,1]} \frac{k(x-y)}{p_0(x)} \mathrm{l}\{p_0(x) > \sigma\}.$$

Again, $c_2, c_3, \ldots$ will be suitable, strictly positive constants depending on c_1. It is easy to see that

$$\int G^2\, dP \leq 9 \int_{p_0(x)>\sigma} \frac{1}{p_0(x)}\, dx \leq 9c_2^2 \log\left(\frac{1}{\sigma}\right), \tag{7.47}$$

and

$$\int_{p_0(x)\leq\sigma} p_0(x)\, dx \leq c_3^2\sigma^2.$$

Therefore, we take $\sigma_n = \delta_n/c_3$. Write

$$\tilde{\mathscr{G}} = \left\{\frac{k(\cdot - y)}{p_0} \mathrm{l}\{p_0 > \sigma_n\} \,:\, y \in [0,1]\right\}.$$

On the set $\{\int G^2\, dP_n \leq T^2 \log(1/\delta_n)\}$, we find for $dQ_n = dP_n/(T \log^{1/2}(1/\delta_n))$,

$$N(\delta, \tilde{\mathscr{G}}, Q_n) \leq c_4 \frac{1}{\delta}.$$

So by Theorem 3.14, on $\{\int G^2\, dP_n \leq T^2 \log(1/\delta_n)\}$,

$$H(\delta, \overline{\mathrm{conv}}(\tilde{\mathscr{G}}), Q_n) \leq A\delta^{-\frac{2}{3}},$$

and hence

$$H(\delta, \overline{\mathrm{conv}}(\tilde{\mathscr{G}}), P_n) \leq A\left(\frac{T \log^{1/2}(1/\delta_n)}{\delta}\right)^{2/3}.$$

In view of (7.47), it follows that

$$\sup_{\delta>0} \frac{H(\delta, \mathscr{G}_{\sigma_n}^{(\mathrm{conv})}, P_n)}{\log^{1/3}(1/\delta_n)\delta^{-2/3}} = O_{\mathbf{P}}(1).$$

So we arrive at the rate

$$h(\hat{p}_n, p_0) = O_{\mathbf{P}}(n^{-3/8} \log^{1/8} n).$$

Example 7.4.6. A suboptimal rate Let μ be Lebesgue measure, and
(7.48)
$$\mathscr{P} = \left\{ p : [0,1] \to \mathbf{R},\ \int p(x)\,dx = 1,\ |p(x) - p(\tilde{x})| \le M|x - \tilde{x}|^\alpha,\ x, \tilde{x} \in [0,1] \right\},$$

where $M > 0$ and $\alpha > 0$ are given. Assume that

$$p_0 \ge \eta_0^2 > 0. \tag{7.49}$$

Using the same arguments as in the proof of Lemma 2.3, one finds that for some constant A,

$$H_\infty(\delta, \bar{\mathscr{P}}^{1/2}) \le A\delta^{-1/\alpha}, \quad \text{for all } \delta > 0. \tag{7.50}$$

Hence, by Theorem 7.4, for $\alpha > \frac{1}{2}$, one obtains the rate

$$h(\hat{p}_n, p_0) = O_{\mathbf{P}}(n^{-\alpha/(2\alpha+1)}). \tag{7.51}$$

However, if $\alpha \le \frac{1}{2}$, the entropy integral

$$\int_0^\delta H_B^{1/2}(u, \bar{\mathscr{P}}^{1/2}(\delta), \mu)\,du$$

diverges. To see this, we have to show that the bound $Au^{-1/\alpha}$ for

$$H_B(u, \bar{\mathscr{P}}^{1/2}(\delta), \mu)$$

cannot be improved. Therefore, we introduce the following lemma.

Lemma 7.13 *For* $0 < \alpha \le 1$, *let*

$$\tilde{\mathscr{G}} = \{g : [0,1] \to \mathbf{R},\ |g| \le 1,\ |g(x) - g(\tilde{x})| \le |x - \tilde{x}|^\alpha,\ x, \tilde{x} \in [0,1]\},$$

and $\tilde{\mathscr{G}}(\delta) = \{g \in \tilde{\mathscr{G}} : \|g\|_\mu \le \delta\}$. *Then for some constant* $A_1 > 0$,

$$H\left(\frac{\delta}{8}, \tilde{\mathscr{G}}(\delta), \mu\right) \ge A_1\delta^{-1/\alpha}, \quad \textit{for all } \delta > 0.$$

Proof In this proof, $k = 1, \ldots, N-1$. Take $0 = a_0 < a_1 < \cdots < a_N = 1$, with $a_k = k\delta^{1/\alpha}$, so that $\delta^{-1/\alpha} \le N \le \delta^{-1/\alpha} + 1$. Define

$$\psi_k(x) = \begin{cases} \delta^{-(\frac{1}{\alpha}-1)}(x - a_{k-1}), & a_{k-1} \le x \le a_{k-1} + \frac{1}{2}\delta^{\frac{1}{\alpha}}, \\ -\delta^{-(\frac{1}{\alpha}-1)}(x - a_k), & a_{k-1} + \frac{1}{2}\delta^{\frac{1}{\alpha}} \le x \le a_k, \\ 0, & \text{else.} \end{cases}$$

Then

$$\int \psi_k^2(x)\,dx \le \frac{1}{4}\delta^{2+\frac{1}{\alpha}}, \tag{7.52}$$

$$|\psi_k(x) - \psi_k(\tilde{x})| \le |x-\tilde{x}|^\alpha, \quad x,\tilde{x} \in [0,1], \tag{7.53}$$

and

$$\int \psi_k^2(x)\,dx \ge \frac{1}{8}\delta^{2+\frac{1}{\alpha}}. \tag{7.54}$$

Define for $\tau \in \{0,1\}^N$,

$$g_\tau(x) = \sum_{k=1}^N \tau_k \psi_k(x).$$

Then from (7.52) and (7.53), one sees that $g_\tau \in \tilde{\mathscr{G}}(\delta)$. Moreover, from (7.54), for $\sum_{k=1}^N |\tau_k - \tilde{\tau}_k| \ge N/8$,

$$\|g_\tau - g_{\tilde{\tau}}\|_\mu^2 \ge \left(\frac{\delta}{8}\right)^2.$$

In Birgé (1983, Proposition 3.8), it is shown that for $N \ge 8$, the largest set $\mathscr{T} \subset \{0,1\}^N$ with $\sum_{k=1}^N |\tau_k - \tilde{\tau}_k| \ge N/8$ for all $\tau, \tilde{\tau} \in \mathscr{T}$, $\tau \ne \tilde{\tau}$, satisfies

$$\log(\mathscr{T} - 1) \ge 0.316N. \qquad \square$$

By noting that for $u \le \delta$,

$$H_B\left(u, \bar{\mathscr{P}}^{1/2}(\delta), \mu\right) \ge H\left(u, \bar{\mathscr{P}}^{1/2}(u), \mu\right),$$

one indeed finds from Lemma 7.13 that for $0 < \alpha \le \frac{1}{2}$ and some constant $A_2 > 0$,

$$H_B\left(u, \bar{\mathscr{P}}^{1/2}(\delta), \mu\right) \ge A_2 u^{-1/\alpha}, \quad \text{for all } u < \frac{\delta}{c_1},$$

with c_1 some constant, depending on M. Therefore, the rate of convergence implied by Theorem 7.4 is

$$h(\hat{p}_n, p_0) = \begin{cases} O_{\mathbf{P}}(n^{-\frac{1}{4}} \log^{\frac{1}{2}} n), & \alpha = \frac{1}{2}, \\ O_{\mathbf{P}}(n^{-\frac{\alpha}{2}}), & 0 < \alpha < \frac{1}{2}. \end{cases}$$

Of course, these slow rates could be a consequence of our method of proof, i.e., the actual *exact* rate for the maximum likelihood estimator could be faster. However, Birgé and Massart (1993) have shown for a related problem that this is not the case. Because there do exist other estimators with a better rate of convergence (see Section 10.4), the conclusion is that one should not use the maximum likelihood estimator when the entropy integral diverges.

7.5. Notes

Theorem 7.4 can be found in Wong and Shen (1995), albeit that they impose the entropy conditions on the class $\mathcal{P}$, instead of $\bar{\mathcal{P}}$. The usefulness of $\bar{\mathcal{P}}$ was pointed out by Birgé and Massart (1993). The latter prove Theorem 7.4 under condition (7.1). If $\mathcal{P}$ is indexed by a finite-dimensional parameter, say $\mathcal{P} = \{p_\theta : \theta \in \Theta\}$, with $\Theta \subset \mathbf{R}^d$, Theorem 7.4 can sometimes recover the $n^{-1/2}$-rate (if appropriate). On the other hand, the conditions of Theorem 7.4 are too strong in certain cases. For example, if $d = 1$, and if $h(p_\theta, p_{\tilde{\theta}})$ behaves like $|\theta - \tilde{\theta}|$, then the entropy with bracketing may be replaced by entropy without bracketing. This is because the second moment of $\sqrt{p_\theta / p_0}$ exists. When $d > 1$, one needs higher order moments (see Ibragimov and Has'minskii (1981a) and Section 12.3).

Using local entropy instead of global entropy means that the rate of convergence can depend on p_0. Moreover, in finite-dimensional cases, the rate $n^{-1/2}$ can only follow from local entropy calculations. In infinite-dimensional models, local and global entropy are often of the same order. In fact, it is shown in Yang and Barron (1996) that in such cases there always exists a p_0 at which local and global entropy are of the same order.

In van de Geer (1993), conditions on the entropy with bracketing as well as on the empirical entropy are imposed. This leads to variants of Theorem 7.4 and Theorem 7.6. Theorem 7.7 is from van de Geer (1996).

Minimax rates of convergence for various statistical quantities, are obtained by Ibragimov and Has'minskii (1981b). For smooth densities, these rates coincide with the ones found in Example 7.4.1. Rates of convergence, in, for example, L_1-distance, for the maximum likelihood estimator of a decreasing density (the Grenander estimator) are given in Groeneboom (1985). Interval censored observations are treated in Groeneboom and Wellner (1992).

7.6. Problems and complements

7.1. Let $\mathcal{P}$ be a class of densities with respect to the dominating measure μ, and suppose

$$\int q \, d\mu_0 < \infty,$$

where

$$q = \sup_{p \in \mathscr{P}} p,$$

and $d\mu_0 = 1\{p_0 > 0\}d\mu$. Take $d\tilde{\mu}_0 = q d\mu_0$ and $\tilde{\mathscr{P}} = \{p/q : p \in \mathscr{P}\}$. Verify that $\tilde{\mathscr{P}}$ is a uniformly bounded class of densities, with respect to the finite dominating measure $\tilde{\mu}_0$.

7.2. Show that if $\mathscr{G} = \{\frac{1}{2}\log(p + p_0)/2p_0 : p \in \mathscr{P}\}$ is a P-Donsker class, then $h(\hat{p}_n, p_0) = o_{\mathbf{P}}(n^{-\frac{1}{4}})$.

7.3. It often happens that the entropy with bracketing of $\bar{\mathscr{P}}^{1/2}(\delta)$ is of smaller order for small values of δ, than for large values of δ. Of course, if $H_B(u, \bar{\mathscr{P}}^{1/2}(\delta), \mu_0)$ is finite for all u and δ, then it follows from Theorem 4.3 that $h(\hat{p}_n, p_0) \to 0$, a.s., so that for proving a rate of convergence, it suffices to consider $\bar{\mathscr{P}}^{1/2}(\delta)$ with δ arbitrarily small. One may use Theorem 7.4 here, but only the result on the rate of convergence, and not the exponential bound. However, in many cases, the exponential bound is still valid. Suppose that for some constants A_0, $0 \le \nu \le 1$ and $\eta > 0$,

$$J(\delta, \bar{\mathscr{P}}^{1/2}(\delta), \mu_0) \le \begin{cases} A_0\delta^\nu, & 0 < \delta \le \eta \\ A_0 f(\delta) & \eta < \delta \le 1, \end{cases}$$

where $f(\delta) \ge \delta^\nu$ for all $0 < \delta \le 1$. Invoke Theorem 7.4 to find that for some constant c_1 and for all $T \ge c_1$,

$$\mathbf{P}(h(\hat{p}_n, p_0) > Tn^{-\frac{1}{2(2-\nu)}}) \le c_1 \exp\left[-\frac{T^2 n^{(1-\nu)/(2-\nu)}}{c_1^2}\right].$$

7.4. Let Θ be a bounded subset of $\mathbf{R}^d$, and

$$\mathscr{P} = \{p_\theta : \theta \in \Theta\}.$$

Write $p_0 = p_{\theta_0}$. Suppose that for some $\eta > 0$ and c_0,

$$h(p_\theta, p_{\theta_0}) \ge \frac{|\theta - \theta_0|}{c_0}, \quad |\theta - \theta_0| \le \eta,$$

and that for all $\delta > 0$,

$$\int \sup_{|\theta - \tilde{\theta}| \le \delta} \left(\left(\frac{p_\theta + p_{\theta_0}}{2}\right)^{1/2} - \left(\frac{p_{\tilde{\theta}} + p_{\theta_0}}{2}\right)^{1/2} \right)^2 d\mu_0 \le c_0^2\delta^2.$$

Show that $h(\hat{p}_n, p_0) = O_{\mathbf{P}}(n^{-1/2})$.

7.5. Let μ be Lebesgue measure on the real line, and

$$\mathscr{P} = \left\{ p_\theta(x) = \frac{1}{\theta} 1\{0 < x < \theta\} : \theta > 0 \right\}.$$

Write $p_0 = p_{\theta_0}$. In Problem 4.8, it was shown that $h(\hat{p}_n, p_0) \to 0$, a.s., so in order to arrive at a rate of convergence, it suffices to consider $\bar{\mathscr{P}}(\delta)$, with δ small. Show that Theorem 7.4 produces the rate $h(\hat{p}_n, p_0) = O_{\mathbf{P}}(n^{-1/2})$. This implies $|\hat{\theta}_n - \theta_0| = O_{\mathbf{P}}(n^{-1})$, where $\hat{\theta}_n$ is the maximum likelihood estimator of θ_0.

7.6. Let $X_1, X_2, \ldots$ be i.i.d. copies of

$$X = Y + Z,$$

where Y and Z are independent positive random variables, Z is exponentially distributed, with intensity $\lambda = 1$, and Y has an unknown distribution $F_0 \in \Lambda$, where

$$\Lambda = \{\text{all distributions } F \text{ on } (0, \infty)\}.$$

Suppose that $\int_0^\infty e^y \, dF_0(y) < \infty$. Show that $h(\hat{p}_n, p_0) = O_{\mathbf{P}}(n^{-1/3})$.
(Hint: use the fact that the class of densities with respect to $d\mu(x) = e^{-x} dx$ is

$$\mathscr{P} = \left\{ p(x) = \int_0^x e^y \, dF(y) : F \in \Lambda \right\}.$$

This is a convex class of increasing densities.)

7.7. Let μ be Lebesgue measure on $[0, 2]$ and

$$\mathscr{P} = \left\{ p_F(x) = \int k(x \mid y) \, dF(y) : F \in \Lambda \right\},$$

with Λ the collection of all probability measures on $[0, 1]$, and

$$k(x) = 2x 1\{0 \le x \le 1\}.$$

Write $p_0 = p_{F_0}$, and suppose F_0 has a density with respect to Lebesgue measure, that is bounded away from zero and infinity on $[0, 1]$. Then $h(\hat{p}_n, p_0) = O_{\mathbf{P}}(n^{-3/8} (\log n)^{1/8})$ (see van de Geer (1996)).

8

The Non-I.I.D. Case

The maximal inequality of Lemma 3.2 is completed, in the sense that its assumption of an exponential probability inequality for weighted sums is further investigated. Moreover, the maximal inequality given in Theorem 5.11 is extended from the i.i.d. case to independent, non-identically distributed random variables. This allows one to apply it to weighted sums, so that a comparison with Lemma 3.2 is possible. The difference is that the non-i.i.d. version of Theorem 5.11 needs weaker moment conditions and stronger entropy conditions. Theorem 5.11 is also generalized to dependent random variables. The result is applied to maximum likelihood problems.

8.1. Independent non-identically distributed random variables

This section is splitted into two parts. The first part is confined to weighted sums, whereas the second part treats independent non-identically distributed random variables in general, with weighted sums as a special case. Both subsections set out with an exponential probability inequality, which is then extended to a *uniform* exponential probability inequality.

8.1.1. Maximal inequalities for weighted sums revisited

Let $z_1, \ldots, z_n$ be a set of points in a space $\mathscr{Z}$, $\mathscr{G}$ be a class of real-valued functions on $\mathscr{Z}$ and let $W_1, \ldots, W_n$ be independent, real-valued random variables with expectation zero. In Section 3.3, Lemma 3.2, we proved a maximal inequality for $\frac{1}{n}\sum_{i=1}^{n} W_i g(z_i)$, $g \in \mathscr{G}$, under the assumption that for some C_1 and C_2, and for all $\gamma \in \mathbf{R}^n$ and $a > 0$,

$$(8.1) \qquad \mathbf{P}\left(\left|\sum_{i=1}^{n} W_i\gamma_i\right| \geq a\right) \leq C_1 \exp\left[-\frac{a^2}{C_2^2 \sum_{i=1}^{n} \gamma_i^2}\right].$$

If $|W_i| \leq K$ for all $i \in \{1, \ldots, n\}$, then (8.1) holds, with $C_1 = 2$, $C_2^2 = 2K^2$. This follows from Hoeffding's inequality (see Lemma 3.5). We shall now consider more general W_i. This will allow us to apply Lemma 3.2 in regression problems, with W_i the measurement error in the model $Y_i = g_0(z_i) + W_i$, $i = 1, \ldots, n$ (see Chapter 9). Of course, one does not want to assume that the errors in a regression model are bounded. One often assumes normally distributed errors. Here, we shall consider the more general situation, where

$$(8.2) \qquad \max_{i=1,\ldots,n} K^2 \left(\mathbf{E}e^{|W_i|^2/K^2} - 1\right) \leq \sigma_0^2.$$

Then $W_1, \ldots, W_n$ are called uniformly sub-Gaussian. We show that under (8.2), (8.1) is met for certain C_1 and C_2 depending on K and σ_0.

To verify (8.1), we first need to relate (8.2) to the behaviour of moment generating function of W_i.

Lemma 8.1 *Let W be a random variable with $\mathbf{E}W = 0$ and with*

$$(8.3) \qquad K^2(\mathbf{E}e^{|W|^2/K^2} - 1) \leq \sigma_0^2.$$

Then for all β,

$$(8.4) \qquad \mathbf{E}e^{\beta W} \leq \exp[2(K^2 + \sigma_0^2)\beta^2].$$

Proof Take $L = 1 + (\sigma_0^2/K^2)$. By Chebyshev's inequality, we have for all $t > 0$,

$$\mathbf{P}(|W| > t) \leq L \exp[-t^2/K^2].$$

Hence, for $m \in \{2, 3, \ldots\}$,

$$\mathbf{E}|W|^m \leq L \int_0^\infty \exp[-t^{2/m}/K^2]\, dt = LK^m \Gamma\left(\frac{m}{2} + 1\right).$$

So, for $\beta \geq 0$,

$$\begin{aligned}
\mathbf{E}e^{\beta W} &= 1 + \sum_{m=2}^{\infty} \frac{1}{m!}\beta^m \mathbf{E}W^m \\
&\leq 1 + \sum_{m=2}^{\infty} L \frac{\Gamma(\frac{m}{2}+1)}{\Gamma(m+1)} K^m \beta^m \leq 1 + \sum_{m=2}^{\infty} L \frac{K^m \beta^m}{\Gamma(\frac{m}{2}+1)} \\
&= 1 + \sum_{m=1}^{\infty} L \frac{(K^2\beta^2)^m}{\Gamma(m+1)} + \sum_{m=1}^{\infty} L \frac{(K^2\beta^2)^{m+\frac{1}{2}}}{\Gamma(m+\frac{3}{2})} \\
&\leq 1 + \sum_{m=1}^{\infty} L \frac{(K^2\beta^2)^m (1 + (K^2\beta^2)^{1/2})}{\Gamma(m+1)} \\
&\leq 1 + \sum_{m=1}^{\infty} \frac{(K^2 L \beta^2)^m (1 + (K^2 L \beta^2)^{1/2})}{\Gamma(m+1)} \\
&\leq \exp[2K^2 L \beta^2].
\end{aligned}$$

The result for $\beta \leq 0$ follows by replacing W by $-W$. □

Lemma 8.2 below states that indeed, the exponential probability inequality needed in Lemma 3.2 is met, provided $W_1, \ldots, W_n$ are uniformly sub-Gaussian.

Lemma 8.2 *Suppose that $W_1, \ldots, W_n$ are independent random variables with expectation zero and with*

$$\max_{i=1,\ldots,n} K^2 \left(\mathbf{E}e^{|W_i|^2/K^2} - 1\right) \leq \sigma_0^2. \tag{8.5}$$

Then for all $\gamma \in \mathbf{R}^n$ and $a > 0$,

$$\mathbf{P}\left(\left|\sum_{i=1}^{n} W_i \gamma_i\right| \geq a\right) \leq 2\exp\left[-\frac{a^2}{8(K^2+\sigma_0^2)\sum_{i=1}^{n}\gamma_i^2}\right]. \tag{8.6}$$

Proof By Lemma 8.1, for all β,

$$\mathbf{E}\exp\left[\beta \sum_{i=1}^{n} W_i \gamma_i\right] = \prod_{i=1}^{n} \exp[\beta W_i \gamma_i] \leq \exp\left[2(K^2+\sigma_0^2)\beta^2 \sum_{i=1}^{n} \gamma_i^2\right].$$

Hence, from Chebyshev's inequality, for $\beta > 0$,

$$\mathbf{P}\left(\sum_{i=1}^{n} W_i \gamma_i \geq a\right) \leq \exp\left[2(K^2+\sigma_0^2)\beta^2 \sum_{i=1}^{n} \gamma_i^2 - \beta a\right].$$

Now, take

$$\beta = \frac{a}{4(K^2+\sigma_0^2)\sum_{i=1}^n \gamma_i^2},$$

to obtain that

$$\mathbf{P}\left(\sum_{i=1}^n W_i\gamma_i \geq a\right) \leq \exp\left[-\frac{a^2}{8(K^2+\sigma_0^2)\sum_{i=1}^n \gamma_i^2}\right].$$

Because the same inequality holds for $-\sum_{i=1}^n W_i\gamma_i$, the lemma follows. □

As a corollary, the maximal inequality for weighted sums holds whenever $W_1, \ldots, W_n$ are uniformly sub-Gaussian. To state this corollary, recall the notation used in Section 3.3. Let

$$Q_n = \frac{1}{n}\sum_{i=1}^n \delta_{z_i},$$

and for $g : \{z_1, \ldots, z_n\} \to \mathbf{R}$,

$$\|g\|_{Q_n}^2 = \int g^2\, dQ_n.$$

Moreover, let $H(\delta, \mathcal{G}, Q_n)$ be the δ-entropy of $(\mathcal{G}, \|\cdot\|_{Q_n})$.

Corollary 8.3 *Suppose that* $\sup_{g\in\mathcal{G}} \|g\|_{Q_n} \leq R$ *and that*

(8.7) $$\max_{i=1,\ldots,n} K^2\left(\mathbf{E}e^{|W_i|^2/K^2} - 1\right) \leq \sigma_0^2.$$

Then for some constant C depending only on K and σ_0, and for $\delta > 0$ and $\sigma > 0$ satisfying $R > \delta/\sigma$ and

(8.8) $$\sqrt{n}\delta \geq 2C\left(\int_{\delta/8\sigma}^R H^{1/2}(u, \mathcal{G}, Q_n)\, du \vee R\right),$$

we have

(8.9) $$\mathbf{P}\left(\sup_{g\in\mathcal{G}} \left|\frac{1}{n}\sum_{i=1}^n W_i g(z_i)\right| \geq \delta \wedge \frac{1}{n}\sum_{i=1}^n W_i^2 \leq \sigma^2\right) \leq C\exp\left[-\frac{n\delta^2}{4C^2R^2}\right].$$

As in Section 5.5, one can derive, from Corollary 8.3, the modulus of continuity of the process $\left\{\frac{1}{\sqrt{n}}\sum_{i=1}^n W_i g(z_i) : g \in \mathcal{G}\right\}$. Let us simplify the exposition by assuming that

(8.10) $$H(\delta, \mathcal{G}, Q_n) \leq A\delta^{-\alpha}, \quad \text{for all } \delta > 0,$$

for some constants $0 < \alpha < 2$ and A. Then we may choose $\sigma \to \infty$ in (8.8) and (8.9), because the entropy integral converges:

$$\int_0^1 H^{1/2}(u, \mathcal{G}, Q_n)\, du < \infty.$$

Lemma 8.4 *Suppose that* (8.10) *is met, that* $\sup_{g\in\mathcal{G}} \|g\|_{Q_n} \leq R$ *and that*

$$\max_{i=1,\dots,n} K^2 \left(\mathbf{E}e^{|W_i|^2/K^2} - 1\right) \leq \sigma_0^2.$$

Then for some constant c depending on A, α, R, K and σ_0, *we have for all* $T \geq c$,

$$\mathbf{P}\left(\sup_{g\in\mathcal{G}} \frac{\left|\frac{1}{\sqrt{n}}\sum_{i=1}^n W_i g(z_i)\right|}{\|g\|_{Q_n}^{1-\frac{\alpha}{2}}} \geq T \right) \leq c \exp\left[-\frac{T^2}{c^2}\right]. \tag{8.11}$$

Proof We have

$$\int_0^\delta H^{1/2}(u, \mathcal{G}, Q_n)\, du \leq A_0 \delta^{1-\frac{\alpha}{2}}.$$

The proof is now straightforwardly peeling off the class $\mathcal{G}$. For $C_1 \geq 1$,

$$\mathbf{P}\left(\sup_{g\in\mathcal{G},\ \|g\|_{Q_n}\leq\delta} \left| \frac{1}{\sqrt{n}} \sum_{i=1}^n W_i g(z_i) \right| \geq 2C_1 C A_0 \delta^{1-\frac{\alpha}{2}} \right) \leq C \exp\left[-C_1^2 A_0^2 \delta^{-\alpha}\right].$$

It follows that for $T = 2C_1 C A_0 2^{1-(\alpha/2)}$,

$$\begin{aligned}
&\mathbf{P}\left(\sup_{g\in\mathcal{G}} \frac{\left|\frac{1}{\sqrt{n}}\sum_{i=1}^n W_i g(z_i)\right|}{\|g\|_{Q_n}^{1-\frac{\alpha}{2}}} \geq T \right) \\
&\leq \sum_{s=1}^\infty \mathbf{P}\left(\sup_{g\in\mathcal{G},\ \|g\|_{Q_n}\leq 2^{-s+1}R} \left| \frac{1}{\sqrt{n}} \sum_{i=1}^n W_i g(z_i) \right| \geq T(2^{-s}R)^{1-\frac{\alpha}{2}} \right) \\
&= \sum_{s=1}^\infty \mathbf{P}\left(\sup_{g\in\mathcal{G},\ \|g\|_{Q_n}\leq 2^{-s+1}R} \left| \frac{1}{\sqrt{n}} \sum_{i=1}^n W_i g(z_i) \right| \geq 2C_1 C A_0 (2^{-s+1}R)^{1-\frac{\alpha}{2}} \right) \\
&\leq \sum_{s=1}^\infty C \exp[-C_1^2 A_0^2 (2^{-s+1}R)^{-\alpha}] \leq c \exp\left[-\frac{T^2}{c^2}\right].
\end{aligned}$$

□

We end this section with an extension of Corollary 8.3, which will be useful in Chapter 12. Consider a subset Θ of a metric space (Λ, d). The δ-entropy of this subset is defined in the usual sense, i.e. it is the logarithm of the minimum number of balls with radius δ, necessary to cover Θ. Now, we shall study stochastic processes indexed by $\theta \in \Theta$, which might not be defined for $\theta \notin \Theta$. Therefore, we require that the centers of the balls covering Θ are a subset of Θ. Thus, let $H(\delta, \Theta; d) = \log N(\delta, \Theta; d)$, where $N(\delta, \Theta; d) = \min\{N :$ there exists $\{\theta_1, \ldots, \theta_N\} \subset \Theta$ such that $\min_{j=1,\ldots,N} \max_{\theta \in \Theta} d(\theta, \theta_j) \leq \delta\}$.

Suppose now that the metric is of the form

$$d^2 = \frac{1}{n} \sum_{i=1}^{n} d_i^2,$$

where $d_1, \ldots, d_n$ are also metrics on Θ. Consider real-valued random variables $U_{i,\theta}$, $i = 1, \ldots, n$, $\theta \in \Theta$. We assume that the processes $\{U_{i,\theta} : \theta \in \Theta\}$, $i = 1, \ldots, n$, are independent and centered, and that

$$|U_{i,\theta} - U_{i,\tilde{\theta}}| \leq |V_i| d_i(\theta, \tilde{\theta}), \quad i = 1, \ldots, n, \ \theta, \tilde{\theta} \in \Theta, \tag{8.12}$$

where $V_1, \ldots, V_n$ are uniformly sub-Gaussian:

$$\max_{i=1,\ldots,n} K^2 \left(\mathbf{E} e^{|V_i|^2/K^2} - 1 \right) \leq \sigma_0^2. \tag{8.13}$$

Lemma 8.5 *Assume* (8.12) *and* (8.13) *and that* $\sup_{\theta \in \Theta} d(\theta, \theta_0) \leq R$. *Then for some constant* $\tilde{C}$ *depending only on* K *and* σ_0, *and for all* $\delta > 0$ *and* $\sigma > 0$ *satisfying*

$$\sqrt{n}\delta \geq \tilde{C} \left(\int_{\delta/8\sigma}^{R} H^{1/2}(u, \Theta; d)\, du \vee R \right), \tag{8.14}$$

we have

(8.15)

$$\mathbf{P}\left(\sup_{\theta \in \Theta} \left| \frac{1}{n} \sum_{i=1}^{n} (U_{i,\theta} - U_{i,\theta_0}) \right| \geq \delta \wedge \frac{1}{n} \sum_{i=1}^{n} V_i^2 \leq \sigma^2 \right) \leq \tilde{C} \exp\left[-\frac{n\delta^2}{\tilde{C}^2 R^2} \right].$$

Proof Since

$$U_{i,\theta} - U_{i,\tilde{\theta}} = W_i(\theta, \tilde{\theta}) d_i(\theta, \tilde{\theta}),$$

with

$$W_i(\theta, \tilde{\theta}) = \frac{U_{i,\theta} - U_{i,\tilde{\theta}}}{d_i(\theta, \tilde{\theta})},$$

satisfying

$$\mathbf{E} W_i(\theta, \tilde{\theta}) = 0,$$

and

$$|W_i(\theta, \tilde{\theta})| \leq |V_i|,$$

for $i = 1, \ldots, n$, we know from Lemma 8.2 that for all $a > 0$,

$$\mathbf{P}\left(\left|\sum_{i=1}^{n}(U_{i,\theta} - U_{i,\tilde{\theta}})\right| \geq a\right) \leq 2\exp\left[-\frac{a^2}{8(K^2 + \sigma_0^2)nd^2(\theta, \tilde{\theta})}\right].$$

Now, apply the same arguments as in the proof of Lemma 3.2. □

8.1.2. Maximal inequalities based on Bernstein's inequality

Consider independent, non-identically distributed random variables $X_1, \ldots, X_n$ with values in $(\mathscr{X}, \mathscr{A})$. Denote the distribution of X_i by $P^{(i)}$, $i = 1, \ldots, n$, and let $\bar{P} = \sum_{i=1}^{n} P^{(i)}/n$. Moreover, let $P_n = \sum_{i=1}^{n} \delta_{X_i}/n$ be the empirical distribution. For $\mathscr{G}$ a class of real-valued functions on $\mathscr{X}$, consider the empirical process

(8.16) $$\{v_n(g) = \sqrt{n}\int g\, d(P_n - \bar{P}),\ g \in \mathscr{G}\}.$$

It turns out that an inequality of exactly the same form as the one in Theorem 5.11 holds for the empirical process, i.e. Theorem 5.11 directly extends to the non-i.i.d. case. The proof of this fact is based on Bernstein's inequality, which we present here for completeness.

Lemma 8.6 (Bernstein's inequality) *Consider independent random variables $Z_1, \ldots, Z_n$ with expectation zero, and with*

(8.17) $$\frac{1}{n}\sum_{i=1}^{n} \mathbf{E}|Z_i|^m \leq \frac{m!}{2}K^{m-2}R^2, \quad m = 2, 3, \ldots.$$

Then for all $a > 0$,

(8.18) $$\mathbf{P}\left(\left|\frac{1}{n}\sum_{i=1}^{n} Z_i\right| \geq a\right) \leq 2\exp\left[-\frac{na^2}{2(aK + R^2)}\right].$$

See for example Shorack and Wellner (1986). The inequality for general discrete time martingales is given in Lemma 8.9.

We avoid repetitions (the final extension of Theorem 5.11 is given in Theorem 8.13), we now only give the recipe. Define

$$\bar{\rho}_K^2(g) = 2K^2\int\left(e^{|g|/K} - 1 - |g|/K\right) d\bar{P}.$$

The generalized entropy with bracketing $\mathscr{H}_{B,K}(\delta, \mathscr{G}, \bar{P})$ is defined as in Definition 5.1, with ρ_K replaced by $\bar{\rho}_K$. Assuming that

$$\sup_{g \in \mathscr{G}} \bar{\rho}_K(g) \leq R,$$

one has the same result for the empirical process $\{\nu_n(g) : g \in \mathscr{G}\}$ as the one given in Theorem 5.11, provided that in (5.33), one replaces P by $\bar{P}$.

These findings can be applied to the special case of weighted sums. Let $W_1, \ldots, W_n$ be independent random variables with expectation zero, and $\{z_1, \ldots, z_n\}$ be (non-random) elements of a space $\mathscr{Z}$. Furthermore, let $\mathscr{G}$ be a class of real-valued functions on $\mathscr{Z}$. We shall consider the process

$$\left\{ \frac{1}{\sqrt{n}} \sum_{i=1}^n W_i g(z_i) : g \in \mathscr{G} \right\}. \tag{8.19}$$

Taking $\mathscr{F} = \{f(w,z) = wg(z),\ g \in \mathscr{G}\}$, we see that we may express (8.19) as $\{\nu_n(f) : f \in \mathscr{F}\}$. As before, we use the notation $Q_n = \sum_{i=1}^n \delta_{z_i}/n$ and $\|\cdot\|_{Q_n}^2 = \int (\cdot)^2 \, dQ_n$. Write $H_B(\delta, \mathscr{G}, Q_n)$ for the δ-entropy with bracketing of $(\mathscr{G}, \|\cdot\|_{Q_n})$. If the random variables $W_1, \ldots, W_n$ possess an exponential moment, and if moreover $\mathscr{G}$ is uniformly bounded, then the generalized entropy with bracketing of $\mathscr{F}$ is bounded by the entropy with bracketing of $(\mathscr{G}, \|\cdot\|_{Q_n})$. This is shown in Lemma 8.7 below.

Lemma 8.7 *Suppose that*

$$\max_{i=1,\ldots,n} 2K_1^2 \mathbf{E}(e^{|W_i|/K_1} - 1 - |W_i|/K_1) \le \sigma_0^2, \tag{8.20}$$

and that

$$\sup_{g \in \mathscr{G}} |g|_\infty \le K_2. \tag{8.21}$$

Then for $K = 4K_1K_2$ *and for all* $\delta > 0$,

$$\mathscr{H}_{B,K}(\sqrt{2}\sigma_0\delta, \mathscr{F}, \bar{P}) \le H_B(\delta, \mathscr{G}, Q_n). \tag{8.22}$$

Proof Take $-K_2 \le g^L \le g \le g^U \le K_2$, $\|g^U - g^L\|_{Q_n} \le \delta$, and

$$f^L(w,z) = wg^L(z)\mathrm{l}\{w \ge 0\} + wg^U(z)\mathrm{l}\{w < 0\}.$$

Similarly, take

$$f^U(w,z) = wg^U(z)\mathrm{l}\{w \ge 0\} + wg^L(z)\mathrm{l}\{w < 0\}.$$

Clearly, we then have for $f(w,z) = wg(z)$,

$$f^L \le f \le f^U.$$

Inequality (8.20) implies that

$$\max_{i=1,\dots,n} \mathbf{E}|W_i|^m \le \frac{m!}{2} K_1^{m-2}\sigma_0^2, \quad m = 2, 3, \dots.$$

Therefore,

$$\frac{1}{n}\sum_{i=1}^{n} \mathbf{E}|f^U(W_i, z_i) - f^L(W_i, z_i)|^m \le \frac{m!}{2}(2K_1K_2)^{m-2}\sigma_0^2\delta^2, \quad m = 2, 3, \dots.$$

But then, for $K = 4K_1K_2$,

$$\begin{aligned}\bar{\rho}_K^2(f^U - f^L) &= \frac{1}{n}\sum_{i=1}^{n}\sum_{m=2}^{\infty} \frac{2}{m!K^{m-2}} \mathbf{E}|f^U(W_i, z_i) - f^L(W_i, z_i)|^m \\ &\le \sum_{m=2}^{\infty} \frac{1}{2^{m-2}}\sigma_0^2\delta^2 = 2\sigma_0^2\delta^2. \qquad \square\end{aligned}$$

The application of the non-i.i.d. version of Theorem 5.11 to weighted sums now yields the following corollary.

Corollary 8.8 *Suppose that* $\sup_{g\in\mathcal{G}} \|g\|_{Q_n} \le R$ *and that*

(8.23) $$\max_{i=1,\dots,n} 2K_1^2\mathbf{E}\left(e^{|W_i|/K_1} - 1 - |W_i|/K_1\right) \le \sigma_0^2.$$

Moreover, assume that

(8.24) $$\sup_{g\in\mathcal{G}} |g|_\infty \le K_2.$$

Then for some universal constant C and for

(8.25) $$K = 4K_1K_2,$$

(8.26) $$\delta \le C_1 2R^2\sigma_0^2/K,$$

(8.27) $$\delta \le 8\sqrt{2}R\sigma_0,$$

(8.28) $$\sqrt{n}\delta \ge C_0\left(\int_{\delta/2^6}^{\sqrt{2}R\sigma_0} H_B^{1/2}\left(\frac{u}{\sqrt{2}\sigma_0}, \mathcal{G}, Q_n\right) du \vee \sqrt{2}R\sigma_0\right),$$

(8.29) $$C_0^2 \geq C^2(C_1+1),$$

we have

(8.30) $$\mathbf{P}\left(\sup_{g\in\mathcal{G}}\left|\frac{1}{n}\sum_{i=1}^{n} W_i g(z_i)\right| \geq \delta\right) \leq C\exp\left[-\frac{n\delta^2}{C^2(C_1+1)2R^2\sigma_0^2}\right].$$

A comparison of Corollary 8.3 and Corollary 8.8 reveals that the condition of sub-Gaussian W_i may be replaced by the assumption of exponential tails. But there is a price to pay: Corollary 8.8 uses entropy *with bracketing*, and moreover requires a uniform bound on the class of functions.

From Corollary 8.8, one can deduce the modulus of continuity of the process $\left\{\frac{1}{\sqrt{n}}\sum_{i=1}^{n} W_i g(z_i) : g \in \mathcal{G}\right\}$. The result is similar to Lemma 5.13 and Lemma 8.4. But note however that, in contrast to Lemma 8.4, the modulus of continuity cannot be derived for arbitrary small values of $\|g\|_{Q_n}$, due to condition (8.26).

Remark Suppose now that $(W_1,Z_1),\ldots,(W_n,Z_n)$ are i.i.d. copies of (W,Z), with $W \in \mathbf{R}$ and $Z \in \mathcal{Z}$. Let P be the distribution of (W,Z), Q be the distribution of Z, and let $\mathcal{G}$ be a class of functions on $\mathcal{Z}$. Suppose that

(8.31) $$\mathbf{E}(W \mid Z) = 0, \quad \text{a.s.},$$

(8.32) $$2K_1^2\mathbf{E}\left(e^{|W|/K_1} - 1 - |W|/K_1 \mid Z\right) \leq \sigma_0^2, \quad \text{a.s.},$$

and

(8.33) $$\sup_{g\in\mathcal{G}} |g|_\infty \leq K_2.$$

Then, from the same arguments as those in the proof of Lemma 8.7, one finds for $\mathcal{F} = \{wg(z) : g \in \mathcal{G}\}$ and $K = 4K_1K_2$,

(8.34) $$\mathcal{H}_{B,K}(\sqrt{2}\sigma_0\delta, \mathcal{F}, P) \leq H_B(\delta, \mathcal{G}, Q), \quad \text{for all } \delta > 0.$$

Thus, Theorem 5.11 also leads to an i.i.d. version of Corollary 8.8, with Q_n replaced by Q.

8.2. Martingales

Consider a probability space $(\Omega, \mathcal{F}, \mathbf{P})$ and let $\mathcal{F}_0 \subset \mathcal{F}_1 \ldots$ be an increasing sequence of sub-sigma algebras of $\mathcal{F}$. For $i \geq 1$, let Z_i be $\mathcal{F}_i$-measurable random variables with conditional expectation $\mathbf{E}(Z_i \mid \mathcal{F}_{i-1}) = 0$. Write $M_0 = 0$, and for $n \geq 1$, $M_n = \sum_{i=1}^n Z_i$. Thus $\{M_n\}$ is a martingale. We establish an exponential probability inequality in Subsection 8.2.1, and then extend this in Subsection 8.2.2 to a uniform inequality, in the situation where the martingale depends on a parameter. The result will be applied to the maximum likelihood problem in Section 8.3.

8.2.1. Bernstein's inequality for martingales

In the proof of Lemma 8.9 below, we use the notation

$$\rho_{i,K}^2 = 2K^2 \mathbf{E}\big(e^{|Z_i|/K} - 1 - |Z_i|/K \mid \mathcal{F}_{i-1}\big), \quad i = 1, 2, \ldots,$$

and

$$\bar{\rho}_{n,K}^2 = \frac{1}{n} \sum_{i=1}^n \rho_{i,K},$$

where $K > 0$ is some constant. This notation will also be used later on.

Lemma 8.9 *Let n be fixed. Suppose that for some $\mathcal{F}_{n-1}$ measurable random variable R_n^2,*

$$\frac{1}{n} \sum_{i=1}^n \mathbf{E}(|Z_i|^m \mid \mathcal{F}_{i-1}) \leq \frac{m!}{2} K^{m-2} R_n^2, \quad m = 2, 3, \ldots. \tag{8.35}$$

Then for all $a > 0$, $R > 0$,

$$\mathbf{P}(M_n \geq a \wedge R_n \leq R) \leq \exp\left[-\frac{a^2}{2(aK + nR^2)}\right]. \tag{8.36}$$

Proof Take $L > K$ and define for $i = 1, 2, \ldots,$

$$U_{i,L} = \frac{M_i}{L} - \frac{\sum_{j=1}^i \rho_{j,L}^2}{2L^2},$$

and

$$W_{i,L} = e^{|Z_i|/L} - 1 - |Z_i|/L.$$

Note that $\left\{ \sum_{j=1}^i \rho_{j,L}^2/(2L^2) \right\}$ is the compensator of $\left\{ \sum_{j=1}^i W_{j,L} \right\}$. We have

$$e^{U_{i,L}} - e^{U_{i-1,L}} = e^{U_{i-1,L}} \left(\exp\left[\frac{Z_i}{L} - \frac{\rho_{i,L}^2}{2L^2}\right] - 1 \right),$$

and

$$\exp\left[\frac{Z_i}{L} - \frac{\rho_{i,L}^2}{2L^2}\right] - 1 \leq \frac{\exp[Z_i/L]}{1 + \rho_{i,L}^2/(2L^2)} - 1 = \frac{W_{i,L} + 1 + Z_i/L}{1 + \rho_{i,L}^2/(2L^2)} - 1$$
$$= \frac{W_{i,L} - \rho_{i,L}^2/(2L^2) + Z_i/L}{1 + \rho_{i,L}^2/(2L^2)}.$$

It follows that

$$\mathbf{E}\left(\exp\left[\frac{Z_i}{L}-\frac{\rho_{i,L}^2}{2L^2}\right]-1 \mid \mathscr{F}_{i-1}\right)\leq 0, \quad i=1,2\ldots.$$

Hence,

$$\mathbf{E}(e^{U_{i,L}} \mid \mathscr{F}_{i-1}) \leq e^{U_{i-1,L}}, \quad i=1,2,\ldots,$$

i.e., $\{e^{U_{i,L}}\}$ is a supermartingale. Therefore,

$$\mathbf{E}e^{U_{n,L}} \leq 1.$$

Note now that

$$U_{n,L}=\frac{M_n}{L}-n\frac{\bar{\rho}_{n,L}^2}{2L^2},$$

and under condition (8.35), for $L>K$,

$$\bar{\rho}_{n,L}^2 \leq \frac{LR_n^2}{L-K}.$$

Therefore, on the set $A_n=\{M_n\geq a,\ R_n\leq R\}$,

$$e^{U_{n,L}}\geq \exp\left[\frac{a}{L}-\frac{nR^2}{2L(L-K)}\right].$$

So,

$$\mathbf{P}(A_n)\leq \exp\left[-\frac{a}{L}+\frac{nR^2}{2L(L-K)}\right].$$

Take

$$L=K+\frac{nR^2}{a}$$

to complete the proof. □

If $|Z_i|\leq K$ for all i, we may take nR_n^2 as the predictable variation

$$\sum_{i=1}^{n}\mathbf{E}(|Z_i|^2 \mid \mathscr{F}_{i-1}).$$

Moreover, condition (8.35) is automatically fulfilled if one chooses $R_n^2=\bar{\rho}_{n,K}^2$.

Corollary 8.10 *We have for all n, $K>0$, $a>0$ and $R>0$,*

$$\mathbf{P}(M_n\geq a \wedge \bar{\rho}_{n,K}\leq R)\leq \exp\left[-\frac{a^2}{2(aK+nR^2)}\right]. \tag{8.37}$$

If the conditions of Lemma 8.9 hold uniformly in n, the exponential probability inequality also holds uniformly in n. We shall not use this result, but we state it for completeness.

Lemma 8.11 *Suppose that for some predictable increasing sequence* $\{nR_n^2\}$,

$$\sum_{i=1}^{n} \mathbf{E}(|Z_i|^m \mid \mathscr{F}_{i-1}) \leq \frac{m!}{2} K^{m-2} nR_n^2, \quad m = 2, 3, \ldots, \ n \geq 1. \tag{8.38}$$

Then for all $a > 0$, $b > 0$,

$$\mathbf{P}\big(M_n \geq a \wedge nR_n^2 \leq b^2 \textit{ for some } n\big) \leq \exp\left[-\frac{a^2}{2(aK + b^2)}\right]. \tag{8.39}$$

Proof Define $U_{i,L}$, $i = 1, 2, \ldots$, $L > K$, as in the proof of Lemma 8.9. We showed there that

$$\{X_{n,L}\} = \{e^{U_{n,L}}\}$$

is a supermartingale. Hence, for any stopping time σ,

$$\mathbf{E}X_{\sigma,L}\{\sigma < \infty\} \leq 1.$$

Let

$$A = \{M_n \geq a \wedge nR_n^2 \leq b^2 \text{ for some } n\},$$

and $\sigma = \inf\{n : M_n \geq a\}$. Because $A \subset \{\sigma < \infty\}$,

$$\int_A X_{\sigma,L} d\mathbf{P} \leq 1.$$

On A, we have that $M_\sigma \geq a$ and $\sigma R_\sigma^2 \leq b^2$. The result now follows from the same arguments as in the proof of Lemma 8.9. □

Since (8.38) is met for all n when $R_n^2 = \bar{\rho}_{n,K}^2$, we know that Corollary 8.10 in fact holds uniformly in n.

Corollary 8.12 *We have for all* $K > 0$, $a > 0$, $b > 0$,

$$\mathbf{P}\big(M_n \geq a \wedge n\bar{\rho}_{n,K}^2 \leq b^2 \textit{ for some } n\big) \leq \exp\left[-\frac{a^2}{2(aK + b^2)}\right]. \tag{8.40}$$

8.2.2. *A uniform inequality*

Consider a probability triple $(\Omega, \mathscr{F}, \mathbf{P})$, and sub-sigma algebras $\mathscr{F}_0 \subset \mathscr{F}_1 \subset \cdots \subset \mathscr{F}_n \subset \cdots \subset \mathscr{F}$.

Let $Z_{i,\theta}$ be $\mathscr{F}_i$-measurable random variables, $i = 1, \ldots, n, \ldots$, depending on a parameter $\theta \in \Theta$. Denote the average of the first n by

$$\bar{Z}_{n,\theta} = \frac{1}{n} \sum_{i=1}^{n} Z_{i,\theta},$$

and define

(8.41)
$$\bar{A}_{n,\theta} = \frac{1}{n}\sum_{i=1}^{n} \mathbf{E}(Z_{i,\theta} \mid \mathscr{F}_{i-1}).$$

Then $\{M_{n,\theta} = n(\bar{Z}_{n,\theta} - \bar{A}_{n,\theta})\}$ is a martingale, indexed by a parameter $\theta \in \Theta$. Now, in this subsection, n will be fixed, and we write $\bar{Z}_\theta = \bar{Z}_{n,\theta}$ and $\bar{A}_\theta = \bar{A}_{n,\theta}$.

Define

$$\rho_K(Z_{i,\theta}) = 2K^2\mathbf{E}\left(e^{|Z_{i,\theta}|/K} - 1 - |Z_{i,\theta}|/K \mid \mathscr{F}_{i-1}\right), \quad i = 1, \ldots, n,$$

and for $\mathbf{Z}_\theta = (Z_{1,\theta}, \ldots, Z_{n,\theta})$, write

(8.42)
$$\bar{\rho}_K^2(\mathbf{Z}_\theta) = \frac{1}{n}\sum_{i=1}^{n} \rho_K(Z_{i,\theta}).$$

We now introduce an extension of Definition 5.1 to the case of dependent variables. We measure the 'closeness' of two random vectors, say $\mathbf{Z}$ and $\tilde{\mathbf{Z}}$ (with ith component $\mathscr{F}_i$-measurable), by the random quantity $\bar{\rho}_K(\mathbf{Z} - \tilde{\mathbf{Z}})$. However, our definition does not give rise to random covering numbers. Instead, we look for a measurable set $\mathbf{F} \in \mathscr{F}$, such that a certain collection of random vectors forms a bracketing set for all $\omega \in \mathbf{F}$. Therefore, our entropy concept will also depend on $\mathbf{F}$. In applications, one should try to choose $\mathbf{F}$ in such a way that it has large probability.

Definition 8.1 (Generalized entropy with bracketing) For $0 < \delta \le R$ and $\mathbf{F} \in \mathscr{F}$, let $\{[\mathbf{Z}_j^L, \mathbf{Z}_j^U]\}_{j=1}^N$ be a collection of pairs of random vectors $\mathbf{Z}_j^L = (Z_{1,j}^L, \ldots, Z_{n,j}^L)$ and $\mathbf{Z}_j^U = (Z_{1,j}^U, \ldots, Z_{n,j}^U)$, with $[Z_{i,j}^L, Z_{i,j}^U]$ $\mathscr{F}_i$-measurable, $i = 1, \ldots, n$, $j = 1, \ldots, N$, such that for all $\theta \in \Theta$, there is a $j = j(\theta) \in \{1, \ldots, N\}$, with $\theta \mapsto j(\theta)$ non-random, such that

(i) $\bar{\rho}_K^2(\mathbf{Z}_j^U - \mathbf{Z}_j^L) \le \delta^2$ on $\{\bar{\rho}_K(\mathbf{Z}_\theta) \le R\} \cap \mathbf{F}$,

(ii) $Z_{i,j}^L \le Z_{i,\theta} \le Z_{i,j}^U$, $i = 1, \ldots, n$, on $\{\bar{\rho}_K(\mathbf{Z}_\theta) \le R\} \cap \mathbf{F}$.

Let $\mathscr{N}_{B,K}(\delta, R, \mathbf{F})$ be the smallest non-random value of N for which such a collection $\{[\mathbf{Z}_j^L, \mathbf{Z}_j^U]\}_{j=1}^N$ exists. Then $\mathscr{H}_{B,K}(\delta, R, \mathbf{F}) = \log \mathscr{N}_{B,K}(\delta, R, \mathbf{F})$ is called the *generalized δ-entropy with bracketing*.

The generalized entropy with bracketing might be hard to calculate. On the other hand, in the following fairly general situation, it can be bounded by the entropy for the supremum norm. Let $Z_{i,\theta} = g_\theta(X_i)$, $\theta \in \Theta$, where $X_i \in \mathscr{X}$ is $\mathscr{F}_i$-measurable and where $g_\theta : \mathscr{X} \to \mathbf{R}$ is some non-random function. Assume that

$$\sup_{\theta \in \Theta} |g_\theta|_\infty \le K.$$

Then, using the idea of Lemma 5.8,

$$\bar{\rho}_{2K}(\mathbf{Z}_\theta)^2 \le 2K^2.$$

Moreover, for all $\delta > 0$,

$$\mathscr{H}_{B,4K}(\sqrt{2}\delta, \sqrt{2}K, \Omega) \le H_\infty\left(\frac{\delta}{2}, \mathscr{G}\right),$$

where $\mathscr{G} = \{g_\theta : \theta \in \Theta\}$ (invoke the arguments of Lemma 2.1 and 5.10).

Here is a uniform inequality for martingales.

Theorem 8.13 *Take*

$$a \le C_1 \sqrt{n} R^2 / K, \tag{8.43}$$

$$a \le 8\sqrt{n} R, \tag{8.44}$$

$$a \ge C_0 \left(\int_{a/2^6\sqrt{n}}^{R} \mathscr{H}_{B,K}^{1/2}(u, R, \mathbf{F})\, du \vee R \right), \tag{8.45}$$

and

$$C_0^2 \ge C^2(C_1 + 1). \tag{8.46}$$

Then

$$\mathbf{P}\left((\sqrt{n}|\bar{Z}_\theta - \bar{A}_\theta| \ge a \wedge \bar{\rho}_K(\mathbf{Z}_\theta) \le R \text{ for some } \theta \in \Theta) \cap \mathbf{F} \right) \le C \exp\left[-\frac{a^2}{C^2(C_1 + 1)R^2} \right]. \tag{8.47}$$

The proof is given in the appendix.

For example, if $Z_i = g_\theta(X_i)$, $\theta \in \Theta$, where $X_i \in \mathscr{X}$ is $\mathscr{F}_i$-measurable, and $g_\theta : \mathscr{X} \to \mathbf{R}$ are non-random functions, then we may invoke the entropy of $\mathscr{G} = \{g_\theta : \theta \in \Theta\}$, for the supremum norm, in (8.45). Note moreover that Theorem 8.13 reduces to Theorem 5.11 when $X_1, \ldots, X_n$ are i.i.d.

8.3. Application to maximum likelihood

Suppose one observes $(X_1, \ldots, X_n)$ with distribution $\mathbf{P}$. Let f_{θ_0} be the density of $(X_1, \ldots, X_n)$, with respect to a σ-finite dominating measure μ. The parameter θ_0 is assumed to be a member of a given parameter space Θ. The maximum likelihood estimator of θ_0 is

$$\hat{\theta}_n = \arg\max_{\theta\in\Theta} f_\theta(X_1,\dots,X_n). \tag{8.48}$$

Let $\mathscr{F}_0 = \emptyset$, $\mathscr{F}_i = \sigma(X_1,\dots,X_i)$, $i = 1,\dots,n$, and suppose that the conditional densities $p_{i,\theta}$ given $\mathscr{F}_{i-1}$ exist for $i = 1,\dots,n$. These conditional densities $p_{i,\theta}$ are supposed to be defined with respect to a $\mathscr{F}_{i-1}$-measurable, σ-finite measure μ_i, independent of θ.

The Hellinger process is

$$h^2\left(\mathbf{p}_{\theta_1},\mathbf{p}_{\theta_2}\right) = \frac{1}{2}\sum_{i=1}^{n}\int\left(p_{i,\theta_1}^{1/2} - p_{i,\theta_2}^{1/2}\right)^2 d\mu_i. \tag{8.49}$$

We write

$$\bar{h}^2\left(\mathbf{p}_{\theta_1},\mathbf{p}_{\theta_2}\right) = \frac{1}{n}h^2\left(\mathbf{p}_{\theta_1},\mathbf{p}_{\theta_2}\right). \tag{8.50}$$

We have seen that in the i.i.d. case the generalized entropy with bracketing of the set of log-likelihood ratios (involving the convex combinations) is bounded by the entropy with bracketing of the class of densities, endowed with Hellinger metric. The same is true in the dependent case.

Definition 8.2 (Hellinger entropy with bracketing) Let $\mathbf{N}_B(\delta,R,\mathbf{F})$ be the smallest (non-random) value of N such that there exists a collection $\left\{\left[\mathbf{p}_j^L,\mathbf{p}_j^U\right]\right\}_{j=1}^N$, with $\mathbf{p}_j^L = (p_{1,j}^L,\dots,p_{n,j}^L)$ and $\mathbf{p}_j^U = (p_{1,j}^U,\dots,p_{n,j}^U)$, and with $\left[p_{i,j}^L, p_{i,j}^U\right]$ a pair of $\mathscr{F}_{i-1}$-measurable functions, $i = 1,\dots,n$, $j = 1,\dots,N$, such that for all $\theta\in\Theta$, there is a (non-random) $j = j(\theta)\in\{1,\dots,N\}$, such that

(i) $\bar{h}\left(\frac{\mathbf{p}_j^L+\mathbf{p}_{\theta_0}}{2}, \frac{\mathbf{p}_j^U+\mathbf{p}_{\theta_0}}{2}\right) \le \delta$ on $\left\{\bar{h}(\mathbf{p}_\theta,\mathbf{p}_{\theta_0})\le R\right\}\cap\mathbf{F}$,

(ii) $p_{i,j}^L \le p_{i,\theta} \le p_{i,j}^U$, $i = 1,\dots,n$, on $\{\bar{h}(\mathbf{p}_\theta,\mathbf{p}_{\theta_0})\le R\}\cap\mathbf{F}$.

Then $\{\mathbf{H}_B(\delta,R,\mathbf{F}) = \log\mathbf{N}_B(\delta,R,\mathbf{F}) : \delta > 0\}$ is called *the Hellinger entropy with bracketing.*

Now, let c be an appropriate universal constant, and denote the entropy integral by

$$\mathbf{J}_B(\delta,\mathbf{F}) = \int_{\frac{\delta^2}{c}}^{\delta}\mathbf{H}_B^{1/2}(u,\delta,\mathbf{F})\,du \vee \delta, \quad 0<\delta\le 1.$$

Theorem 8.14 *Take $\Psi(\delta) \geq \mathbf{J}_B(\delta, \mathbf{F})$ in such a way that $\Psi(\delta)/\delta^2$ is a non-increasing function of δ. Then for*

$$\sqrt{n}\delta_n^2 \geq c\Psi(\delta_n), \tag{8.51}$$

we have for all $\delta \geq \delta_n$,

$$\mathbf{P}\left(\bar{h}(\mathbf{p}_{\hat{\theta}_n}, \mathbf{p}_{\theta_0}) > \delta\right) \leq c\exp\left[-\frac{n\delta^2}{c^2}\right] + \mathbf{P}(\mathbf{F}^c). \tag{8.52}$$

Proof Take for $\theta \in \Theta$,

$$Z_{i,\theta} = \frac{1}{2}\log\left(\frac{p_{i,\theta}(X_i) + p_{i,\theta_0}(X_i)}{2p_{i,\theta_0}(X_i)}\right), \quad i = 1, \ldots, n,$$

$$\bar{Z}_\theta = \frac{1}{n}\sum_{i=1}^n Z_{i,\theta},$$

and

$$\bar{A}_\theta = \frac{1}{n}\sum_{i=1}^n \mathbf{E}(Z_{i,\theta} \mid \mathscr{F}_{i-1}).$$

The proof is now along the lines of the proof of Theorem 7.4. We have

$$\mathbf{P}\left(\bar{h}(\mathbf{p}_{\hat{\theta}_n}, \mathbf{p}_{\theta_0}) > \delta\right) \leq \mathbf{P}\left(\bar{h}\left(\frac{\mathbf{p}_{\hat{\theta}_n} + \mathbf{p}_{\theta_0}}{2}, \mathbf{p}_{\theta_0}\right) > \frac{\delta}{4}\right)$$

$$\leq \sum_{s=0}^{S}\mathbf{P}\left(\left(\sqrt{n}|\bar{Z}_\theta - \bar{A}_\theta| \geq \sqrt{n}2^{2s}\left(\frac{\delta}{4}\right)^2 \wedge \bar{h}\left(\frac{\mathbf{p}_\theta + \mathbf{p}_{\theta_0}}{2}, \mathbf{p}_{\theta_0}\right)\right.\right.$$

$$\left.\left.\leq 2^{s+1}\frac{\delta}{4} \text{ for some } \theta\right) \cap \mathbf{F}\right) + \mathbf{P}(\mathbf{F}^c)$$

$$= \sum_{s=0}^{S}\mathbf{P}_s + \mathbf{P}(\mathbf{F}^c),$$

with $S = \min\{\, s : \, 2^{s+1}\frac{\delta}{4} > 1\,\}$. It is not difficult to verify that

$$\bar{\rho}_1(\mathbf{Z}_\theta) \leq 4h\left(\frac{\mathbf{p}_\theta + \mathbf{p}_{\theta_0}}{2}, \mathbf{p}_{\theta_0}\right),$$

i.e., we have a generalization of Lemma 7.2. Also, Lemma 7.3 can be extended to this situation. Application of Theorem 8.13 to each $\mathbf{P}_s$ gives the required result. □

If $\{p_{i,\theta} : \theta \in \Theta\}$ is convex for all $i \in \{1,\ldots,n\}$, one can use the ideas of the first part of Section 7.3 to relax the entropy conditions.

8.4. Examples

Example 8.4.1. Estimating the error distribution in an auto-regression model Let Φ be the distribution function of the standard normal distribution on $\mathbf{R}$, with density ϕ with respect to Lebesgue measure. Let $X_0 = 0$, and $X_1,\ldots,X_n$ be real-valued observations, with conditional densities

$$p_{i,\theta_0}(x) = \gamma_0(\Phi(x-\alpha_0 X_{i-1}))\phi(x-\alpha_0 X_{i-1}), \quad i = 1,\ldots,n, \tag{8.53}$$

with respect to Lebesgue measure on $\mathbf{R}$, where $\theta_0 = (\alpha_0, \gamma_0)$, with

$$\alpha_0 \in \{|\alpha| < 1\},$$

and
(8.54)

$$\gamma_0 \in \Gamma = \left\{\gamma : [0,1] \to [0,\infty) : \int_0^1 \gamma(z)\,dz = 1,\ \int_0^1 \frac{(\gamma'(z))^2}{\gamma(z)}\,dz \le M^2\right\}.$$

So we assume that the Fisher information for location of the density γ_0 is known to be bounded by a given constant M^2. If $\gamma_0 \equiv 1$ were known, the model is the classical auto-regression model with known mean and variance. We suppose however that γ_0 is unknown, but known to lie in the Sobolev space Γ.

Notice now that for $\gamma \in \Gamma$,

$$\gamma(x) = 1 + \int(\gamma(x)-\gamma(u))\,du = 1 + \int\int_u^x \gamma'(v)\,dv\,du \le 1 + M.$$

Because

$$\frac{d\gamma^{1/2}(z)}{dz} = \frac{\gamma'(z)}{2\gamma^{1/2}(z)},$$

we can apply Theorem 2.4 to the class $\Gamma^{1/2} = \{\gamma^{1/2} : \gamma \in \Gamma\}$ to obtain

$$H_\infty(\delta, \Gamma^{1/2}) \le A\frac{1}{\delta}, \quad \text{for all } \delta > 0.$$

We have for $0 \le \tilde{x} < x \le 1$, $\gamma \in \Gamma$,

$$\gamma(x) - \gamma(\tilde{x}) = \int_{\tilde{x}}^x \gamma'(u)\,du \le |x-\tilde{x}|^{1/2}\frac{M}{2}.$$

Therefore, if $|\alpha - \alpha_j| \leq \delta^2$, we obtain

$$\left|\gamma^{1/2}(\Phi(x - \alpha_j X_{i-1})) - \gamma^{1/2}(\Phi(x - \alpha X_{i-1}))\right| \leq \frac{1}{2(2\pi)^{\frac{1}{4}}} \delta |X_{i-1}|^{1/2}.$$

It is now not hard to see that for

$$\mathbf{F} = \left\{ \frac{1}{n} \sum_{i=1}^{n} |X_{i-1}| \leq c_1^2 \right\}. \tag{8.55}$$

and for a constant A_1 depending only on M and c_1, we find

$$\mathbf{H}_B(\delta, 1, \mathbf{F}) \leq A_1 \frac{1}{\delta}, \quad \text{for all } \delta > 0.$$

In view of Theorem 8.14, we conclude that for all $T \geq c$, where c depends on M and c_1,

$$\mathbf{P}\left(\bar{h}(\mathbf{p}_{\hat{\theta}_n}, \mathbf{p}_{\theta_0}) > T n^{-1/3}\right) \leq c \exp\left[-\frac{T^2 n^{1/3}}{c^2}\right] + \mathbf{P}\left(\frac{1}{n}\sum_{i=1}^{n} |X_{i-1}| > c_1^2\right).$$

Example 8.4.2. A binary choice model for dependent observations Let $X_0 = (y_0, z_0)$, $y_0 \in \{0,1\}$ and $z_0 \in \mathbf{R}$ fixed, $X_i = (Y_i, Z_i)$, with $Y_i \in \{0,1\}$, $Z_i \in \mathbf{R}$, $i = 1, \ldots, n$. The conditional densities with respect to μ_i = (counting measure on $\{0,1\}$) $\times \tilde{Q}_i$, where $\tilde{Q}_i$ is the distribution of Z_i given $\mathscr{F}_{i-1}$, are supposed to be of the form

$$p_{i,F}(y,z) = F^y(z)\left(1 - F(z)\right)^{1-y},$$

where

$$F \in \{0 \leq F(z) \leq 1, \quad F \text{ increasing}\}.$$

Suppose now that the value of the covariate at time i depends only on whether or not a success occurred at time $i-1$, i.e. $\tilde{Q}_i = Q_{Y_{i-1}}$. Then it suffices to consider the non-random entropy with bracketing with respect to the non-random measure $Q = Q_0 + Q_1$. Application of Theorem 2.4 now yields that for some constant A,

$$\mathbf{H}_B(\delta, 1, \Omega) \leq A \frac{1}{\delta}, \quad \text{for all } \delta > 0,$$

so that by Theorem 8.14, for all $T \geq c$,

$$\mathbf{P}\left(\bar{h}(\mathbf{p}_{\hat{F}_n}, \mathbf{p}_{F_0}) > T n^{-1/3}\right) \leq c \exp\left[-\frac{T^2 n^{1/3}}{c^2}\right].$$

8.5. Notes

Lemma 8.2 is from Kuelbs (1978). An extension of Lemma 8.9 to continuous time martingales is given in van de Geer (1995a). There, a version of Theorem 8.13 for counting processes is presented. The proof of Theorem 8.13 is along the lines of Birgé and Massart (1993), employing a truncation device introduced by Bass (1985).

Theorem 8.14 relates the likelihood ratio to the Hellinger process itself, instead of to its expectation. As a consequence, the rates are in a random metric, which may be difficult to interpret.

For finite-dimensional models, the conditions of Theorem 8.14 may be too strong. We refer to Dzhaparidze and Valkeila (1990), who present the modulus of continuity of the likelihood process in terms of the higher order moments of the Hellinger process.

8.6. Problems and complements

8.1. Let W be a real-valued random variable with $EW = 0$, $\text{var}(W) = \tau_0^2$ and $|W| \leq K_0 < \infty$. Show that

$$\lim_{K\to\infty} K^2 \left(Ee^{|W|^2/K^2} - 1\right) = \lim_{K\to\infty} 2K \left(Ee^{|W|/K} - 1 - |W|/K\right) = \tau_0^2.$$

8.2. Consider a continuous time martingale $\{M_t\}_{t\geq 0}$ with respect to the filtration $\{\mathscr{F}_t\}_{t\geq 0}$. Let $\Delta M_t = M_t - M_{t-}$ denote its jumps, and suppose that for some constant $0 \leq K < \infty$,

$$|\Delta M_t| \leq K, \ t > 0.$$

Let $\{V_t\}$ be the predictable variation of $\{M_t\}$. Then for all $a > 0$ and $b > 0$,

$$\mathbf{P}(M_t \geq a \text{ and } V_t \leq b \text{ for some } t) \leq \exp\left[-\frac{a^2}{2(aK + b^2)}\right]$$

(see Shorack and Wellner (1986)). The condition $|\Delta M_t| \leq K$ for all t can be relaxed to conditions on the higher order variational processes (see van de Geer (1995a)).

8.3. For the result below, we need to calculate the entropy of differentiable functions on higher-dimensional Euclidean space. Let $\kappa = (\kappa_1, \ldots, \kappa_r)$ be a multi-index, i.e., $\kappa_j \in \{0, 1, \ldots\}$, $j = 1, \ldots, r$, and write $|\kappa| = \sum_{j=1}^r \kappa_j$. For a function $g : \mathbf{R}^r \to \mathbf{R}$, we let

$$\mathscr{D}^\kappa g(v) = \frac{\partial^{|\kappa|}}{\partial v_1^{\kappa_1} \ldots \partial v_r^{\kappa_r}} g(v_1, \ldots, v_r).$$

Kolmogorov and Tikhomirov (1959) prove that for

$$\mathscr{G} = \left\{ g : [0,1]^r \to \mathbf{R} : \sum_{|\kappa| \le m} |\mathscr{D}^\kappa g|_\infty \le M \right\},$$

one has

$$H_\infty(\delta, \mathscr{G}) \le A\delta^{-r/m}, \quad \text{for all } \delta > 0,$$

where the constant A depends on m, M, and r.

Suppose now that $(X_0, \ldots, X_{1-t}) = (x_0, \ldots, x_{1-t}) \in [0,1]^t$ is fixed, and that $X_1, \ldots, X_n$ are real-valued observations in $[0,1]$, with conditional density

$$p_{i,\theta_0}(x) = \theta_0(x; X_{i-1}, \ldots, X_{i-t}),$$

where

$$\theta_0(\cdot\,;\cdot) \in \Theta = \left\{ \theta : [0,1]^{t+1} \to [0,\infty) : \int_0^1 \theta(x;z)\,dx = 1, \text{ for all } z \in [0,1]^t, \right.$$

$$\left. \sum_{|\kappa| \le m} \sup_{v \in [0,1]^{t+1}} |\mathscr{D}^\kappa \theta^{1/2}(v_1; v_2, \ldots, v_{t+1})| \le M \right\}.$$

Show that if $m > (t+1)/2$,

$$\bar{h}\big(\mathbf{p}_{\hat{\theta}_n}, \mathbf{p}_{\theta_0}\big) = O_{\mathbf{P}}\big(n^{-m/(2m+t+1)}\big).$$

8.4. (The binary choice model for dependent observations, revisited.) Let $X_0 = (y_0, z_0)$, $y_0 \in \{0,1\}$, and $z_0 \in [0,1]$ fixed, and $X_i = (Y_i, Z_i)$, with $Y_i \in \{0,1\}$, $Z_i \in [0,1]$, $i = 1, \ldots, n$. Let $\tilde{Q}_i$ denote the distribution of Z_i given $\mathscr{F}_{i-1}$, and suppose that the conditional density with respect to $\mu_i =$ (counting measure on $\{0,1\}$) $\times \tilde{Q}_i$ is

$$p_{i,F_0}(y,z) = F_0^y(z)\big(1 - F_0(z)\big)^{1-y},$$

where

$$F_0 \in \left\{ 0 \le F(z) \le 1 : \left| \frac{d}{dz} F^{1/2}(z) \right| \le M, \text{ for all } z \in [0,1] \right\}.$$

Show that

$$\bar{h}(\mathbf{p}_{\hat{F}_n}, \mathbf{p}_{F_0}) = O_{\mathbf{P}}(n^{-1/3}).$$

8.5. Let $X_0 = x_0 \in [0,1]$ be fixed, and let $X_1, \ldots, X_n$ be real-valued observations satisfying the auto-regression model

$$X_i = \theta_0(X_{i-1}) + W_i, \quad i = 1, \ldots, n,$$

where for $i = 1, \ldots, n$, $\mathbf{E}(W_i \mid \mathscr{F}_{i-1}) = 0$, a.s., and

$$2K_1^2 \mathbf{E}\left(e^{|W_i|/K_1} - 1 - |W_i|/K_1 \mid \mathscr{F}_{i-1}\right) \leq \sigma_0^2, \quad \text{a.s.},$$

and where

$$\theta_0 \in \Theta = \left\{\theta : [0,1] \to [-K_2, K_2] : \int_0^1 \left(\theta^{(m)}(x)\right)^2 dx \leq M^2\right\}.$$

Let $\hat{\theta}_n$ be the least squares estimator

$$\hat{\theta}_n = \arg\min_{\theta \in \Theta} \sum_{i=1}^n \left(X_i - \theta(X_{i-1})\right)^2.$$

Then

$$\left(\frac{1}{n}\sum_{i=1}^n \left(\hat{\theta}_n(X_{i-1}) - \theta_0(X_{i-1})\right)^2\right)^{1/2} = O_{\mathbf{P}}\left(n^{-m/(2m+1)}\right).$$

This follows from Theorem 8.13, using the same arguments as in Theorem 8.14.

9

Rates of Convergence for Least Squares Estimators

Probability inequalities for the least squares estimator are obtained, under conditions on the entropy of the class of regression functions. A comparison is made between sub-Gaussian errors and errors with exponential tails. In the examples, we study smooth regression functions, functions of bounded variation, concave functions, analytic functions, and image restoration.

We return to the regression model considered in Sections 4.3 and 4.4. There, it was proved that the least squares estimator is consistent if the covering numbers of the class of possible regression functions do not grow exponentially fast. Here, we establish the link between these covering numbers and the rate of convergence of the least squares estimator. Corollaries 8.3 and 8.8 are tailored for this.

We start by recalling the model and notation used in Section 4.3. Let $Y_1, \dots, Y_n$ be real-valued observations satisfying

$$Y_i = g_0(z_i) + W_i, \quad i = 1, \dots, n,$$

with $z_1, \dots, z_n$ (fixed) covariates in a space $\mathscr{Z}$, $W_1, \dots, W_n$ independent errors with expectation zero, and with the unknown regression function g_0 in a given class $\mathscr{G}$ of regression functions. The least squares estimator is

$$\hat{g}_n = \arg\min_{g \in \mathscr{G}} \sum_{i=1}^{n} (Y_i - g(z_i))^2 .$$

Throughout, we assume that a minimizer $\hat{g}_n \in \mathscr{G}$ of the sum of squares exists, but it need not be unique.

The following notation will be used. The empirical measure of the covariates is

$$Q_n = \frac{1}{n}\sum_{i=1}^{n} \delta_{z_i}.$$

For g a function on $\mathscr{Z}$, we denote its $L_2(Q_n)$-norm by

$$\|g\|_n^2 = \|g\|_{Q_n}^2 = \frac{1}{n}\sum_{i=1}^{n} g^2(z_i).$$

The empirical inner product between error and regression function is written as

$$(w, g)_n = \frac{1}{n}\sum_{i=1}^{n} W_i g(z_i).$$

Moreover, with some abuse of notation, we write

$$\|y - g\|_n^2 = \frac{1}{n}\sum_{i=1}^{n} \left(Y_i - g(z_i)\right)^2,$$

and

$$\|w\|_n^2 = \frac{1}{n}\sum_{i=1}^{n} W_i^2.$$

Finally, we let

$$\mathscr{G}_n(R) = \{g \in \mathscr{G} : \|g - g_0\|_n \le R\}$$

denote a ball around g_0 with radius R, intersected with $\mathscr{G}$.

The main idea to arrive at rates of convergence for $\hat{g}_n$ is as in Chapter 7. The Basic Inequality of Lemma 4.7 says that

$$\|\hat{g}_n - g_0\|_n^2 \le 2(w, \hat{g}_n - g_0)_n, \tag{9.1}$$

and the modulus of continuity of the process $\{(w, g - g_0) : g \in \mathscr{G}_n(R)\}$ can be derived from the entropy of $\mathscr{G}_n(R)$.

9.1. Sub-Gaussian errors

As pointed out in Subsection 8.1.1, we have a maximal inequality at our disposal for $\{(w, g - g_0)_n : g \in \mathscr{G}_n(R)\}$, provided that the errors W_i are uniformly sub-Gaussian. Therefore, the latter assumption plays a key role in our considerations and we shall refer to it as condition (A):

$$\max_{i=1,\ldots,n} K^2\left(\mathbf{E}e^{|W_i|^2/K^2} - 1\right) \le \sigma_0^2. \tag{A}$$

In the next section however, we shall relax (A). The price we shall pay is stronger entropy conditions. Note that (A) is met if $W_1, \ldots, W_n$ are i.i.d. $\mathcal{N}(0, \tau_0^2)$-distributed. Then one can take $K = 2\tau_0$ and $\sigma_0 = 3\tau_0$.

Fix $\sigma > 0$, and define, for $0 < \delta < 2^6\sigma$,

$$J(\delta, \mathcal{G}_n(\delta), Q_n) = \int_{\delta^2/2^6\sigma}^{\delta} H^{1/2}(u, \mathcal{G}_n(\delta), Q_n)\, du \vee \delta. \tag{9.2}$$

Theorem 9.1 *Suppose that* (A) *is met. Take* $\Psi(\delta) \geq J(\delta, \mathcal{G}_n(\delta), Q_n)$ *in such a way that* $\Psi(\delta)/\delta^2$ *is a non-decreasing function of* δ, $0 < \delta < 2^6\sigma$. *Then for a constant c depending on K and* σ_0 *(but not on* σ*), and for*

$$\sqrt{n}\delta_n^2 \geq c\Psi(\delta_n) \tag{9.3}$$

we have for all $\delta \geq \delta_n$,

$$\mathbf{P}\big(\|\hat{g}_n - g_0\|_n > \delta\big) \leq c\exp\left[-\frac{n\delta^2}{c^2}\right] + \mathbf{P}\big(\|w\|_n > \sigma\big). \tag{9.4}$$

Proof By the Cauchy-Schwarz inequality, we may deduce from the Basic Inequality (9.1) that $\|\hat{g}_n - g_0\|_n \leq 2\sigma$ on $\{\|w\|_n \leq \sigma\}$. Therefore, it suffices to prove (9.4) for $\delta \leq 2\sigma$. Let $S = \min\{s \in \{0, 1, \ldots\} : 2^s\delta > 2\sigma\}$. Then, using (9.1) once more, and applying the peeling device, we obtain

$$\begin{aligned}
&\mathbf{P}\big(\|\hat{g}_n - g_0\|_n > \delta \wedge \|w\|_n \leq \sigma\big) \\
&\leq \sum_{s=0}^{S} \mathbf{P}\left(\sup_{g \in \mathcal{G}_n(2^{s+1}\delta)} (w, g - g_0)_n \geq 2^{2s-1}\delta^2 \wedge \|w\|_n \leq \sigma\right) \\
&= \sum_{s=0}^{S} \mathbf{P}_s.
\end{aligned}$$

Now, if

$$\sqrt{n}\delta_n^2 \geq 16C\Psi(\delta_n),$$

then also for all $2^{s+1}\delta > \delta_n$, $2^{s+1}\delta < 2^6\sigma$,

$$\sqrt{n}2^{2s+2}\delta^2 \geq 16C\Psi(2^{s+1}\delta).$$

So we may apply Corollary 8.3 to each $\mathbf{P}_s$. This gives

$$\sum_{s=0}^{S} \mathbf{P}_s \leq \sum_{s=0}^{S} C\exp\left[-\frac{n2^{4s-2}\delta^4}{4C^2 2^{2s+2}\delta^2}\right] \leq c\exp\left[-\frac{n\delta^2}{c^2}\right]. \qquad \square$$

Under condition (A), $\mathbf{P}(\|w\|_n > \sigma)$ is small for all $\sigma > \sigma_0$, since it implies $\mathbf{E}\|w\|_n^2 \le \sigma_0^2$. Inserting Bernstein's inequality gives for instance

$$\mathbf{P}\big(\|w\|_n^2 > 2\sigma_0^2\big) \le \exp\left[-\frac{n\sigma_0^2}{12K^2}\right].$$

Moreover, if the entropy integral converges, one may take $\sigma \to \infty$, in which case the second term on the right-hand side of (9.4) vanishes.

9.2. Errors with exponential tails

In Corollary 8.8, we also proved a maximal inequality for $\{(w, g - g_0)_n : g \in \mathcal{G}_n(R)\}$. The assumption there is that the tails of the error distribution decrease exponentially fast. We refer to this as assumption (B):

$$\text{(B)} \qquad \max_{i=1,\dots,n} 2K_1^2 \mathbf{E}\left(e^{|W_i|/K_1} - 1 - |W_i|/K_1\right) \le \sigma_0^2.$$

Clearly, (A) is stronger than (B).

Note however that Corollary 8.8 needs the assumption that the functions involved are uniformly bounded, which is in general not true for a class of regression functions. To handle this problem we introduce a class of renormalized functions. Define for $L_n \ge 0$,

$$\mathcal{F}_n = \left\{ \frac{g - g_0}{1 + L_n\|g - g_0\|_n} : g \in \mathcal{G} \right\},$$

and

$$\mathcal{F}_n(\delta) = \{ f \in \mathcal{F}_n : \|f\|_n \le \delta \}.$$

Moreover, let

$$J_B\big(\delta, \mathcal{F}_n(\delta), Q_n\big) = \int_{\delta^2/c_1}^{\delta} H_B^{1/2}\big(u, \mathcal{F}_n(\delta), Q_n\big)\, du \vee \delta. \tag{9.5}$$

where c_1 is a constant depending on σ_0 and K_1 (appropriately chosen) and where $\delta < c_1$.

Theorem 9.2 *Suppose that* (B) *holds. Assume moreover that*

$$\sup_{f \in \mathcal{F}_n} |f|_\infty \le K_2. \tag{9.6}$$

Take $\Psi(\delta) \ge J_B\big(\delta, \mathcal{F}_n(\delta), Q_n\big)$ in such a way that $\Psi(\delta)/\delta^2$ is a non-increasing function of δ, $0 < \delta < c_1$. Then for some constant c_2 depending on σ_0, K_1 and K_2, and for $\delta_n \le 1/(2L_n)$ and

$$\sqrt{n}\delta_n^2 \ge c_2 \Psi(\delta_n), \tag{9.7}$$

we have for all $1/(2L_n) \geq \delta \geq \delta_n$,

$$\text{(9.8)} \qquad \mathbf{P}\big(\|\hat{g}_n - g_0\|_n > 2\delta\big) \leq c_2 \exp\left[-\frac{n\delta^2}{c_2^2}\right].$$

Proof Since

$$\|\hat{g}_n - g_0\|_n^2 \leq 2(w, \hat{g}_n - g_0)_n,$$

we have for all $0 \leq \alpha \leq 1$,

$$\alpha^2 \|\hat{g}_n - g_0\|_n^2 \leq \alpha \|\hat{g}_n - g_0\|_n^2 \leq 2\alpha(w, \hat{g}_n - g_0)_n.$$

Choose $\alpha = 1/(1 + L_n\|\hat{g}_n - g_0\|_n)$ to find the Basic Inequality

$$\|\hat{f}_n\|_n^2 \leq 2(w, \hat{f}_n)_n,$$

where $\hat{f}_n = (\hat{g}_n - g_0)/(1 + L_n\|\hat{g}_n - g_0\|_n)$. As in the proof of Theorem 9.1, using Corollary 8.8 instead of Corollary 8.3, we obtain that for all $\delta \geq \delta_n$,

$$\mathbf{P}\big(\|\hat{f}_n\|_n > \delta\big) \leq c_2 \exp\left[-\frac{n\delta^2}{c_2^2}\right].$$

The condition $\delta \leq 1/(2L_n)$ implies a similar probability inequality for $\|\hat{g}_n - g_0\|_n$. To see this, note that if $\delta \leq 1/(2L_n)$, and $\|\hat{f}_n\|_n \leq \delta$, then

$$\|\hat{g}_n - g_0\|_n \leq \delta + \delta L_n \|\hat{g}_n - g_0\|_n \leq \delta + \frac{1}{2}\|\hat{g}_n - g_0\|_n,$$

so that

$$\|\hat{g}_n - g_0\|_n \leq 2\delta. \qquad \square$$

Thus, Theorem 9.2 yields the rate $\|\hat{g}_n - g_0\|_n = O_{\mathbf{P}}(\delta_n)$, if $L_n\delta_n \to 0$. In the proof, we used the convexity of the least squares loss function, so that we could make the step from $\mathcal{G}$ to the class of renormalized functions $\mathcal{F}_n$. This idea is also employed in, for example, Subsection 10.2.2, and Section 12.2: convexity of the loss function makes a renormalization possible.

In Theorem 9.1, the class $\mathcal{G}$ can be replaced by the class of renormalized functions $\mathcal{F}_n$ as well. However, there appears to be little merit in doing so.

9.3. Examples

Of course, if in a particular application, the entropy with bracketing is of the same order as the entropy without bracketing, Theorem 9.2 is better, because it needs weaker conditions on the errors. In Example 9.3.1, we derive exponential probability inequalities for the case of sub-Gaussian errors as well as the case of errors with exponential tails. Examples 9.3.2, 9.3.3 and 9.3.5 use Theorem 9.2 (errors with exponential tails), whereas Examples 9.3.4, 9.3.6 and 9.3.7 are applications of Theorem 9.1 (sub-Gaussian errors).

Example 9.3.1. Linear regression Let

$$\mathcal{G} = \{g(z) = \theta_1\psi_1(z) + \cdots + \theta_d\psi_d(z) : \theta \in \mathbf{R}^d\}.$$

Let us first apply Theorem 9.1. By Corollary 2.5,

$$H(u, \mathcal{G}_n(\delta), Q_n) \leq d \log\left(\frac{\delta + 4u}{u}\right), \quad \text{for all } 0 < u < \delta, \ \delta > 0.$$

So

$$\int_0^\delta H^{\frac{1}{2}}(u, \mathcal{G}_n(\delta), Q_n)\, du \leq d^{\frac{1}{2}} \int_0^\delta \log^{\frac{1}{2}}\left(\frac{\delta + 4u}{\delta}\right) du = d^{\frac{1}{2}}\delta \int_0^1 \log^{\frac{1}{2}}(1 + 4v)\, dv$$
$$= A_0 d^{\frac{1}{2}} \delta.$$

So (9.3) is met for

$$\delta_n \geq c A_0 \sqrt{\frac{d}{n}}.$$

Theorem 9.1 yields that under condition (A), for some constant c and for all $T \geq c$,

$$\text{(9.9)} \qquad \mathbf{P}\left(\|\hat{g}_n - g_0\|_n > T\sqrt{\frac{d}{n}}\right) \leq c \exp\left[-\frac{T^2 d}{c^2}\right].$$

Note that we made extensive use here of the fact that it suffices to calculate the *local* entropy of $\mathcal{G}$.

Let us now see what happens if we relax the assumptions on the errors. The exponential inequality (9.9) implies that

$$\mathbf{E}\|\hat{g}_n - g_0\|_n^2 \leq \text{const.} \frac{d}{n}.$$

This result can also be obtained using only conditions on the second moment of the errors, due to the linearity of the problem. We define the vector valued function $\psi(z) = (\psi_1(z), \ldots, \psi_d(z))^T$ and

$$\Sigma_n = \int \psi\psi^T \, dQ_n.$$

For simplicity, let us assume that Σ_n is non-singular (if this is not the case, the result can be improved by replacing d by the rank of Σ_n). Then we may assume without loss of generality that $\Sigma_n = I$, the $d \times d$ identity matrix. In that case

$$\|\hat{g}_n - g_0\|_n^2 = (\hat{\theta}_n - \theta_0)^T(\hat{\theta}_n - \theta_0) = |\hat{\theta}_n - \theta_0|^2,$$

and by the Cauchy–Schwarz inequality,

$$(9.10) \quad |(\hat{g}_n - g_0, w)_n|^2 = |(\hat{\theta}_n - \theta_0)^T(\psi, w)_n|^2 \leq |\hat{\theta}_n - \theta_0|^2 \left(\sum_{k=1}^{d}(\psi_k, w)_n\right)^2.$$

We now drop assumption (A), but still require that the variance of the error remains bounded:

$$(9.11) \quad \max_{i=1,\ldots,n} \mathbf{E}W_i^2 \leq \tau_0^2.$$

Since $\|\psi_k\|_n = 1$, $k = 1, \ldots, d$, (9.11) implies

$$\mathbf{E}\left(\sum_{k=1}^{d}(\psi_k, w)_n\right)^2 \leq \frac{d\tau_0^2}{n}.$$

In view of the Basic Inequality (9.1), we arrive at

$$(9.12) \quad \mathbf{E}\|\hat{g}_n - g_0\|_n^2 \leq 4\frac{d\tau_0^2}{n}.$$

Clearly, (9.11) is much weaker than condition (A), but the result (9.12) is also much weaker than the exponential inequality (9.9).

Condition (B) is intermediate between condition (A) and (9.11). Let us first see what Theorem 9.2 tells us. We now have to calculate the entropy with bracketing. We assume that

$$|\psi_k|_\infty \leq 1, \quad k = 1, \ldots, d.$$

The assumption that Σ_n is non-singular is now more important. Denote its smallest eigenvalue by $\lambda_{n,d}^2 > 0$, and take $L_n = \sqrt{d}/\lambda_{n,d}$ to meet condition (9.6). For some constant A,

$$H_B(u, \mathcal{F}_n(\delta), Q_n) \leq Ad\left(\log\left(\frac{\delta}{u}\right) + \log\left(\frac{d}{\lambda_{n,d}^2}\right)\right),$$

so that

$$\int_0^\delta H_B^{1/2}(u, \mathcal{F}_n(\delta), Q_n)\, du \leq A_0 d^{1/2}\delta \log^{1/2}\left(\frac{d}{\lambda_{n,d}^2}\right).$$

Under condition (B), for some constant c_2 depending on σ, and for all $T \geq c_2$, $T\sqrt{\lambda_{n,d}^2/n}\log^{1/2}(d/\lambda_{n,d}) \leq 1$,

$$\mathbf{P}\left(\|\hat{g}_n - g_0\|_n > T\sqrt{\frac{d}{n}}\log^{1/2}\left(\frac{d}{\lambda_{n,d}}\right)\right) \leq c_2 \exp\left[-\frac{nT^2 d\log(d/\lambda_{n,d})}{c_2^2}\right].$$

Thus, we are unable to recover the optimal rate $\sqrt{d/n}$ from Theorem 9.2, whenever $d \to \infty$ and/or $\lambda_{n,d} \to 0$. Of course, condition (B) *does* allow us to improve on (9.12), by applying it directly to (9.10). We then find that for some constant c_3 depending on K_1 and σ_0, and for $T \geq c_3$,

$$\text{(9.13)} \qquad \mathbf{P}\left(\|\hat{g}_n - g_0\|_n \geq T\sqrt{\frac{d}{n}}\right) \leq c_3 \exp\left[-\frac{T^2 d}{c_3^2(T\sqrt{d/n}+1)}\right].$$

A comparison with (9.9) reveals the unsurprising fact that condition (B) only gives rise to exponential tails for the least squares estimator, instead of sub-Gaussian tails.

Example 9.3.2. Smooth functions Let

$$\mathscr{G} = \left\{ g : [0,1] \to \mathbf{R},\ \int \left(g^{(m)}(z)\right)^2 dz \leq M^2 \right\}.$$

In this example we assume throughout that condition (B) is met. We shall apply Theorem 9.2.

Let $\psi_k(z) = z^{k-1}$, $k = 1, \ldots, m$, $\psi(z) = (\psi_1(z), \ldots, \psi_m(z))^T$ and $\Sigma_n = \int \psi\psi^T\, dQ_n$. Denote the smallest eigenvalue of Σ_n by λ_n, and assume that

$$\lambda_n \geq \lambda > 0, \quad \text{for all } n \geq n_0.$$

Lemma 3.9 states that for $L_n = 1$,

$$H_\infty(\delta, \mathscr{F}_n) \leq A\delta^{-1/m}, \quad \text{for all } \delta > 0,$$

where the constant A depends on R and λ. Hence, then (9.6) is fulfilled, and we find from Theorem 9.2 that for $T \geq c_2$, $Tn^{-m/(2m+1)} \leq 1$,

$$\mathbf{P}\left(\|\hat{g}_n - g_0\|_n > Tn^{-\frac{m}{2m+1}}\right) \leq c_2 \exp\left[-\frac{T^2 n^{\frac{1}{2m+1}}}{c_2^2}\right].$$

In general, the rate of convergence for certain aspects of $\hat{g}_n$ can be much faster than the global rate. We illustrate this here, for the projection of $\hat{g}_n$ on the polynomials of degree $m-1$. (Related results are obtained in Sections 10.1 and 11.1. There, the constant M may be unknown, and the roughness of the least squares estimator is penalized.) Consider the Taylor expansion of a function $g \in \mathcal{G}$:

$$g(z) = g(0) + zg^{(1)}(0) + \cdots + \int_0^z \frac{(z-u)^{m-1}}{(m-1)!} g^{(m)}(u)\, du$$
$$= \sum_{k=1}^m \theta_k \psi_k(z) + \int_0^1 \beta_u \phi_u(z)\, du,$$

where

$$\theta_k = \frac{g^{(k-1)}(0)}{(k-1)!}, \quad k = 1, \ldots, m,$$
$$\beta_u = g^{(m)}(u), \quad 0 < u \le 1,$$

and

$$\phi_u(z) = \frac{(z-u)^{m-1}}{(m-1)!} \mathrm{l}\{u \le z\}, \quad 0 < u \le 1.$$

Define $\mathbf{z} = (z_1, \ldots, z_n)$, let $\psi_k(\mathbf{z}) = (\psi_k(z_1), \ldots, \psi_k(z_n))^T$, $k = 1, \ldots, m$, $\phi_u(\mathbf{z}) = (\phi_u(z_1), \ldots, \phi_u(z_n))^T$, $0 < u \le 1$. Let

$$\bar{\phi}_u(\mathbf{z}) = \sum_{k=1}^m \gamma_{k,u} \psi_k(\mathbf{z})$$

be the projection in $L_2(Q_n)$ of $\phi_u(\mathbf{z})$ on the linear space spanned by $\psi_1(\mathbf{z}), \ldots, \psi_m(\mathbf{z})$, and

$$\tilde{\phi}_u(\mathbf{z}) = \phi_u(\mathbf{z}) - \bar{\phi}_u(\mathbf{z}),$$

$0 < u \le 1$. Then for $g(\mathbf{z}) = \big(g(z_1), \ldots, g(z_n)\big)^T$,

$$g(\mathbf{z}) = g_1(\mathbf{z}) + g_2(\mathbf{z}), \tag{9.14}$$

where

$$g_1(\mathbf{z}) = \sum_{k=1}^m \alpha_k \psi_k(\mathbf{z}), \tag{9.15}$$

$$g_2(\mathbf{z}) = \int_0^1 \beta_u \tilde{\phi}_u(\mathbf{z})\, du, \tag{9.16}$$

and

$$\alpha_k = \theta_k + \int \beta_u \gamma_k du, \quad k = 1, \ldots, m.$$

Let $g_0 = g_{1,0} + g_{2,0}$ be the corresponding expression for g_0. Because $(\psi_k, \tilde{\phi}_u)_n = 0$ for all $k \in \{1,\dots,m\}$ and $u \in (0,1]$, it it easy to see that the least squares estimator $\hat{g}_n = \hat{g}_{1,n} + \hat{g}_{2,n}$ satisfies

$$\hat{g}_{1,n} = \arg\min\left((-2(w, g_1 - g_{1,0})_n + \|g_1 - g_{1,0}\|_n^2 \, : \, g_1 = \sum_{k=1}^m \alpha_k \psi_k\right),$$

$$\hat{g}_{2,n} = \arg\min\left(-2(w, g_2 - g_{2,0})_n + \|g_2 - g_{2,0}\|_n^2 \, : \, g_2 = \int_0^1 \beta_u \tilde{\phi}_u \, du\right).$$

So the estimate $\hat{\alpha}_n$ of $\alpha_0 = (\alpha_{1,0},\dots,\alpha_{m,0})^T$ is the classical least squares estimator in the linear model

$$\tilde{Y}_i = \sum_{k=1}^m \alpha_{k,0}\psi_k(z_i) + W_i, \quad i = 1,\dots,n.$$

It follows that $|\hat{\alpha}_n - \alpha_0|$ (as well as $\|\hat{g}_{1,n} - g_{1,0}\|_n$) converges with rate $n^{-1/2}$.

Example 9.3.3. Functions of bounded variation in R Let

$$\mathscr{G} = \left\{g : \mathbf{R} \to \mathbf{R}, \int |g'(z)|\, dz \le M\right\}.$$

Without loss of generality, we may assume that $z_1 \le \cdots \le z_n$. The derivative should be understood in the generalized sense:

$$\int |g'(z)|\, dz = \sum_{i=2}^n |g(z_i) - g(z_{i-1})|.$$

We assume condition (B) holds, and apply Theorem 9.2. Define for $g \in \mathscr{G}$,

$$\alpha = \int g \, dQ_n.$$

Then it is easy to see that

$$\max_{i=1,\dots,n} |g(z_i)| \le \alpha + M.$$

Thus, (9.6) holds, with $L_n = 1$. In view of (2.5), we obtain

$$H_B(\delta, \mathscr{F}_n, Q_n) \le A\delta^{-1}, \quad \text{for all } \delta > 0,$$

and therefore, for all $T \ge c_2$, $Tn^{-1/3} \le 1$,

$$\mathbf{P}\left(\|\hat{g}_n - g_0\|_n > Tn^{-1/3}\right) \le c_2 \exp\left[-\frac{T^2 n^{1/3}}{c_2^2}\right]. \tag{9.17}$$

Note that for $g \in \mathcal{G}$,

$$g = g_1 + g_2,$$

with $g_1(z) = \alpha\psi(z)$, $\psi(z) \equiv 1$, and $(g_2, \psi)_n = 0$. So, from the same arguments as in the previous example, we know that $|\hat{\alpha}_n - \alpha_0| = |\int(\hat{g}_n - g_0)\, dQ_n|$ converges with rate $n^{-1/2}$.

Example 9.3.4. Functions of bounded variation in $\mathbf{R}^2$ Suppose that $z_i = (u_k, v_l)$, $i = kl$, $k = 1, \ldots, n_1$, $l = 1, \ldots, n_2$, $n = n_1 n_2$, with $u_1 \le \cdots \le u_{n_1}$, $v_1 \le \cdots \le v_{n_2}$. Consider the class

$$\mathcal{G} = \{g : \mathbf{R}^2 \to \mathbf{R},\ I(g) \le M\}$$

where

$$I(g) = I_0(g) + I_1(g_{1\cdot}) + I_2(g_{\cdot 2}),$$

$$I_0(g) = \sum_{k=2}^{n_2} \sum_{l=2}^{n_2} |g(u_k, v_l) - g(u_{k-1}, v_l) - g(u_k, v_{l-1}) + g(u_{k-1}, v_{l-1})|,$$

$$g_{1\cdot}(u) = \frac{1}{n_2} \sum_{l=1}^{n_2} g(u, v_l), \qquad g_{\cdot 2}(v) = \frac{1}{n_1} \sum_{k=1}^{n_1} g(u_k, v),$$

$$I_1(g_{1\cdot}) = \sum_{k=2}^{n_1} |g_{1\cdot}(u_k) - g_{1\cdot}(u_{k-1})|,$$

and

$$I_2(g_{\cdot 2}) = \sum_{l=2}^{n_2} |g_{\cdot 2}(v_l) - g_{\cdot 2}(v_{l-1})|.$$

Thus, each $g \in \mathcal{G}$ as well as its marginals have total variation bounded by M. We assume condition (A) is met and apply Theorem 9.1 and the result of Ball and Pajor on convex hulls. Let

$$\Lambda = \{\text{all distribution functions } F \text{ on } \mathbf{R}^2\},$$

and

$$\mathcal{K} = \{1_{(y,\infty]} : y \in \mathbf{R}^2\}.$$

Clearly, $\Lambda = \overline{\text{conv}(\mathcal{K})}$, and

$$N(\delta, \mathcal{K}, Q_n) \le c\frac{1}{\delta^4}, \quad \text{for all } \delta > 0.$$

From Theorem 3.14, we now deduce that

$$H(\delta, \Lambda, Q_n) \le A\delta^{-4/3}, \quad \text{for all } \delta > 0.$$

The same bound holds therefore for any uniformly bounded subset of $\mathcal{G}$. Now, any function $g \in \mathcal{G}$ can be expressed as

$$g(u,v) = \tilde{g}(u,v) + \tilde{g}_{1\cdot}(u) + \tilde{g}_{\cdot 2}(v) + \alpha,$$

where

$$\alpha = \frac{1}{n_1 n_2} \sum_{k=1}^{n_1} \sum_{l=1}^{n_2} g(u_k, v_k),$$

and where

$$\sum_{k=1}^{n_1} \sum_{l=1}^{n_2} \tilde{g}(u_k, v_l) = 0, \qquad \sum_{k=1}^{n_1} \tilde{g}_{1\cdot}(u_k) = 0 \qquad \sum_{l=1}^{n_2} \tilde{g}_{\cdot 2}(v_l) = 0.$$

It is easy to see that

$$|\tilde{g}(u_k, v_l)| \le I_0(\tilde{g}) = I_0(g), \quad k = 1, \dots, n_1, \ l = 1, \dots, n_2,$$

$$|\tilde{g}_{1\cdot}(u_k)| \le I_1(\tilde{g}_{1\cdot}) = I_1(g_{1\cdot}), \quad k = 1, \dots, n_1,$$

and

$$|g_{\cdot 2}(v_l)| \le I_2(\tilde{g}_{\cdot 2}) = I_2(g_{\cdot 2}), \quad l = 1, \dots, n_2.$$

Whence

$$\{\tilde{g} + \tilde{g}_{1\cdot} + \tilde{g}_{\cdot 2} : g \in \mathcal{G}\}$$

is a uniformly bounded class, for which the entropy bound $A_1 \delta^{-4/3}$ holds, with A_1 depending on M. It follows that

$$H\big(u, \mathcal{G}_n(\delta), Q_n\big) \le A_1 u^{-4/3} + A_2 \log\left(\frac{\delta}{u}\right), \quad 0 < u \le \delta.$$

From Theorem 9.1, we find for all $T \ge c$,

$$\text{(9.18)} \qquad \mathbf{P}\big(\|\hat{g}_n - g_0\|_n > T n^{-3/10}\big) \le c \exp\left[-\frac{T^2 n^{2/5}}{c^2}\right].$$

Orthogonality arguments show moreover that the marginals $g_{1\cdot,0}$ and $g_{\cdot 2,0}$ are estimated with rate $n^{-1/3}$, and the constant term

$$\alpha_0 = \sum_{k=1}^{n_1} \sum_{l=1}^{n_2} g_0(u_k, v_l)/n_1 n_2$$

is again estimated with rate $n^{-1/2}$.

The result can be extended to functions of bounded variation in $\mathbf{R}^r$. Then one finds the rate $\|\hat{g}_n - g_0\|_n = O_{\mathbf{P}}(n^{-(1+r)/(2+4r)})$.

Example 9.3.5. Concave functions Let

$$\mathcal{G} = \{g : [0,1] \to \mathbf{R},\ 0 \le g' \le M,\ g' \text{ decreasing}\}.$$

Then $\mathcal{G}$ is a subset of

$$\left\{g : [0,1] \to \mathbf{R},\ \int_0^1 |g''(z)|\, dz \le 2M\right\}.$$

Birman and Solomjak (1967) prove that for all $m \in \{2, 3, \dots\}$,

$$H_\infty\left(\delta, \left\{g : [0,1] \to [0,1] : \int_0^1 |g^{(m)}(z)|\, dz \le 1\right\}\right) \le A\delta^{-1/m}, \quad \text{for all } \delta > 0.$$

Again, our class $\mathcal{G}$ is not uniformly bounded, but we can write for $g \in \mathcal{G}$,

$$g = g_1 + g_2,$$

with $g_1(z) = \theta_1 + \theta_2 z$ and $|g_2|_\infty \le 2M$. Assume now that $\sum_{i=1}^n (z_i - \bar{z})^2/n$ stays away from 0. Using the same arguments as in Example 9.3.2, we obtain for $T \ge c_2$, $Tn^{-2/5} \le 1$,

$$\text{(9.19)} \qquad \mathbf{P}\big(\|\hat{g}_n - g_0\|_n > Tn^{-2/5}\big) \le c_2 \exp\left[-\frac{T^2 n^{1/5}}{c_2^2}\right].$$

Example 9.3.6. Analytic functions Let

$$\mathcal{G} = \big\{g : [0,1] \to \mathbf{R} :\ g^{(k)} \text{ exists for all } k \ge 0,\ |g^{(k)}|_\infty \le M \text{ for all } k \ge m\big\}.$$

We assume condition (A) holds, and apply Theorem 9.1.

Lemma 9.3 *We have*

$$\text{(9.20)} \quad H\big(u, \mathcal{G}_n(\delta), Q_n\big) \le \left(\left(\frac{\log\left(\frac{3M}{u}\right)}{\log 2} + 1\right) \vee m\right) \log\left(\frac{3\delta + 6u}{u}\right), \quad 0 < u < \delta.$$

Proof Take

$$d = \left(\left\lfloor \frac{\log\left(\frac{M}{u}\right)}{\log 2} \right\rfloor + 1\right) \vee m,$$

where $\lfloor x \rfloor$ is the integer part of $x \geq 0$. For each $g \in \mathscr{G}$, we can find a polynomial f of degree $d-1$ such that

$$|g(z) - f(z)| \leq M \left| z - \frac{1}{2} \right|^d \leq M \left(\frac{1}{2} \right)^d \leq u.$$

Now, let $\mathscr{F}$ be the collection of all polynomials of degree $d-1$, and let $f_0 \in \mathscr{F}$ be the approximating polynomial of g_0, with $|g_0 - f_0|_\infty \leq u$.

If $\|g - g_0\|_n \leq \delta$, we find $\|f - f_0\|_n \leq \delta + 2u$. From Corollary 2.5, we know that

$$H\big(u, \mathscr{F}_n(\delta + 2u), Q_n\big) \leq d \log\left(\frac{\delta + 6u}{u} \right), \quad u > 0,\ \delta > 0.$$

If $\|f - \tilde{f}\|_n \leq u$ and $|g - f|_\infty \leq u$ as well as $|\tilde{g} - \tilde{f}|_\infty \leq u$, we obtain $\|g - \tilde{g}\|_n \leq 3u$. So,

$$H\big(3u, \mathscr{G}_n(\delta), Q_n\big) \leq \left(\left(\frac{\log(\frac{M}{u})}{\log 2} + 1 \right) \vee m \right) \log\left(\frac{\delta + 6u}{u} \right), \quad u > 0,\ \delta > 0.$$

□

From Theorem 9.1, it follows that for $T \geq c$,

$$\mathbf{P}\big(\|\hat{g}_n - g_0\|_n > T n^{-1/2} \log^{1/2} n\big) \leq c \exp\left[-\frac{T^2 \log n}{c^2} \right].$$

Example 9.3.7. Image restoration

Case (i) Let $\mathscr{Z} \subset \mathbf{R}^2$ be some subset of the plane. Each site $z \in \mathscr{Z}$ has a certain grey-level $g_0(z)$, which is expressed as a number between 0 and 1, i.e., $g_0(z) \in [0,1]$. We have noisy data on a set of $n = n_1 n_2$ pixels $\{z_{kl} : k = 1, \ldots, n_1,\ l = 1, \ldots, n_2\} \subset \mathscr{Z}$:

$$Y_{kl} = g_0(z_{kl}) + W_{kl},$$

where the measurement errors $\{W_{kl} : k = 1, \ldots, n_1,\ l = 1, \ldots, n_2\}$ are independent centered random variables. We assume that they are uniformly sub-Gaussian (condition (A)). Now, each patch of a certain gray-level is a mixture of certain amounts of black and white. Let

$$\mathscr{G} = \overline{\mathrm{conv}}(\mathscr{K}),$$

where

$$\mathscr{K} = \{1_{\mathscr{D}} : D \in \mathscr{D}\}.$$

Assume that

$$N(\delta, \mathscr{K}, Q_n) \leq c\delta^{-w}, \quad \text{for all } \delta > 0.$$

Then by Theorem 3.14,

$$H(\delta, \mathscr{G}, Q_n) \leq A\delta^{-2w/(2+w)}, \quad \text{for all } \delta > 0.$$

It follows from Theorem 9.1 that for $T \geq c$,

$$\mathbf{P}\left(\|\hat{g}_n - g_0\|_n > Tn^{-\frac{2+w}{4+4w}}\right) \leq c \exp\left[-\frac{T^2 n^{\frac{w}{2+2w}}}{c^2}\right].$$

For example, if $\mathscr{D}$ is the collection of all half-spaces, then the picture could look star-shaped, dark in the center, and gradually fading out when one moves away from the center. In that case, $w = 6$ (see Theorem 3.11), so that the rate of convergence is $O_{\mathbf{P}}(n^{-2/7})$.

Case (ii) Consider a black-and-white image observed with noise. Let $\mathscr{Z} = [0, 1]^2$ be the unit square, and

$$g_0(z) = \begin{cases} 1, & \text{if } z \text{ is black,} \\ 0, & \text{if } z \text{ is white}. \end{cases}$$

The black part of the image is

$$D_0 = \{z \in [0, 1]^2 : g_0(z) = 1\}.$$

We observe

$$Y_{kl} = g(z_{kl}) + W_{kl},$$

with $z_{kl} = (u_k, v_l)$, $u_k = k/m$, $v_l = l/m$, $k, l \in \{1, \ldots, m\}$. The total number of pixels is thus $n = m^2$.

Suppose that

$$D_0 \in \mathscr{D} = \{\text{all convex subsets of } [0, 1]^2\},$$

and write

$$\mathscr{G} = \{1_D : D \in \mathscr{D}\}.$$

Dudley (1984) shows that for all $\delta > 0$ sufficiently small

$$H(\delta, \mathscr{G}, Q_n) \leq A\delta^{-1/2},$$

so that under condition (A) on the errors, for $T \geq c$,

$$\mathbf{P}(\|\hat{g}_n - g_0\|_n > Tn^{-2/5}) \leq c \exp\left[-\frac{T^2 n^{1/5}}{c^2}\right].$$

Let $\hat{D}_n$ be the estimate of the black area, so that $\hat{g}_n = 1_{\hat{D}_n}$. For two sets D_1 and D_2, denote the symmetric difference by

$$D_1 \Delta D_2 = (D_1 \cap D_2^c) \cup (D_1^c \cap D_2).$$

Since $Q_n(D) = \|1_D\|_n^2$, we find

$$Q_n(\hat{D}_n \Delta D_0) = O_{\mathbf{P}}(n^{-4/5}).$$

Remark In higher dimensions, say $\mathscr{X} = [0,1]^r$, $r \geq 2$, the class $\mathscr{G}$ of indicators of convex sets has entropy

$$H(\delta, \mathscr{G}, Q_n) \leq A\delta^{-(r-1)/2}, \quad \delta \downarrow 0,$$

provided that the pixels are on a regular grid (see Dudley (1984)). So the rate is then

$$Q_n(\hat{D}_n \Delta D_0) = \begin{cases} O_{\mathbf{P}}(n^{-4/(r+3)}), & \text{if } r \in \{2,3,4\}, \\ O_{\mathbf{P}}(n^{-1/2}\log n), & \text{if } r = 5, \\ O_{\mathbf{P}}(n^{-2/(r-1)}), & \text{if } r \geq 6\,. \end{cases}$$

For $r \geq 5$, the least squares estimator converges with suboptimal rate (see also Example 10.4.1).

9.4. Notes

Theorem 9.1 was proved by van de Geer (1990), and Theorem 9.2 essentially by Birgé and Massart (1993). The rates of convergence in Example 9.3.2 coincide with known minimax rates (Stone (1982)). Mammen (1991) studies, for example, the least squares estimator of a concave function. In Ibragimov and Has'minskii (1984), one finds minimax rates for the estimation of an analytic function. Minimax rates for image reconstruction are given in Korostelev and Tsybakov (1993a).

9.5. Problems and complements

9.1. Let $Y_1, \ldots, Y_n$ be independent, uniformly sub-Gaussian random variables, with $EY_i = \alpha_0$ for $i = 1, \ldots, \lfloor n\gamma_0 \rfloor$, and $EY_i = \beta_0$ for $i = \lfloor n\gamma_0 \rfloor + 1, \ldots, n$, where α_0, β_0 and the change point γ_0 are completely unknown. Write $g_0(i) = g(i; \alpha_0, \beta_0, \gamma_0) = \alpha_0 1\{1 \leq i \leq \lfloor n\gamma_0 \rfloor\} + \beta_0 1\{\lfloor n\gamma_0 \rfloor + 1 \leq i \leq n\}$. We call the parameter $(\alpha_0, \beta_0, \gamma_0)$ identifiable if $\alpha_0 \neq \beta_0$ and $\gamma_0 \in (0,1)$. Let $\hat{g}_n = g(\cdot; \hat{\alpha}_n, \hat{\beta}_n, \hat{\gamma}_n)$ be the least squares estimator. Show that if $\alpha_0, \beta_0, \gamma_0$

is identifiable, then $\|\hat{g}_n - g_0\|_n = O_{\mathbf{P}}(n^{-1/2})$, and $|\hat{\alpha}_n - \alpha_0| = O_{\mathbf{P}}(n^{-1/2})$, $|\hat{\beta}_n - \beta_0| = O_{\mathbf{P}}(n^{-1/2})$, and $|\hat{\gamma}_n - \gamma_0| = O_{\mathbf{P}}(n^{-1})$. If $(\alpha_0, \beta_0, \gamma_0)$ is not identifiable, show that $\|\hat{g}_n - g_0\|_n = O_{\mathbf{P}}(n^{-1/2}(\log\log n)^{1/2})$.

9.2. Let $z_i = i/n$, $i = 1, \ldots, n$, and let $\mathcal{G}$ consist of the functions

$$g(z) = \begin{cases} \alpha_1 + \alpha_2 z, & \text{if } z \leq \gamma, \\ \beta_1 + \beta_2 z, & \text{if } z > \gamma. \end{cases}$$

Suppose g_0 is continuous, but does have a kink at γ_0: $\alpha_{1,0} = \alpha_{2,0} = 0$, $\beta_{1,0} = -\frac{1}{2}$, $\beta_{2,0} = 1$, and $\gamma_0 = \frac{1}{2}$. Show that under condition (B), $\|\hat{g}_n - g_0\|_n = O_{\mathbf{P}}(n^{-1/2})$, and that $|\hat{\alpha}_n - \alpha_0| = O_{\mathbf{P}}(n^{-1/2})$, $|\hat{\beta}_n - \beta_0| = O_{\mathbf{P}}(n^{-1/2})$ and $|\hat{\gamma}_n - \gamma_0| = O_{\mathbf{P}}(n^{-1/3})$.

9.3. Let us assume condition (B) is met. If $\mathcal{G}$ is a uniformly bounded class of increasing functions, then it follows from Theorem 9.2 that $\|\hat{g}_n - g_0\|_n = O_{\mathbf{P}}(n^{-1/3})$. Now, let

$$\mathcal{G} = \{g : \mathbf{R} \to \mathbf{R},\ g \text{ increasing}\}.$$

Suppose that $|g_0|_\infty \leq K$ (with K unknown). We shall show that the rate $O_{\mathbf{P}}(n^{-1/3})$ is valid on large subintervals.

Take $L > K$, and define for $g \in \mathcal{G}$,

$$g_L(z) = \begin{cases} g_0(z), & \text{if } |g(z)| \leq L, \\ g(z), & \text{if } |g(z)| > L\,. \end{cases}$$

Then also $g_L \in \mathcal{G}$, so that

$$\|y - \hat{g}_n\|_n \leq \|y - \hat{g}_{n,L}\|_n.$$

Rewrite this as the Basic Inequality

$$\|(\hat{g}_n - g_0)\mathrm{l}\{|\hat{g}_n| \leq L\}\|_n^2 \leq 2(w, (\hat{g}_n - g_0)\mathrm{l}\{|\hat{g}_n| \leq L\})_n.$$

Using the same arguments as in Theorem 9.2, this implies

$$\|(\hat{g}_n - g_0)\mathrm{l}\{|\hat{g}_n| \leq L\}\|_n = O_{\mathbf{P}}(n^{-1/3}).$$

9.4. Consider the regression model

$$Y_i = g_0(Z_i) + W_i, \quad i = 1, \ldots, n,$$

where $(Z_1, W_1), \ldots, (Z_n, W_n)$ are i.i.d. copies of (Z, W). Assume that $E(W \mid Z) = 0$, a.s., and

$$2K_1^2 E\left(e^{|W|/K_1} - 1 - |W|/K_1 \mid Z\right) \leq \sigma_0^2, \quad \text{a.s.}$$

It is known that $g_0 \in \mathcal{G}$. Now, let Q be the distribution of Z, and suppose that for some $0 < \alpha < 2$,

$$H_B(\delta, \mathcal{G}, Q) \leq A\delta^{-\alpha}, \quad \text{for all } \delta > 0.$$

Show that

$$\|\hat{g}_n - g_0\| = O_{\mathbf{P}}(n^{-1/(2+\alpha)}).$$

(Hint: use Lemma 5.16 and the remark following Corollary 8.8.)

9.5. Let $W_1, \ldots, W_n$ be $\mathcal{N}(0, \tau_0^2)$-distributed. If the entropy integral converges, Theorem 9.1 essentially gives the minimax rate of convergence. This follows from Lemma VII.1.1 in Ibragimov and Has'minskii (1981a). A variant of this lemma is in Birgé (1983): let $\mathbf{P}_{g_1}, \ldots, \mathbf{P}_{g_N}$ be $N \geq 2$ probability measures on $(\Omega, \mathcal{F})$, and let $g^* \in \{g_1, \ldots, g_N\}$ be an estimator; then

$$\sup_{1 \leq j \leq N} \mathbf{P}_{g_j}(g^* \neq g_j) \geq 1 - \frac{\frac{1}{N^2}\sum_{j_1, j_2} K(\mathbf{P}_{g_{j_1}}, \mathbf{P}_{g_{j_2}}) + \log 2}{\log(N-1)}.$$

Here, $K(\cdot, \cdot)$ denotes the Kullback–Leibler information.

In the regression model, let $g_1, \ldots, g_N$ be a maximal subset of $\mathcal{G}(\delta)$, such that

$$\|g_{j_1} - g_{j_2}\|_n > u, \quad \text{for all } j_1 \neq j_2.$$

Let $\mathbf{P}_g$ correspond to the model

$$Y_i = g(z_i) + W_i, \quad i = 1, \ldots, n.$$

Then

$$K\left(\mathbf{P}_{g_{j_1}}, \mathbf{P}_{g_{j_2}}\right) = \frac{n}{2}\|g_{j_1} - g_{j_2}\|_n^2/\tau_0^2 \leq \frac{n\delta^2}{\tau_0^2}.$$

Let $\tilde{g}_n$ be any estimator, and define g^* by

$$\|\tilde{g}_n - g^*\|_n = \min_{j=1,\ldots,N} \|\tilde{g}_n - g_j\|_n.$$

Note now that $g^* \neq g_j$ implies $\|\tilde{g}_n - g_j\|_n > u/2$. We thus find

$$\sup_{g \in \mathcal{G}(\delta)} \mathbf{P}_g\left(\|\tilde{g}_n - g\|_n > \frac{u}{2}\right) \geq \max_{j=1,\ldots,N} \mathbf{P}_{g_j}\left(\|\tilde{g}_n - g_j\|_n > \frac{u}{2}\right) \geq 1 - \frac{n\delta^2/\tau_0^2 + \log 2}{\log(N-1)}.$$

Now, use the fact that $\log N \geq H(u, \mathcal{G}(\delta), Q_n)$, and $\log(N-1) \geq \log N/2$ for $N \geq 3$. Then for

$$\tau_0^2 \log 2 \leq n\delta_n^2 \leq \frac{\tau_0^2}{8} H(u_n, \mathcal{G}(\delta_n), Q_n),$$

one has

$$\sup_{g \in \mathcal{G}(\delta_n)} \mathbf{P}_g \left(\|\tilde{g}_n - g\|_n > \frac{u_n}{2} \right) \geq \frac{1}{2}.$$

10

Penalties and Sieves

The least squares estimator $\hat{g}_n$, with a penalty on its complexity (or roughness), is studied. The rate of convergence follows from the modulus of continuity of the empirical process, expressed in terms of the distance between $\hat{g}_n$ and the true underlying regression function g_0, as well as in terms of the complexity of $\hat{g}_n$ and g_0. A similar approach is used to prove rates of convergence for the penalized maximum likelihood estimator. Here, the attention is restricted to a penalty on the L_2-norm of the higher order derivative of the density or the log-density.

A sieve is an approximation of the (possibly too rich) parameter space, by a space with smaller entropy. Because the approximating space will in general not contain the true parameter, the Basic Inequalities we used earlier are no longer valid. In the regression context however, the Basic Inequality of Lemma 4.7 has a straightforward extension to the sieved case. For the maximum likelihood estimator with sieve, we need a further condition. We assume that the ratio of the true density and its approximation in the sieved space is bounded. In some examples, we show that a sieve can overcome the problem of suboptimal rates, which occur when the empirical process, indexed by a too rich set, is not asymptotically equicontinuous.

10.1. Penalized least squares

We consider again the regression model of the previous chapter. For each $n \geq 1$, let $Y_1, \ldots, Y_n$ be independent real-valued response variables satisfying

$$Y_i = g_0(z_i) + W_i, \quad i = 1, \ldots, n,$$

with g_0 an unknown regression function which is known to lie in a class $\mathcal{G}$ of regression functions, $z_1, \ldots, z_n$ given covariates in some space $\mathcal{Z}$, and with $\mathbf{E}W_i = 0$, $i = 1, \ldots, n$. We have established (rates of) convergence for the least squares estimator, under the assumption that a ball around g_0, intersected with $\mathcal{G}$, is totally bounded. This assumption means that quite a lot is known a priori about g_0. For instance, in Example 9.3.3. we have imposed an a priori bound on the total variation, etc. In this section, we assume that g_0 possesses certain smoothness properties, but we do not require an a priori bound on its roughness. Instead, we let its roughness be estimated by the data by adding a penalty term in the least squares loss function.

Let $I : \mathcal{G} \to [0, \infty)$ be a pseudo-norm on $\mathcal{G}$. We think of $I(g)$ as a measure for the complexity or roughness of g. For example, if $\mathcal{Z} = [0, 1]$, one could take the Sobolev norm $I(g) = \left(\int_0^1 (g^{(m)}(z))^p \, dz\right)^{1/p}$, with $m \geq 1$ and $p \geq 1$ given.

We assume that $I(g_0) < \infty$, but that a bound for $I(g_0)$ is unknown. The least squares estimator $\hat{g}_n$ with roughness penalty is defined as

$$\hat{g}_n = \arg\min_{g \in \mathcal{G}} \left(\frac{1}{n} \sum_{i=1}^n (Y_i - g(z_i))^2 + \lambda_n^2 I^v(g) \right). \tag{10.1}$$

Here, $v > 0$ and $\lambda_n > 0$. The quantity λ_n is called the smoothing parameter. If λ_n is large, the resulting estimator will be smooth. In practice, a choice for λ_n could be obtained by, for example, cross-validation, or by inspecting the various curves for different values of λ_n. This means that in most situations, λ_n will depend on the data. The parameter v is in principle free, but there is usually a natural choice, related to the pseudo-norm I. For example, if I is the above Sobolev norm, the natural choice is $v = p$.

A rate of convergence for the penalized least squares estimator can be derived in very much the same way as in the previous chapter: write down a Basic Inequality, and study the increments of the empirical process occurring in this Basic Inequality. The Basic Inequality is presented in Lemma 10.1. Here, we use the same notation as in Chapter 9:

$$Q_n = \frac{1}{n} \sum_{i=1}^n \delta_{z_i}, \quad \|g\|_n = \|g\|_{Q_n}, \quad (w, g)_n = \frac{1}{n} \sum_{i=1}^n W_i g(z_i),$$

and

$$\|y - g\|_n^2 = \frac{1}{n} \sum_{i=1}^n (Y_i - g(z_i))^2.$$

10.1. (Basic Inequality) *We have*

$$\|\hat{g}_n - g_0\|_n^2 + \lambda_n^2 I^v(\hat{g}_n) \leq 2(w, \hat{g}_n - g_0)_n + \lambda_n^2 I^v(g_0). \tag{10.2}$$

Proof This is simply rewriting the inequality

$$\|y - \hat{g}_n\|_n^2 + \lambda_n^2 I^v(\hat{g}_n) \leq \|y - g_0\|_n^2 + \lambda_n^2 I^v(g_0). \quad \square$$

Lemma 10.1 reveals that we need the modulus of continuity of the empirical process $\{(w, g - g_0)_n\}$, both in terms of $\|g - g\|_n$, as well as in terms of the roughness of g and g_0. Lemma 8.4 supplies one with this modulus of continuity, under entropy conditions. Let us suppose that $\mathcal{G}$ is a cone, and that

$$H\big(\delta, \{g \in \mathcal{G} : I(g) \leq 1\}, Q_n\big) \leq A\delta^{-\alpha}, \quad \text{for all } \delta > 0,\ n \geq 1, \tag{10.3}$$

for some constants $A > 0$ and $0 < \alpha < 2$. Then the same entropy bound holds for the normalized functions $(g - g_0)/(I(g) + I(g_0))$, $g \in \mathcal{G}$, whenever $I(g) + I(g_0) > 0$:

$$H\left(\delta, \left\{\frac{g - g_0}{I(g) + I(g_0)} : g \in \mathcal{G},\ I(g) + I(g_0) > 0\right\}, Q_n\right) \leq \tilde{A}\delta^{-\alpha}, \quad \text{for all } \delta > 0,\ n \geq 1. \tag{10.4}$$

In order to be able to apply Lemma 8.4, we require sub-Gaussian errors:

$$\sup_n \max_{i=1,\ldots,n} K^2\big(\mathbf{E}e^{|W_i|^2/K^2} - 1\big) \leq \sigma_0^2. \tag{10.5}$$

This allows us to conclude that for $I(g_0) > 0$,

$$\sup_{g \in \mathcal{G}} \frac{|(w, g - g_0)_n|}{\|g - g_0\|_n^{1-\frac{\alpha}{2}}(I(g) + I(g_0))^{\frac{\alpha}{2}}} = O_{\mathbf{P}}(n^{-\frac{1}{2}}). \tag{10.6}$$

Thus, the modulus of continuity of the empirical process also depends on the complexity of the functions involved. A similar, but technically more complicated bound holds if we replace (10.5) by the weaker assumption of errors with exponential tails

$$\sup_n \max_{i=1,\ldots,n} 2K\mathbf{E}\big(e^{|W_i|/K} - 1 - |W_i|/K\big) \leq \sigma_0^2, \tag{10.7}$$

provided that (10.4) is replaced by the corresponding bound on the entropy with bracketing. However, to avoid technicalities here, we impose throughout this section the condition (10.5) of sub-Gaussian errors. In the next section, we shall re-encounter this issue.

At first reading, it may be helpful to take $I(g_0) = 1$ in Theorem 10.2 below. One then finds that (apart from (10.8)), the rate of convergence does not depend on v. Moreover, the choice $\lambda_n = n^{-1/(2+\alpha)}$ then yields the fastest rate.

Now, in principle, g_0 and hence $I(g_0)$ is allowed to change with n, for example $I(g_0) \to 0$ or $I(g_0) \to \infty$, as $n \to \infty$. From Theorem 10.2 one may, for example, conclude that if $I(g_0) = O(n^{-1/2})$ (for instance $I(g_0) = 0$), the choice $\lambda_n = n^{-(2-v)/4}$ yields the rate $O_{\mathbf{P}}(n^{-1/2})$ for both $\|\hat{g}_n - g_0\|_n$ as well as $I(\hat{g}_n)$. Of course, in practice $I(g_0)$ is not known, so we do not know a priori how to choose λ_n.

Theorem 10.2 *Assume* (10.4) *and* (10.5). *Then for*

$$v > \frac{2\alpha}{2+\alpha}, \tag{10.8}$$

$$I(g_0) > 0, \tag{10.9}$$

and

$$\lambda_n^{-1} = O_{\mathbf{P}}(n^{1/(2+\alpha)}) I^{(2v-2\alpha+v\alpha)/2(2+\alpha)}(g_0), \tag{10.10}$$

we have

$$\|\hat{g}_n - g_0\|_n = O_{\mathbf{P}}(\lambda_n) I^{v/2}(g_0), \tag{10.11}$$

and

$$I(\hat{g}_n) = O_{\mathbf{P}}(1) I(g_0). \tag{10.12}$$

Moreover, if (10.8) *holds and*

$$I(g_0) = 0, \tag{10.13}$$

then

$$\|\hat{g}_n - g_0\|_n = O_{\mathbf{P}}\left(n^{-\frac{v}{2v-2\alpha+v\alpha}}\right) \lambda_n^{-\frac{2\alpha}{2v-2\alpha+v\alpha}}, \tag{10.14}$$

and

$$I(\hat{g}_n) = O_{\mathbf{P}}\left(n^{-\frac{2}{2v-2\alpha+v\alpha}}\right) \lambda_n^{-\frac{2(2+\alpha)}{2v-2\alpha+v\alpha}}. \tag{10.15}$$

Proof The result is a straightforward consequence of the behaviour of the empirical process, as stated in (10.6).

Case (i) Suppose that $I(\hat{g}_n) > I(g_0)$. Then by Lemma 10.1, combined with (10.6),

$$\|\hat{g}_n - g_0\|_n + \lambda_n^2 I^{\nu}(\hat{g}_n) \leq O_{\mathbf{P}}(n^{-\frac{1}{2}})\|\hat{g}_n - g_0\|_n^{1-\frac{\alpha}{2}} I^{\frac{\alpha}{2}}(\hat{g}_n) + \lambda_n^2 I^{\nu}(g_0). \tag{10.16}$$

Therefore, either

$$\|\hat{g}_n - g_0\|_n^2 + \lambda_n^2 I^{\nu}(\hat{g}_n) \leq O_{\mathbf{P}}(n^{-\frac{1}{2}})\|\hat{g}_n - g_0\|_n^{1-\frac{\alpha}{2}} I^{\frac{\alpha}{2}}(\hat{g}_n), \tag{10.17}$$

or

$$\|\hat{g}_n - g_0\|_n^2 + \lambda_n^2 I^{\nu}(\hat{g}_n) \leq 4\lambda_n^2 I^{\nu}(g_0). \tag{10.18}$$

Clearly, (10.18) implies (10.11) and (10.12). Solving (10.17) yields

$$\|\hat{g}_n - g_0\|_n = O_{\mathbf{P}}\left(n^{-\frac{\nu}{2\nu-2\alpha+\nu\alpha}}\right)\lambda_n^{-\frac{2\alpha}{2\nu-2\alpha+\nu\alpha}}, \tag{10.19}$$

and

$$I(\hat{g}_n) = O_{\mathbf{P}}\left(n^{-\frac{2}{2\nu-2\alpha+\nu\alpha}}\right)\lambda_n^{-\frac{2(2+\alpha)}{2\nu-2\alpha+\nu\alpha}}. \tag{10.20}$$

This proves (10.14) and (10.15) for the case $I(g_0) = 0$. If $I(g_0) > 0$, (10.19) and (10.20) also yield (10.11)and (10.12), due to assumption (10.10).

Case (ii) Suppose that $I(\hat{g}_n) \leq I(g_0)$ and $I(g_0) > 0$. Then from Lemma 10.1, combined with (10.6),

$$\|\hat{g}_n - g_0\|_n^2 \leq O_{\mathbf{P}}\left(n^{-\frac{1}{2}}\right)\|\hat{g}_n - g_0\|_n^{1-\frac{\alpha}{2}} I^{\frac{\alpha}{2}}(g_0) + \lambda_n I^{\nu}(g_0).$$

It follows that either

$$\|\hat{g}_n - g_0\|_n = O_{\mathbf{P}}\left(n^{-\frac{1}{2+\alpha}}\right) I^{\frac{\alpha}{2+\alpha}}(g_0) = O_{\mathbf{P}}(\lambda_n) I^{\frac{\nu}{2}}(g_0),$$

or

$$\|\hat{g}_n - g_0\|_n^2 \leq 4\lambda_n^2 I^{\nu}(g_0). \qquad \square$$

10.1.1. Penalizing a function in a Sobolev space

In this subsection, we assume that $\mathscr{Z} = [0, 1]$, and take the Sobolev norm $I(g) = \left(\int_0^1 (g^{(m)}(z))^2 \, dz\right)^{1/2}$, with $m \in \{1, 2, \ldots\}$, as a measure of complexity of the function g. The estimator $\hat{g}_n$ given in (10.1) with moreover $\nu = 2$ is

$$\hat{g}_n = \arg\min_{g \in \mathscr{G}} \left(\|y - g\|_n^2 + \lambda_n^2 \int \left(g^{(m)}(z)\right)^2 dz \right). \tag{10.21}$$

Here,

$$\mathcal{G} = \left\{ g : [0,1] \to \mathbf{R}, \ \int_0^1 \left(g^{(m)}(z)\right)^2 dz < \infty \right\}. \tag{10.22}$$

The estimator is known as the penalized smoothing spline.

The entropy result (10.3) does not hold, but an orthogonality argument can fix this. It goes as follows. In Example 9.3.2, we showed that one can write any $g \in \mathcal{G}$ as

$$g = g_1 + g_2,$$

where

$$g_1 = \sum_{k=1}^m \alpha_k \psi_k, \quad g_2 = \int_0^1 \beta_u \tilde{\phi}_u,$$

and where

$$(\psi_k, \tilde{\phi}_u)_n = \frac{1}{n} \sum_{i=1}^n \psi_k(z_i) \tilde{\phi}_u(z_i) = 0, \quad k = 1, \ldots, m, \ 0 < u \leq 1. \tag{10.23}$$

Due to the orthogonality (10.23), and the fact that the penalty is only on the coefficients $\{\beta_u : 0 < u \leq 1\}$, and not on the $\{\alpha_k : 1 \leq k \leq m\}$, one has

$$\hat{g}_{1,n} = \arg\min \left(-2(w, g_1 - g_{1,0})_n + \|g_1 - g_{1,0}\|_n^2 : g_1 = \sum_{k=1}^m \alpha_k \psi_k \right),$$

$$\hat{g}_{2,n} = \arg\min \left(-2(w, g_2 - g_{2,0})_n + \|g_2 - g_{2,0}\|_n^2 + \lambda_n^2 I^2(g_2) : g_2 = \int_0^1 \beta_u \tilde{\phi}_u \, du \right).$$

Therefore, $\|\hat{g}_{1,n} - g_{1,0}\|_n$ converges with rate $n^{-1/2}$. Now, define

$$\begin{aligned} \mathcal{G}_2 &= \left\{ g_2 = \int_0^1 \beta_u \tilde{\phi}_u, \ \int_0^1 \beta_u^2 \, du < \infty \right\} \\ &= \left\{ g_2 : \int (g_2^{(m)}(z))^2 \, dz < \infty, \ (g_2, \psi_k)_n = 0, \ k = 1, \ldots, m \right\}. \end{aligned}$$

Define $\psi = (\psi_1, \ldots, \psi_m)^T$ and $\Sigma_n = \int \psi \psi^T \, dQ_n$. Assuming that the smallest eigenvalue of Σ_n stays away from zero, one has that (10.3) *does* hold for $\mathcal{G}_2$, with $\alpha = 1/m$ (see also Mammen (1991)).

We now may apply Theorem 10.2 to $\hat{g}_{2,n}$, but we prefer to treat bias and variance separately. Define

$$g_n^* = \arg\min_{g \in \mathcal{G}} \left(\|g_0 - g\|_n^2 + \lambda_n^2 I^2(g) \right).$$

We refer to $\|g_n^* - g_0\|_n$ as the approximation error, and to $\|\hat{g}_n - g_n^*\|_n$ as the estimation error.

Lemma 10.3 *We have*

$$(10.24)\qquad \hat{g}_n - g_n^* = \arg\min_{g\in\mathcal{G}} \left(-2(w,g)_n + \|g\|_n^2 + \lambda_n^2 I^2(g)\right).$$

Proof The estimator $\hat{g}_n$ minimizes $\left(\|y-g\|_n^2 + \lambda_n^2 I^2(g)\right)$ iff for all $h \in \mathcal{G}$, and for $\hat{g}_{n,t} = \hat{g}_n + th$,

$$\frac{d}{dt}\left(\|y - \hat{g}_{n,t}\|_n^2 + \lambda_n^2 I^2(\hat{g}_{n,t})\right)\Big|_{t=0} = 0,$$

i.e., iff for all $h \in \mathcal{G}$,

$$(10.25)\qquad -(w,h)_n + (\hat{g}_n - g_0, h)_n + \lambda_n^2 I(\hat{g}_n, h) = 0,$$

where

$$(g,h)_n = \int gh\, dQ_n,$$

and

$$I(g,h) = \int_0^1 g^{(m)}(z) h^{(m)}(z)\, dz.$$

Similarly, for all $h \in \mathcal{G}$,

$$(10.26)\qquad (g_n^* - g_0, h)_n + \lambda_n^2 I(g_n^*, h) = 0.$$

Subtracting (10.26) from (10.25) yields that for all $h \in \mathcal{G}$,

$$-(w,h)_n + (\hat{g}_n - g_n^*, h)_n + \lambda_n^2 I(\hat{g}_n - g_n^*, h) = 0.$$

This is equivalent to (10.24). □

We summarize the result in a corollary.

Corollary 10.4 *For the approximation error, one has*

$$\|g_n^* - g_0\|_n^2 + \lambda_n^2 I^2(g_n^*) \le \lambda_n^2 I^2(g_0).$$

For the estimation error, one has

$$\|\hat{g}_{1,n} - g_{1,0}\|_n = O_{\mathbf{P}}\left(n^{-\frac{1}{2}}\right),$$

and, if the smallest eigenvalue of Σ_n stays away from zero,

$$\|\hat{g}_{2,n} - g_{2,n}^*\|_n = O_{\mathbf{P}}\left(n^{-\frac{1}{2}}\right)\lambda_n^{-\frac{1}{2m}},$$

and

$$I(\hat{g}_n - g_n^*) = O_{\mathbf{P}}(n^{-\frac{1}{2}})\lambda_n^{-\frac{2m+1}{2m}}.$$

Hence, if $I(g_0) > 0$ *and*

$$\lambda_n^{-1} = O_{\mathbf{P}}(n^{\frac{m}{2m+1}})I(g_0)^{\frac{2m}{2m+1}}, \tag{10.27}$$

then

$$\|\hat{g}_{2,n} - g_{2,0}\|_n = O_{\mathbf{P}}(\lambda_n)I(g_0),$$

and

$$I(\hat{g}_n) = O_{\mathbf{P}}(1)I(g_0).$$

The smoothing parameter λ_n is often estimated from the data. We conjecture that cross-validation will give the optimal order $n^{-m/(2m+1)}$ for the smoothing parameter, for the case $I(g_0) \asymp O(1)$.

Remark If λ_n is non-random, the estimator is a linear function of the data (and hence, $\mathbf{E}\hat{g}_n(z_i) = g_n^*(z_i)$, $i = 1, \ldots, n$). It can be shown that in that case, the assumption of sub-Gaussian errors can be relaxed: only finite second moments are needed.

10.1.2. Penalizing a function of bounded variation

Suppose

$$\mathscr{G} = \{ g : [0,1] \to \mathbf{R},\ TV(g^{(m-1)}) < \infty \}.$$

with $m \in \{1, 2, \ldots\}$ given. Here $TV(g^{(m-1)})$ is the total variation of the $(m-1)$th derivative:

$$TV(g^{(m-1)}) = \sum_{i=1}^{n} \left| g^{(m-1)}(z_i) - g^{(m-1)}(z_{i-1}) \right|, \quad z_1 \leq \cdots \leq z_n.$$

We take this total variation as a measure of roughness:

$$I(g) = TV(g^{(m-1)}).$$

Let Σ_n be defined as in the previous subsection, and assume that the smallest eigenvalue of Σ_n stays away from zero. Application of Theorem 10.2 gives for $I(g_0) > 0$, and for

$$\lambda_n^{-1} = O_{\mathbf{P}}(n^{\frac{m}{2m+1}})I^{\frac{2m-1}{2(2m+1)}}(g_0),$$

the rates

$$\|\hat{g}_n - g_0\|_n = O_{\mathbf{P}}(\lambda_n I^{\frac{1}{2}}(g_0) \vee n^{-\frac{1}{2}}),$$

and

$$I(\hat{g}_n) = O_{\mathbf{P}}(1)I(g_0).$$

Thus, the rates are essentially the same as in the previous subsection. However, the estimator is much more flexible (it adjusts to local non-smoothness), but also more difficult to compute.

Note that for the case $m = 1$, the assumption of bounded support can be dropped.

10.1.3. Penalizing a function in a Besov space

We now consider a penalty on the Besov norm defined in (10.28) below. The s-th order difference of a function g on $[0, 1]$ is defined as

$$\Delta_h^s(g, z) = \sum_{k=0}^{s} \binom{s}{k} (-1)^{s-k} g(z + kh), \quad z \in [0, 1],\ z + kh \in [0, 1].$$

The L_p-modulus of smoothness is

$$w_s^p(g, y) = \sup_{0<h\leq y} \int_0^{1-sh} |\Delta_h^s(g, z)|^p \, dz, \ y > 0.$$

For $s = \lfloor \gamma \rfloor + 1$, with $\lfloor \gamma \rfloor$ the integer part of γ, we define the Besov space $B_{p,\infty}^\gamma$ as the set of all functions g on $[0, 1]$ with $\int_0^1 g^2(z)\, dz < \infty$, and with

$$I_\gamma(g) = \sup_{y>0} y^{-\gamma} w_s(g, y) < \infty. \tag{10.28}$$

It is shown in Birgé and Massart (1996) that

$$H_\infty\left(\delta, \left\{ g \in B_{p,\infty}^\gamma : I_\gamma(g) \leq 1 \right\}\right) \leq A \left(\frac{1}{\delta}\right)^{1/\gamma}, \quad \delta > 0, \tag{10.29}$$

provided $\gamma > 1/p$. Thus, (10.3) is met. If moreover $\gamma > 1/2$, Theorem 10.2 supplies us with rates of convergence for the least squares estimator with penalty on the Besov norm given in (10.28).

The question arises whether one can adapt to the smoothness γ. It is tempting to use

$$\hat{\gamma}_n = \arg \min_{\gamma > (2\wedge p)^{-1}} \min_{g \in B_{p,\infty}^\gamma} \left(\|y - g\|_n^2 + n^{-\frac{2\gamma}{2\gamma+1}} I_\gamma^p(g) \right)$$

as an estimator of γ. We shall not pursue this here.

An alternative is to use the number of non-zero coefficients in the wavelet decomposition as a penalty. The arguments to prove that then the estimator adapts to the amount of smoothness γ when γ is unknown (in the sense of obtaining the rate $O_{\mathbf{P}}(n^{-\gamma/(2\gamma+1)})$), are similar to the ones used in Theorem 10.2. We shall not present any details here. Lemma 10.12 sketches the idea in an undressed context.

10.2. Penalized maximum likelihood

For each $n \geq 1$, let $X_1, \ldots, X_n$ be independent random variables with distribution P on $[0, 1]$ and let $p_0 = dP/d\mu$ be the density with respect to the Lebesgue measure μ. The distribution P (and hence the density p_0) may depend on n, but we assume throughout that for some η_0 independent of n,

$$p_0 \geq \eta_0^2. \tag{10.30}$$

We also assume that p_0 is a member of the class

$$\mathscr{P} = \left\{ p : [0,1] \to [0, \infty) : \int p(x)\, dx = 1,\ I^2(p) < \infty \right\},$$

where $I(p)$ measures the complexity of the density p. The penalized maximum likelihood estimator $\hat{p}_n$ is defined as

$$\hat{p}_n = \arg\max_{p \in \mathscr{P}} \left(\int \log p \, dP_n - \lambda_n^2 I^2(p) \right). \tag{10.31}$$

Here, λ_n is called the smoothing parameter.

Recall the notation

$$g_p = \frac{1}{2} \log \left(\frac{p + p_0}{2p_0} \right), \quad p \in \mathscr{P}.$$

Lemma 10.5. (Basic Inequality) *Let $\hat{p}_n$ be given by (10.31). We have*

$$h^2(\hat{p}_n, p_0) + 4\lambda_n^2 I^2(\hat{p}_n) \leq 16 \int g_{\hat{p}_n}\, d(P_n - P) + 4\lambda_n^2 I^2(p_0). \tag{10.32}$$

Proof

$$4 \int g_{\hat{p}_n}\, dP_n - \lambda_n^2 I^2(\hat{p}_n) \geq \int \log \left(\frac{\hat{p}_n}{p_0} \right) dP_n - \lambda_n^2 I^2(\hat{p}_n) \geq -\lambda_n^2 I^2(p_0),$$

so

$$\begin{aligned} 16 \int g_{\hat{p}_n}\, d(P_n - P) - 4\lambda_n^2 I^2(\hat{p}_n) &\geq -16 \int g_{\hat{p}_n}\, dP - 4\lambda_n^2 I^2(p_0) \\ &\geq 16 h^2 \left(\frac{\hat{p}_n + p_0}{2}, p_0 \right) - 4\lambda_n^2 I^2(p_0) \\ &\geq h^2(\hat{p}_n, p_0) - 4\lambda_n^2 I^2(p_0). \end{aligned}$$

□

In Subsections 4.2.2 and 7.4.1, we assumed that the density p_0 is known to lie in a Sobolev class. In the literature on penalized maximum likelihood estimation, it is more common to assume that the log-density lies in a Sobolev class (see for example, Barron and Sheu (1991)). We shall consider both situations here.

10.2.1. Roughness penalty on the density

In this subsection, we take

$$I^2(p) = \int_0^1 \left(p^{(m)}(x)\right)^2 dx, \tag{10.33}$$

where $m \geq 1$ is a fixed integer.

It follows from Theorem 2.4 and (10.30) that for all $M \geq 1$,

$$H_B\big(\delta, \{ g_p : p \in \mathscr{P},\ I(p) + I(p_0) \leq M \}\big) \leq A\left(\frac{M}{\delta}\right)^{1/m},\ \delta > 0.$$

To see this, note first of all that g_p is a Lipschitz function of p. Moreover, densities integrate to 1, which implies that for some constant c,

$$\sup_{p \in \mathscr{P},\ I(p)+I(p_0) \leq M} |p|_\infty \leq cM, \quad M \geq 1.$$

Apply Lemma 5.14 with $\alpha = 1/m$, $\beta = 0$, and $d(g_p, g_{p_0}) = h(p, p_0)$. Then we find

$$\sup_{h(p,p_0) \geq n^{-\frac{m}{2m+1}}(1+I(p)+I(p_0))} \frac{\int g_p \, d(P_n - P)}{h^{1-\frac{1}{2m}}(p, p_0)\big(1 + I(p) + I(p_0)\big)^{\frac{1}{2m}}} = O_{\mathbf{P}}(n^{-\frac{1}{2}}), \tag{10.34}$$

and

$$\sup_{h(p,p_0) \leq n^{-\frac{m}{2m+1}}(1+I(p)+I(p_0))} \frac{\int g_p \, d(P_n - P)}{1 + I(p) + I(p_0)} = O_{\mathbf{P}}\big(n^{-\frac{2m}{2m+1}}\big). \tag{10.35}$$

Note that (10.34) only presents the modulus of continuity for $h(p, p_0)$ sufficiently far way from zero. This is a consequence of the fact that we do not have sub-Gaussian tail behaviour.

Combination of (10.34) and (10.35) with Lemma 10.5 gives a rate of convergence for the penalized maximum likelihood estimator.

Theorem 10.6 *Let $\hat{p}_n$ be defined by* (10.31), *with $I(p)$ given in* (10.33). *Then*

$$h(\hat{p}_n, p_0) = O_{\mathbf{P}}(\lambda_n)\big(1 + I(p_0)\big), \tag{10.36}$$

and

$$I(\hat{p}_n) = O_{\mathbf{P}}(1)\big(1 + I(p_0)\big). \tag{10.37}$$

provided that

$$\lambda_n^{-1} = O_{\mathbf{P}}\big(n^{\frac{m}{2m+1}}\big)\big(1 + I(p_0)\big)^{\frac{1}{2}} \tag{10.38}$$

Proof Let us use the shorthand notation

$$\hat{h} = h(\hat{p}_n, p_0),$$

and

$$\hat{I} = I(\hat{p}_n), \ I_0 = I(p_0).$$

Case (i) Suppose $\hat{h} > n^{-\frac{m}{2m+1}}(1 + \hat{I} + I_0)$, and $\hat{I} > 1 + I_0$. Then from (10.34), either

$$\hat{h}^2 + \lambda_n^2 \hat{I}^2 \leq O_{\mathbf{P}}(n^{-\frac{1}{2}})\, \hat{h}^{1-\frac{1}{2m}}\, \hat{I}^{\frac{1}{2m}}$$

or

$$\hat{h}^2 + \lambda_n^2 \hat{I}^2 \leq 4\lambda_n^2 I_0^2.$$

Solving this gives either

$$\hat{h} \leq \lambda_n^{-\frac{1}{2m}} O_{\mathbf{P}}\big(n^{-\frac{1}{2}}\big), \qquad \hat{I} \leq \lambda_n^{-\frac{2m+1}{2m}} O_{\mathbf{P}}\big(n^{-\frac{1}{2}}\big), \tag{10.39}$$

or

$$\hat{h} \leq 2\lambda_n I_0, \ \hat{I} \leq 2I_0. \tag{10.40}$$

Clearly, (10.40) implies (10.36) and (10.37). Because

$$\lambda_n^{-1} = O_{\mathbf{P}}\big(n^{\frac{m}{2m+1}}\big)(1 + I_0)^{\frac{2m}{2m+1}}.$$

the same is true for (10.39).

Case (ii) Suppose that $\hat{h} \leq n^{-\frac{m}{2m+1}}(1 + \hat{I} + I_0)$ and $\hat{I} > 1 + I_0$. Then from (10.35), either

$$\hat{h}^2 + \lambda_n^2 \hat{I}^2 \leq O_{\mathbf{P}}\big(n^{-\frac{2m}{2m+1}}\big)\hat{I},$$

or

$$\hat{h}^2 + \lambda_n^2 \hat{I}^2 \leq 4\lambda_n^2 I_0^2.$$

Solving this gives that either

$$\hat{h} \leq \lambda_n^{-1} O_{\mathbf{P}}\big(n^{-\frac{2m}{2m+1}}\big) = O_{\mathbf{P}}(\lambda_n)(1 + I_0), \quad \hat{I} \leq \lambda_n^{-2} O_{\mathbf{P}}\big(n^{-\frac{2m}{2m+1}}\big) = O_{\mathbf{P}}(1)(1 + I_0),$$

or

$$\hat{h} \leq 2\lambda_n I_0, \quad \hat{I} \leq 2I_0.$$

Here, we invoked assumption (10.38).

Case (iii) Suppose that $\hat{h} > n^{-\frac{m}{2m+1}}(1+\hat{I}+I_0)$ and $\hat{I} \leq 1+I_0$. Then we find that either

$$\hat{h}^2 \leq O_{\mathbf{P}}\big(n^{-\frac{1}{2}}\big)\,\hat{h}^{1-\frac{1}{2m}}(1+I_0)^{\frac{1}{2m}},$$

or

$$\hat{h}^2 \leq 4\lambda_n^2 I_0^2.$$

Again, since

$$\lambda_n^{-1} = O_{\mathbf{P}}\big(n^{\frac{m}{2m+1}}\big)(1+I_0)^{\frac{2m}{2m+1}},$$

this implies (10.36).

Case (iv) Suppose that $\hat{h} \leq n^{-\frac{m}{2m+1}}(1+\hat{I}+I_0)$ and $\hat{I} \leq 1+I_0$. Then (10.36) and (10.37) are trivial. □

Remark Note the difference between (10.27) and (10.38). One could improve (10.38) by taking $\beta > 0$ in our application of Lemma 5.14 (see also the next subsection).

Corollary 10.7 *Suppose* $I(p_0) = O(1)$ *and* $\lambda_n \asymp n^{-\frac{m}{2m+1}}$. *Then it follows from Theorem* 10.6 *and Lemma* 10.9 *below that for integer* $1 \leq k \leq m$,

$$\|\hat{p}_n^{(k)} - p_0^{(k)}\|_\mu = O_{\mathbf{P}}\big(n^{-\frac{m-k}{2m+1}}\big). \tag{10.41}$$

For small values of $I(p_0)$, the rates may be improved. To avoid technicalities, we shall not consider this matter here. The next subsection does look at this issue, albeit in a somewhat different setting.

10.2.2. Roughness penalty on the log-density

We shall now consider the penalty

$$I^2(p) = \int \big(\gamma^{(m)}(x)\big)^2\,dx = J^2(\gamma), \tag{10.42}$$

with

$$\gamma = \log p - \int \log p\,dP.$$

The main ingredient of Subsection 10.2.1 is that g_p is a Lipschitz function of p, because this allows us to conclude that for any subset $\mathscr{P}_1$ of $\mathscr{P}$, the entropy of $\{g_p : p \in \mathscr{P}_1\}$ can be bounded by the entropy of $\mathscr{P}_1$. Since g_p is also Lipschitz in $\log p$, this argument carries over immediately. On the other hand, densities integrate to one, but we do not assume a bound for the (square-)integrated log-densities. This is why we shall need to modify the Basic Inequality of Lemma 10.5.

Define now

$$b(\gamma) = \log \int_0^1 e^{\gamma(x)}\, dx,$$

so that

$$\log p = \gamma - b(\gamma), \quad \int \gamma \, dP = 0,$$

and

$$b(\gamma) - b(\gamma_0) = K(p, p_0),$$

where $K(p, p_0)$ is the Kullback–Leibler information.

For the case $J(\gamma_0) = o(1)$, we shall need the interpolation inequality of Agmon (1965):

Lemma 10.8 *We have for some constant c_1,*

$$\int_0^1 \left(\gamma^{(k)}(x)\right)^2 dx \le c_1^2 \vartheta^{-2k} \int \gamma^2(x)\, dx + c_1^2 \vartheta^{2(m-k)} \int_0^1 \left(\gamma^{(m)}(x)\right)^2 dx, \tag{10.43}$$

for all $\vartheta > 0$ and all $1 \le k \le m$.

Application of this lemma gives

Lemma 10.9 *Suppose $\int_0^1 \gamma^2(x)\, dx \le \delta^2$ and $\int_0^1 \left(\gamma^{(m)}(x)\right)^2 dx \le 1$. Then for $1 \le k \le m$*

$$\int_0^1 (\gamma^{(k)}(x))^2\, dx \le 2c_1^2 \delta^{2(1-\frac{k}{m})}. \tag{10.44}$$

Furthermore,

$$|\gamma|_\infty \le (32)^{\frac{1}{4}} \sqrt{c_1}\, \delta^{1-\frac{1}{2m}}. \tag{10.45}$$

Proof Inequality (10.44) immediately follows from Lemma 10.8, by taking $\vartheta = \delta^{1/m}$. In particular

$$\int_0^1 \left(\gamma^{(1)}(x)\right)^2 dx \le 2c_1^2 \delta^{2(1-\frac{1}{m})}.$$

Now, suppose that $|\gamma(x_0)| \ge s$ for some x_0. Then

$$|\gamma(x)| \ge s - \left| \int_{x_0}^x \gamma^{(1)}(u)\, du \right| \ge s - \sqrt{2} c_1 |x - x_0|^{\frac{1}{2}} \delta^{1-\frac{1}{m}} \ge \frac{s}{2}$$

for all $|x - x_0| \le (s^2/8c_1^2)\delta^{-2(1-(1/m))}$. So,

$$\int_0^1 \gamma^2(x)\,dx \ge \frac{s^4}{32c_1^2}\delta^{-2(1-\frac{1}{m})}.$$

On the other hand, $\int_0^1 \gamma^2(x)\,dx \le \delta^2$, so we must have $s \le (32)^{\frac{1}{4}}\sqrt{c_1}\,\delta^{1-(1/2m)}$. □

Now, if p_0 is rather smooth, in the sense that $J(\gamma_0) = o(1)$, on can obtain a rate faster than $n^{-m/(2m+1)}$, because one needs less smoothing. However, the faster rate requires knowledge on the order of magnitude of $J(\gamma_0)$, because the smoothing parameter depends on it. Hopefully, cross-validation automatically picks the right order.

We use the result of Agmon (Lemma 10.9) to state the best order for the smoothing parameter. Due to (10.45), one may choose $\beta = 1 - (1/2m)$ in the application of Lemma 5.14, and this leads us to condition (10.48) on the smoothing parameter.

Theorem 10.10 *Let $\hat{p}_n$ be given by* (10.31), *with $I(p) = J(\gamma)$ defined in* (10.42). *If $0 < J(\gamma_0) = O(1)$, then*

$$\|\hat{\gamma}_n - \gamma_0\| = O_{\mathbf{P}}\big(\lambda_n J(\gamma_0) \vee n^{-\frac{1}{2}}\big), \tag{10.46}$$

$$J(\hat{\gamma}_n) = O_{\mathbf{P}}\big(J(\gamma_0) \vee n^{-\frac{1}{2}}\lambda_n^{-1}\big), \tag{10.47}$$

provided that

$$\lambda_n^{-1} = O_{\mathbf{P}}\big(n^{\frac{m}{2m+1}}\big)J^{\frac{2m}{2m+1}}(\gamma_0), \quad \lambda_n = o_{\mathbf{P}}(1). \tag{10.48}$$

Proof The first part of the proof supplies us with a principal rate of convergence. Define

$$\alpha = \alpha(\gamma) = \frac{1}{1 + \|\gamma\|},$$

and

$$\gamma_c = \alpha\gamma + (1-\alpha)\gamma_0 = \gamma_0 + f,$$

where

$$f = \frac{\gamma - \gamma_0}{1 + \|\gamma\|}.$$

Let $\log p_c = \gamma_c - b(\gamma_c)$. Use the concavity of the log-function, and the convexity of $b(\gamma)$ and $J^2(\gamma)$ to find

$$\begin{aligned}
&\int \log \frac{\hat{p}_{n,c} + p_0}{2p_0}\, dP_n - \frac{1}{2}\lambda_n^2 \left(J^2(\hat{\gamma}_{n,c}) - J^2(\gamma_0)\right) \\
&\geq \frac{1}{2} \int \log \frac{\hat{p}_{n,c}}{p_0}\, dP_n - \frac{1}{2}\lambda_n^2 \left(J^2(\hat{\gamma}_{n,c}) - J^2(\gamma_0)\right) \\
&= \frac{1}{2}\hat{\alpha}_n \int (\hat{\gamma}_n - \gamma_0)\, dP_n - \frac{1}{2}\left(b(\hat{\gamma}_{n,c}) - b(\gamma_0)\right) - \frac{1}{2}\lambda_n^2 \left(J^2(\hat{\gamma}_{n,c}) - J^2(\gamma_0)\right) \\
&\geq \frac{1}{2}\hat{\alpha}_n \left(\int (\hat{\gamma}_n - \gamma_0)\, dP_n - \left(b(\hat{\gamma}_n) - b(\gamma_0)\right) - \lambda_n^2 \left(J^2(\hat{\gamma}_n) - J^2(\gamma_0)\right) \right) \\
&\geq 0.
\end{aligned}$$

Therefore, the following Basic Inequality holds:

$$h^2(\hat{p}_{n,c}, p_0) + 4\lambda_n^2 J^2(\hat{\gamma}_{n,c}) \leq 8 \int \log \frac{\hat{p}_{n,c} + p_0}{2p_0}\, d(P_n - P) + 4\lambda_n^2 J^2(\gamma_0).$$

Observe now that for $f = (\gamma - \gamma_0)/(1 + \|\gamma\|)$, we have

$$\|f\| \leq \|\gamma_0\| + 1,$$

and

$$J(f) \leq J(\gamma_0 + f) + J(\gamma_0).$$

Hence, for some constant c_1,

$$|f|_\infty \leq c_1 M,$$

whenever $J(\gamma_0 + f) \leq M$, $M \geq 1$. But then also

$$b(\gamma_0 + f) \leq c_2 M.$$

Therefore,

$$H_\infty\left(\delta, \{\log p_c = \gamma_0 + f - b(\gamma_0 + f) : J(\gamma_0 + f) \leq M\}\right) \leq A \left(M/\delta\right)^{1/m}, \quad \text{for all } \delta > 0,\ M \geq 1.$$

But then, also, for all $\delta > 0$, $M \geq 1$,

$$\begin{aligned}
&H_\infty\left(\delta, \left\{\log \frac{p_c + p_0}{2p_0} : \log p_c = \gamma_0 + f - b(\gamma_0 + f),\ J(\gamma_0 + f) \leq M\right\}\right) \\
&\leq A_1 \left(\frac{M}{\delta}\right)^{1/m}.
\end{aligned}$$

From the Basic Inequality, we therefore obtain in the same way as in Theorem 10.6, that

$$h(\hat{p}_{n,c}, p_0) = O_{\mathbf{P}}(\lambda_n),$$

and

$$J(\hat{\gamma}_{n,c}) = O_{\mathbf{P}}(1).$$

But then also

$$J(\hat{f}_n) = O_{\mathbf{P}}(1),$$

which in turn implies

$$|\hat{f}_n|_\infty = O_{\mathbf{P}}(1).$$

But this means that

$$h(\hat{p}_{n,c}, p_0) \geq \frac{1}{c_3} \| \log \hat{p}_{n,c} - \log p_0 \|.$$

But then,

$$\|\hat{f}_n - f_0\| = O_{\mathbf{P}}(\lambda_n).$$

So

$$\|\hat{\gamma}_n - \gamma_0\| = O_{\mathbf{P}}(\lambda_n).$$

So also

$$J(\hat{\gamma}_n) = O_{\mathbf{P}}(1).$$

This proves the theorem when $J(\gamma_0)$ stays away from zero.

Since we have now already proved consistency, we may apply a more refined Basic Inequality to arrive at the result when $J(\gamma_0) = o(1)$. The results $\|\hat{\gamma}_n - \gamma_0\| = o_{\mathbf{P}}(1)$ and $J(\hat{\gamma}_n) = O_{\mathbf{P}}(1)$, imply

$$|\hat{\gamma}_n - \gamma_0|_\infty = o_{\mathbf{P}}(1),$$

which gives

$$b(\hat{\gamma}_n) - b(\gamma_0) = \|\hat{\gamma}_n - \gamma_0\|^2/(1 + O_{\mathbf{P}}(1)).$$

Another modification of the Basic Inequality in Lemma 10.5 now reads

$$\frac{\|\hat{\gamma}_n - \gamma_0\|^2}{1 + O_{\mathbf{P}}(1)} + \lambda_n^2 J^2(\hat{\gamma}_n) \leq \int (\hat{\gamma}_n - \gamma_0)\, dP_n + \lambda_n^2 J^2(\gamma_0).$$

Write

$$\gamma = \sum_{k=2}^{m} \alpha_k \psi_k + \int \beta_u \tilde{\phi}_u \, du = \gamma_1 + \gamma_2,$$

with

$$\psi_k(x) = x^{k-1} - \int x^{k-1} p_0(x)\, dx, \quad k = 2, \ldots, m,$$

and with

$$\int \tilde{\phi}_u(x)\psi_k(x)p_0(x)\, dx = 0, \quad k = 2, \ldots, m, \ u \in (0, 1],$$

and

$$\int \tilde{\phi}_u(x)p_0(x)\, dx = 0, \quad u \in (0, 1].$$

Then it follows that

$$\text{(10.49)} \quad \frac{\|\hat{\gamma}_{1,n} - \gamma_{1,0}\|^2 + \|\hat{\gamma}_{2,n} - \gamma_{2,0}\|^2}{1 + O_{\mathbf{P}}(1)} + \lambda_n^2 J^2(\hat{\gamma}_n)$$
$$\leq \|\hat{\gamma}_{1,n} - \gamma_{1,0}\| O_{\mathbf{P}}(n^{-\frac{1}{2}}) + \int (\hat{\gamma}_{2,n} - \gamma_{2,0})\, dP_n + \lambda_n^2 J^2(\gamma_0).$$

Now, let

$$\Gamma_2 = \left\{ \gamma_2 : \int (\gamma_2^{(m)}(x))^2\, dx \leq 1, \ \int \gamma_2\, dP = 0, \int \gamma_2 \psi_k\, dP = 0, \ k = 2, \ldots, m \right\}.$$

Then Γ_2 is uniformly bounded (see Mammen (1991)), so that

$$H_\infty(\delta, \Gamma_2) \leq A\delta^{-1/m}, \quad \text{for all } \delta > 0.$$

So by Lemma 5.14 (with $\alpha = 1/m$, $\beta = 1 - 1/2m$), if $\|\hat{\gamma}_{2,n} - \gamma_{2,0}\| > n^{-m/2}(J(\hat{\gamma}_n) + J(\gamma_0))$, we have

$$\int (\hat{\gamma}_{2,n} - \gamma_{2,0})\, dP_n = O_{\mathbf{P}}(n^{-\frac{1}{2}}) \|\hat{\gamma}_{2,n} - \gamma_{2,0}\|^{1 - \frac{1}{2m}} \left(J(\hat{\gamma}_n) + J(\gamma_0)\right)^{\frac{1}{2m}},$$

and if $\|\hat{\gamma}_{2,n} - \gamma_{2,0}\| \leq n^{-m/2}(J(\hat{\gamma}_n) + J(\gamma_0))$, we have

$$\int (\hat{\gamma}_{2,n} - \gamma_{2,0})\, dP_n = O_{\mathbf{P}}\left(n^{-\frac{2m+1}{4}}\right)(J(\hat{\gamma}_n) + J(\gamma_0)).$$

Insert this in (10.49). The rest of the proof is straightforward manipulation, as in Theorem 10.6. □

10.3. Least squares on sieves

Recall the regression model

$$Y_i = g_0(z_i) + W_i, \quad i = 1, \ldots, n,$$

with $Y_1, \ldots, Y_n$ real-valued observations, $g_0 \in \mathcal{G}$ an unknown regression function, $z_1, \ldots, z_n$ given covariates in a space $\mathcal{Z}$ and $W_1, \ldots, W_n$ independent errors with expectation zero. For technical reasons, we also impose throughout this section the sub-Gaussian condition

$$\max_{i=1,\ldots,n} K^2 \left(\mathbf{E} e^{|W_i|^2/K^2} - 1 \right) \le \sigma_0^2. \tag{10.50}$$

If the class $\mathcal{G}$ is too large, in the sense that for some δ and R the δ-covering number of $\mathcal{G}_n(R)$ grows exponentially fast in n (see Theorem 4.8), then the least squares estimator obtained by minimizing $\|y - g\|_n$ over all $g \in \mathcal{G}$ may be inconsistent. In Section 10.1, we handled this situation by penalizing the complexity of the functions in $\mathcal{G}$. The approach we take here is to use a sieve. Let $\mathcal{G}_n^*$ be some other class of regression functions, not necessarily a subclass of $\mathcal{G}$, but which is less complex than $\mathcal{G}$, in the sense that it has smaller entropy. Define

$$\hat{g}_n = \arg \min_{g \in \mathcal{G}_n^*} \|y - g\|_n^2. \tag{10.51}$$

We say that $\hat{g}_n$ is a sieved estimator.

Because g_0 is not necessarily a member of $\mathcal{G}_n^*$, the Basic Inequality (9.1) (see also Lemma 4.7), and hence Theorem 9.1 (which presents rates of convergence), cannot be applied. However, the argument used in Theorem 9.1 does carry over. All we need is a modification of (9.1). There appears a bias term

$$\inf_{g \in \mathcal{G}_n^*} \|g - g_0\|_n. \tag{10.52}$$

Let us for simplicity assume that the infimum in (10.52) is attained at $g_n^* \in \mathcal{G}_n^*$. (If the infimum is not attained one can always choose $g_n^* \in \mathcal{G}_n^*$ in such a way that $\|g_n^* - g_0\|_n$ is of the same order as the infimum.)

Define for $0 < \delta < 2^7 \sigma_0$,

$$\mathcal{G}_n^*(\delta) = \left\{ g \in \mathcal{G}_n^* : \ \|g - g_n^*\|_n \le \delta \right\},$$

and

$$J\big(\delta, \mathcal{G}_n^*(\delta), Q_n\big) = \int_{\delta^2/(2^7\sigma_0)}^{\delta} H^{1/2}\big(u, \mathcal{G}_n^*(\delta), Q_n\big)\, du \vee \delta.$$

Theorem 10.11 *Suppose condition* (10.50) *on the errors is met. Take* $\Psi(\delta) \ge J\big(\delta, \mathcal{G}_n^*(\delta), Q_n\big)$ *in such a way that* $\Psi(\delta)/\delta^2$ *is a non-decreasing function of* δ, $0 < \delta < 2^7\sigma_0$. *Then for a constant c depending on K and* σ_0, *and for*

$$\sqrt{n}\delta_n^2 \ge c\Psi(\delta_n), \quad \textit{for all } n, \tag{10.53}$$

we have

(10.54) $$\|\hat{g}_n - g_0\|_n = O_{\mathbf{P}}(\delta_n + \|g_n^* - g_0\|_n).$$

Proof Rewrite the inequality $\|y - \hat{g}_n\|_n \le \|y - g_n^*\|_n$ as

$$\|\hat{g}_n - g_0\|_n^2 \le 2(w, \hat{g}_n - g_n^*)_n + \|g_n^* - g_0\|_n^2.$$

Then we find

$$\|\hat{g}_n - g_n^*\|_n^2 \le 2(\|\hat{g}_n - g_0\|_n^2 + \|g_n^* - g_0\|_n^2) \le 4(w, \hat{g}_n - g_n^*)_n + 4\|g_n^* - g_0\|_n^2.$$

This is our Basic Inequality for the sieved least squares estimator. On the set where $\|g_n^* - g_0\|_n^2 \le (w, \hat{g}_n - g_n^*)_n$, the inequality

$$\|\hat{g}_n - g_n^*\|_n^2 \le 8(w, \hat{g}_n - g_n^*)_n$$

holds, and we can apply the same arguments as in Theorem 9.1. □

In (10.54), we say that δ_n is the estimation error, whereas $\|g_n^* - g_0\|_n$ is the approximation error.

In the remainder of this section, we investigate what happens if $\mathcal{G}_n^*$ is a linear space, and how to choose the dimension.

Suppose that

(10.55) $$\mathcal{G}_n^* = \{ g(z) = \theta_1\psi_1(z) + \cdots + \theta_{d_n}\psi_{d_n}(z) : \theta \in \mathbf{R}^{d_n} \}.$$

Then by Corollary 2.6,

(10.56) $$J(\delta, \mathcal{G}_n^*(\delta), Q_n) \le \text{const.} d_n^{1/2}\delta,$$

so that from Theorem 10.11, we obtain

(10.57) $$\|\hat{g}_n - g_0\|_n = O_{\mathbf{P}}\left(\sqrt{\frac{d_n}{n}} + \|g_n^* - g_0\|_n\right).$$

Remark The assumption (10.50) on the errors can be weakened to

$$\max_{i=1,\ldots,n} \mathbf{E}W_i^2 \le \tau_0^2.$$

One sees this by using the same line of reasoning as in Example 9.3.1. Moreover, g_n^* is now the projection on $\mathcal{G}_n^*$, so that we in fact have $\|\hat{g}_n - g_n^*\|_n = O_{\mathbf{P}}(\sqrt{d_n/n})$.

Generally, the dimension d_n can be chosen in such a way that the estimation error and approximation error are of the same order. However, in some situations the resulting linear estimator converges with a suboptimal rate. For example, if $\mathcal{G}$ is the set of functions of bounded variation

$$\mathcal{G} = \left\{ g : \mathbf{R} \to \mathbf{R} : \int |g'(z)|\, dz < M \right\}, \tag{10.58}$$

then we know from Example 9.3.3 that for $\mathcal{G}_n^* = \mathcal{G}$, $\|\hat{g}_n - g_0\|_n = O_{\mathbf{P}}(n^{-1/3})$. The best d-dimensional linear approximation of $\mathcal{G}$ is given by Kolmogorov's linear d-width

$$K(\mathcal{G}, d, Q_n) = \inf_{\mathcal{G}_n^*\ d\text{-dim. space}} \ \sup_{g_0 \in \mathcal{G}} \ \inf_{g \in \mathcal{G}_n^*} \|g - g_0\|_n.$$

For the class $\mathcal{G}$ given in (10.58), and $z_1, \ldots, z_n$ distinct, we have as $d_n \to \infty$,

$$K(\mathcal{G}, d_n, Q_n) \asymp d_n^{-\frac{1}{2}}$$

(see Pinkus (1985)). Thus, if one takes the optimal choice for $\mathcal{G}_n^*$ in (10.55), we find that generally (except possibly for some a-typical g_0), the rate of convergence is $\|\hat{g}_n - g_0\|_n = O_{\mathbf{P}}(n^{-1/4})$.

As another example, let us suppose that $z_i = i/n$, $i = 1, \ldots, n$, and

$$\mathcal{G} = \left\{ g : [0,1] \to \mathbf{R},\ \int |g^{(m)}(z)|^2\, dz \le M^2 \right\}. \tag{10.59}$$

Then

$$K(\mathcal{G}, d_n, Q_n) \asymp d_n^{-m} \tag{10.60}$$

(Pinkus 1985). So if we take $\mathcal{G}_n^*$ as the best linear approximation, we find the rate $\|\hat{g}_n - g_0\|_n = O_{\mathbf{P}}(n^{-m/(2m+1)})$. This rate is achieved for the choice

$$\mathcal{G}_n^* = \left\{ g(z) = \sum_{j=1}^{N_n} \sum_{k=1}^{m} \alpha_{j,k} z^{k-1} 1\left\{ \frac{j-1}{N_n} < z \le \frac{j}{N_n} \right\},\ \{\alpha_{j,k}\} \in \mathbf{R}^{d_n} \right\},$$

where $d_n = mN_n$ (for the case $m = 1$, see also Problem 2.4). In other words, $\mathcal{G}_n^*$ can be chosen as the set of piecewise polynomials of degree $m - 1$ (or splines of order m). Because this choice is independent of M, the resulting estimator also does not depend on M. Compare this with Example 9.3.2, where the same rate occurs, but where the estimator *does* depend on M. Usually, M is unknown, so that the estimator of Example 9.3.2 has little practical value.

As an alternative procedure for the case when M is unknown, one may think of choosing

$$\mathcal{G}_n^* = \left\{ g : [0,1] \to \mathbf{R}, \int \left(g^{(m)}(z)\right)^2 dz \le M_n^2 \right\},$$

with $M_n \to \infty$ as $n \to \infty$. Then the approximation error is zero from a certain n onwards. However, since

$$H(u, \mathcal{G}_n^*(\delta), Q_n) \asymp M_n^{1/m} u^{-1/m},$$

the estimation error δ_n given in Theorem 10.11 is

$$\delta_n \asymp n^{-\frac{m}{2m+1}} M_n^{\frac{1}{2m+1}}.$$

So we arrive at a suboptimal rate. The phenomenon is real: the $(n^{-\frac{m}{2m+1}} M_n^{\frac{1}{2m+1}})$-rate for this estimator cannot be improved (see van de Geer (1995b)).

Note finally that the penalized least squares estimator of Subsection 10.1.1 achieves the optimal rate $n^{-m/(2m+1)}$ and is also independent of M.

So far, we have not considered Sobolev classes on higher-dimensional Euclidean space, except in Problem 8.3. The theory for this extension is essentially the same, but in higher dimensions, the entropy can be so large that the entropy integral diverges. As a result, the least squares estimator without sieve converges with suboptimal rate. Therefore, let us do the calculations for the higher-dimensional case as well.

For an r-dimensional multi-index $\kappa = (\kappa_1, \ldots, \kappa_r)$ (i.e. all κ_j integer, $\kappa_j \ge 0$), write $|\kappa| = \sum_{j=1}^r \kappa_j$. Let D^κ denote the differential operator

$$D^\kappa g(v) = \frac{\partial^{|\kappa|}}{\partial v_1^{\kappa_1} \ldots \partial v_r^{\kappa_r}} g(v_1, \ldots, v_r).$$

Suppose

$$\mathcal{G} = \left\{ g : [0,1]^r \to \mathbf{R} : \sum_{|\kappa|=m} \int \left(D^\kappa g(z)\right)^2 dz \le M^2 \right\}. \tag{10.61}$$

Suppose the n covariates $\{z_i\}$ are regularly spaced on $[0,1]^r$. Then (see Birman and Solomjak (1967))

$$H(u, \mathcal{G}_n(\delta), Q_n) \asymp \delta^{-r/m}.$$

So if $\mathcal{G}_n^* = \mathcal{G}$ (no sieve), we obtain from Theorem 9.1 the rates

$$\|\hat{g}_n - g_0\|_n = \begin{cases} O_{\mathbf{P}}(n^{-m/(2m+r)}), & \text{if } r < 2m, \\ O_{\mathbf{P}}(n^{-1/4}\sqrt{\log n}), & \text{if } r = 2m, \\ O_{\mathbf{P}}(n^{-m/2r}), & \text{if } r > 2m. \end{cases}$$

Thus, one finds suboptimal rates whenever $r \geq 2m$. The same will be true for the corresponding penalized least squares estimator. On the other hand

$$K(\mathcal{G}, d_n, Q_n) \leq A d_n^{-m/r},$$

so by taking for $\mathcal{G}_n^*$ the best linear approximation, we find the rate $n^{-m/(2m+r)}$ for all values of r and m.

The dimension d_n of the approximating linear space may be chosen to be data-dependent. We shall study a very simple situation, just to sketch the idea. Let $\mathcal{G}_{n,0} \supset \mathcal{G}_{n,1} \supset \cdots \supset \mathcal{G}_{n,s}$ be a decreasing collection of linear spaces, with

$$d_{n,k} = \dim(\mathcal{G}_{n,k}) = n^{1/(2k+1)}, \quad k = 0, \ldots, s. \tag{10.62}$$

Write for $g \in \mathcal{G}_{n,0}$,

$$d_n(g) = \min\{d_{n,k} : g \in \mathcal{G}_{n,k}\}.$$

Suppose that for some fixed m and fixed $g_n^* \in \mathcal{G}_{n,m}$, we have

$$\|g_n^* - g_0\|_n \leq c_0 d_{n,m}^{-m} \tag{10.63}$$

(compare with (10.60)). Let L_0 be a sufficiently large constant, and define

$$\hat{g}_n = \arg \min_{g \in \mathcal{G}_{n,0}} \left(\|y - g\|_n^2 + L_0 \frac{d_n(g)}{n} \right), \tag{10.64}$$

and

$$\hat{d}_n = d_n(\hat{g}_n).$$

Lemma 10.12 *Suppose* (10.50) *and* (10.63) *are met. Let* $\hat{g}_n$ *be given by* (10.64), *with the constant* L_0 *appropriately chosen, depending on* K *and* σ_0. *Then,*

$$\|\hat{g}_n - g_0\|_n = O_{\mathbf{P}}\big(n^{-m/(2m+1)}\big), \tag{10.65}$$

and

$$\hat{d}_n = O_{\mathbf{P}}(d_{n,m}). \tag{10.66}$$

Proof From Corollary 8.3 (or by a simpler direct proof for linear spaces), we know that for some constants T and c depending on K and σ_0,

$$\mathbf{P}\left(\sup_{g\in\mathcal{G}_{n,k}}\sqrt{\frac{n}{d_{n,k}}}\frac{|(w,g-g_n^*)_n|}{\|g-g_n^*\|_n}>T\right)\le c\exp\left[-\frac{d_{n,k}}{c^2}\right].$$

So,

$$\begin{aligned}\mathbf{P}&\left(\sup_{g\in\mathcal{G}_{n,0}}\sqrt{\frac{n}{d_n(g)}}\frac{|(w,g-g_n^*)_n|}{\|g-g_n^*\|_n}>T\right)\\&\le\sum_{k=0}^{s}\mathbf{P}\left(\sup_{g\in\mathcal{G}_{n,k}}\sqrt{\frac{n}{d_{n,k}}}\frac{|(w,g-g_n^*)_n|}{\|g-g_n^*\|_n}>T\right)\\&\le\sum_{k=0}^{s}c\exp\left[-\frac{d_{n,k}}{c^2}\right]\le c(s+1)\exp\left[-\frac{n^{1/(2s+1)}}{c^2}\right].\end{aligned}$$

Let A_n be the event

$$\sqrt{\frac{n}{\hat{d}_n}}\frac{|(w,\hat{g}_n-g_n^*)_n|}{\|\hat{g}_n-g_n^*\|_n}\le T.$$

We have shown above that

$$\mathbf{P}(A_n^c)\to 0,\ \text{as } n\to\infty. \tag{10.67}$$

Take $L_0>8T^2$. Because $\|y-\hat{g}_n\|_n^2+L_0(\hat{d}_n/n)\le\|y-g_n^*\|_n^2+L_0(d_{n,m}/n)$, we find on A_n (using a Basic Inequality),

$$\begin{aligned}\|\hat{g}_n-g_n^*\|_n^2+2L_0\frac{\hat{d}_n}{n}&\le 4(w,\hat{g}_n-g_n^*)_n+2L_0\frac{d_{n,m}}{n}+4\|g_n^*-g_0\|_n^2\\&\le 4T\|\hat{g}_n-g_n^*\|_n\sqrt{\frac{\hat{d}_n}{n}}+(2L_0+4c_0^2)\frac{d_{n,m}}{n}=\mathrm{I}+\mathrm{II}.\end{aligned}$$

Now, it cannot be true that $\mathrm{I}>\mathrm{II}$, because $L_0>8T^2$. But $\mathrm{I}\le\mathrm{II}$ implies that (10.65) and (10.66) hold on A_n. By (10.67), this completes the proof. □

10.4. Maximum likelihood on sieves

Let $X_1,\ldots,X_n$ be i.i.d. random variables with distribution P and density $p_0=(dP/d\mu)\in\mathcal{P}$. For $\mathcal{P}_n^*$ an approximating space of densities w.r.t. μ, the sieved maximum likelihood estimator is

$$\hat{p}_n=\arg\max_{p\in\mathcal{P}_n^*}\int\log p\,dP_n. \tag{10.68}$$

Again, because p_0 is not necessarily a member of $\mathscr{P}_n^*$, we cannot apply Theorem 7.4. Let $p_n^* \in \mathscr{P}_n^*$ be a fixed density satisfying for some constant c_0,

$$\frac{p_0}{p_n^*} \leq c_0^2. \tag{10.69}$$

Under (10.69), we only need to adjust some definitions that were used in Theorem 7.4. Write

$$(\bar{\mathscr{P}}_n^*(\delta))^{1/2} = \left\{ \sqrt{\frac{p + p_n^*}{2}} : p \in \mathscr{P}_n^*,\ h\left(\frac{p+p_n^*}{2}, p_n^*\right) \leq \delta \right\}, \quad 0 < \delta \leq 1.$$

Let $d\mu_n^* = 1\{p_n^* > 0\}d\mu$. For an appropriate universal constant c, define for $\delta \leq 1$,

$$J_B\left(\delta, (\bar{\mathscr{P}}_n^*(\delta))^{1/2}, \mu_n^*\right) = \int_{\delta^2/c}^{\delta} H_B^{1/2}\left(u, (\bar{\mathscr{P}}_n^*(\delta))^{1/2}, \mu_n^*\right) du \vee \delta.$$

Theorem 10.13 *Let $\hat{p}_n$ be given by (10.68). Assume (10.69). Take $\Psi(\delta) \geq J_B\left(\delta, (\bar{\mathscr{P}}_n^*(\delta))^{1/2}, \mu_n^*\right)$ in such a way that $\Psi(\delta)/\delta^2$ is a non-decreasing function of δ. Then for*

$$\sqrt{n}\delta_n^2 \geq c\Psi(\delta_n) \quad \textit{for all } n \geq 1, \tag{10.70}$$

we have

$$h(\hat{p}_n, p_0) = O_{\mathbf{P}}(\delta_n + h(p_n^*, p_0)). \tag{10.71}$$

Proof It suffices to derive an appropriate Basic Inequality for the sieved case. Because $p_n^* \in \mathscr{P}_n^*$, and since by (10.69), $p_n^*/p_0 > 0$ on $\{p_0 > 0\}$, we have

$$\int_{p_n^*>0} \log\left(\frac{\hat{p}_n + p_n^*}{2p_n^*}\right) dP_n \geq 0.$$

On the other hand,

$$\begin{aligned}
&\frac{1}{2}\int_{p_n^*>0} \log\left(\frac{\hat{p}_n + p_n^*}{2p_n^*}\right) dP_n \\
&\quad = \frac{1}{2}\int_{p_n^*>0} \log\left(\frac{\hat{p}_n + p_n^*}{2p_n^*}\right) d(P_n - P) + \frac{1}{2}\int_{p_n^*>0} \log\left(\frac{\hat{p}_n + p_n^*}{2p_n^*}\right) dP, \\
&\quad \leq \frac{1}{2}\int_{p_n^*>0} \log\left(\frac{\hat{p}_n + p_n^*}{2p_n^*}\right) d(P_n - P) - \int_{p_n^*>0} \left(1 - \sqrt{\frac{\hat{p}_n + p_n^*}{2p_n^*}}\right) dP.
\end{aligned}$$

But

$$\int_{p_n^*>0} \left(1 - \sqrt{\frac{\hat{p}_n + p_n^*}{2p_n^*}}\right) dP$$

$$= \int_{p_n^*>0} \left(1 - \sqrt{\frac{\hat{p}_n + p_n^*}{2p_n^*}}\right) p_n^* \, d\mu + \int_{p_n^*>0} \left(1 - \sqrt{\frac{\hat{p}_n + p_n^*}{2p_n^*}}\right) (p_0 - p_n^*) \, d\mu$$

$$= h^2\left(\frac{\hat{p}_n + p_n^*}{2}, p_n^*\right) + \int_{p_n^*>0} \left(\sqrt{p_n^*} - \sqrt{\frac{\hat{p}_n + p_n^*}{2}}\right) (\sqrt{p_0} - \sqrt{p_n^*}) \left(1 + \sqrt{\frac{p_0}{p_n^*}}\right) d\mu$$

$$\geq h^2\left(\frac{\hat{p}_n + p_n^*}{2}, p_n^*\right) - 2(1 + c_0) h\left(\frac{\hat{p}_n + p_n^*}{2}, p_n^*\right) h(p_n^*, p_0).$$

It follows that

$$h^2\left(\frac{\hat{p}_n + p_n^*}{2}, p_n^*\right)$$

$$\leq \frac{1}{2} \int_{p_n^*>0} \log\left(\frac{\hat{p}_n + p_n^*}{2p_n^*}\right) d(P_n - P) + 2(1 + c_0) h\left(\frac{\hat{p}_n + p_n^*}{2}, p_n^*\right) h(p_n^*, p_0)$$

$$= \mathrm{I} + \mathrm{II}.$$

This is our Basic Inequality. If $\mathrm{I} \leq \mathrm{II}$, the result follows immediately. If $\mathrm{I} > \mathrm{II}$, we find

$$h^2\left(\frac{\hat{p}_n + p_n^*}{2}, p_n^*\right) \leq \int_{p_n^*>0} \log\left(\frac{\hat{p}_n + p_n^*}{2p_n^*}\right) d(P_n - P),$$

and we can apply the same arguments as in Theorem 7.4. □

If $\mathcal{P}_n^*$ is convex, one can state a result similar to Theorem 7.6. But let us take a different point of view, which is useful in case of model misspecification. Let

$$p_n^* = \arg\max_{p \in \mathcal{P}_n^*} \int \log p \, dP.$$

We are now only interested in a rate of convergence of $\hat{p}_n$ to p_n^*. Assume (10.69) and define

$$h_n^*(p_1, p_2) = \left(\frac{1}{2} \int (p_1^{1/2} - p_2^{1/2})^2 \frac{p_0}{p_n^*} \, d\mu\right)^{1/2}.$$

Since p_n^* maximizes $\int \log p \, dP$, we have for convex $\mathcal{P}_n^*$,

$$\frac{d}{d\alpha} \int \log\left(\alpha p + (1 - \alpha) p_n^*\right) dP \bigg|_{\alpha=0} \leq 0, \quad p \in \mathcal{P}_n^*.$$

Due to (10.69), differentiation and integration can be interchanged here, so that

$$\int \frac{p - p_n^*}{p_n^*} \, dP \leq 0, \; p \in \mathcal{P}_n^*. \tag{10.72}$$

Lemma 10.14. (Basic Inequality) *Let $\hat{p}_n$ be given by* (10.68) *and suppose $\mathcal{P}_n^*$ is convex and that* (10.69) *holds. Then*

$$(h_n^*(\hat{p}_n, p_0))^2 \leq \int \frac{2\hat{p}_n}{p_n + p_n^*} \, d(P_n - P).$$

Proof Since

$$\begin{aligned} 0 \leq \int \log \frac{2\hat{p}_n}{\hat{p}_n + p_n^*} \, dP_n &\leq \int \frac{2\hat{p}_n}{\hat{p}_n + p_n^*} \, dP_n - 1 \\ &= \int \frac{2\hat{p}_n}{\hat{p}_n + p_n^*} \, d(P_n - P) - \int \frac{p_n^* - \hat{p}_n}{p_n^* + \hat{p}_n} \, dP, \end{aligned}$$

it suffices to show that for all $p \in \mathcal{P}_n^*$,

$$\int \frac{p_n^* - p}{p_n^* + p} \, dP \geq \left(h_n(p, p_n^*)\right)^2.$$

Now, we showed in (10.72) that $\int (p/p_n^*) \, dP \leq 1$ for all $p \in \mathcal{P}_n^*$. Therefore

$$\begin{aligned} \int \frac{p_n^* - p}{p_n^* + p} \, dP = 1 - \int \frac{2p}{p_n^* + p} \, dP &\geq \int \frac{p}{p_n^*} \, dP - \int \frac{2p}{p_n^* + p} \, dP \\ &= \int \frac{(p - p_n^*)^2}{p_n^*(p_n^* + p)} \, dP - \int \frac{p_n^* - p}{p_n^* + p} \, dP. \end{aligned}$$

It follows that

$$\int \frac{p_n^* - p}{p_n^* + p} \, dP \geq \frac{1}{2} \int \frac{(p - p_n^*)^2}{p_n^*(p_n^* + p)} \, dP \geq \left(h_n^*(p, p_n^*)\right)^2. \qquad \square$$

Similar arguments as in Theorem 7.6 now supply us with a rate of convergence for $h_n^*(\hat{p}_n, p_n^*)$, even when p_n^* is not converging to p_0.

We end this section with an illustration of Theorem 10.13.

Example 10.4.1. Estimating the support of a density Suppose that we know that our observations are uniformly distributed on some set $D \in \mathcal{D}$. Let μ be Lebesgue measure on $[0,1]^2$. Our model is

$$p_0 \in \mathcal{P} = \{p_D = 1_D/\mu(D) : D \in \mathcal{D}\}.$$

Note that

$$\|1_{D_1} - 1_{D_2}\|_\mu = \mu^{1/2}(D_1 \Delta D_2),$$

where $D_1 \Delta D_2 = (D_1 \cap D_2^c) \cup (D_1^c \cap D_2)$ is the symmetric difference between the two sets D_1 and D_2.

Case (i) Consider the collection of sets

$$\mathscr{D} = \{ D_g = \{(u, v) : 0 \le v \le g(u)\} : g \in \mathscr{G} \}, \tag{10.73}$$

where

$$\mathscr{G} = \left\{ g : [0,1] \to \mathbf{R}, \int_0^1 (g^{(m)}(u))^2 \, du \le 1 \right\}. \tag{10.74}$$

Thus, we assume that D_0 has a smooth boundary. Using Theorem 2.4, one easily verifies that

$$H_B(\delta, \{1_D : D \in \mathscr{D}\}, \mu) \le A\delta^{-2/m}, \quad \text{for all } \delta > 0, \tag{10.75}$$

which implies that for all $\delta > 0$ sufficiently small

$$H_B\big(u, (\bar{\mathscr{P}}_n^*(\delta))^{1/2}, \mu_n^*\big) \le A' u^{-2/m}, \quad 0 < u \le \delta. \tag{10.76}$$

If $m > 1$, a sieve is not necessary, i.e., one may choose $\mathscr{P}_n^* = \mathscr{P}$. Let $\hat{p}_n = 1_{\hat{D}_n} / \mu(\hat{D}_n)$ be defined by (10.68), with $\mathscr{P}_n^* = \mathscr{P}$. By Theorem 10.13, one finds for the maximum likelihood estimator without sieve

$$\mu(\hat{D}_n \Delta D_0) = O_{\mathbf{P}}(n^{-m/(1+m)}), \quad \text{if } m > 1.$$

In the remainder of Case (i), we study the problem for $m = 1$. We then obtain a suboptimal rate if we do not use a sieve, due to the fact that the entropy integral is not finite. Therefore, we propose a finite approximation of $\mathscr{P}$. Let $\{D_{n,1}, \ldots, D_{n,N_n}\}$ be a collection of sets such that for all $D \in \mathscr{D}$ there is a $j \in \{1, \ldots, N_n\}$ such that

$$D \subset D_{n,j}, \ \mu(D \Delta D_{n,j}) \le n^{-\frac{1}{2}}.$$

Choose

$$\mathscr{P}_n^* = \{ 1_{D_{n,j}} / \mu(D_{n,j}) : j = 1, \ldots, N_n \}. \tag{10.77}$$

In view of (10.75), we may take $\log N_n \le A\sqrt{n}$. Let $D_n^* \in \{D_{n,1}, \ldots, D_{n,N_n}\}$ be the upper approximation of D_0, i.e.,

$$D_0 \subset D_n^*, \ \mu(D_0 \Delta D_n^*) \le n^{-\frac{1}{2}}.$$

Moreover, define $p_n^* = 1_{D_n^*}/\mu(D_n^*)$. Then (10.69) is fulfilled. Let $\hat{p}_n = 1_{\hat{D}_n}/\mu(\hat{D}_n)$ be defined by (10.68), with $\mathscr{P}_n^*$ given in (10.77). We find from Theorem 10.13 that for the sieved maximum likelihood estimator $\hat{D}_n$

$$\mu(\hat{D}_n \Delta D_0) = O_{\mathbf{P}}(n^{-\frac{1}{2}}), \quad \text{if } m = 1.$$

Case (ii) For $r \geq 2$, let

$$\mathscr{D} = \{\text{all convex subsets of } [0,1]^r\}.$$

It is shown in Dudley (1984) that

$$H_B(\delta, \{1_D : D \in \mathscr{D}\}, \mu) \leq A\delta^{-(r-1)/2}, \quad \delta \downarrow 0.$$

Hence, there exists sets $\{D_{n,1}, \ldots, D_{n,N_n}\}$, with $\log N_n \leq An^{(r-1)/(r+3)}$, such that for all $D \in \mathscr{D}$, there is a $j \in \{1, \ldots, N_n\}$, such that $D \subset D_{n,j}$ and $\mu(D_{n,j}\Delta D) \leq n^{-4/(r+3)}$. Let $\mathscr{P}_n^* = \{1_{D_{n,j}}/\mu(D_{n,j}) : j = 1, \ldots, N_n\}$. By Theorem 10.13, the rate of convergence for the sieved maximum likelihood estimator $\hat{D}_n$ is

$$\mu(\hat{D}_n \Delta D_0) = O_{\mathbf{P}}(n^{-4/(r+3)}).$$

The maximum likelihood estimator without sieve converges with a suboptimal rate when $r \geq 5$.

Remark In the same way as in the above example, one can use Theorem 10.13 to repair the suboptimal rates we found in Example 7.4.6. Choose the normalized upper brackets of the densities in $\mathscr{P}$ as approximating space $\mathscr{P}_n^*$.

10.5. Notes

There are quite a few papers on spline smoothing. See for example Silverman (1985) for a discussion. The method of proof employed here is from van de Geer (1990). In Mammen and van de Geer (1997a), the penalized least squares estimator of functions of bounded variation is examined. Papers on penalized maximum likelihood estimation are, for example, Silverman (1982) and Barron and Sheu (1991). Wong and Shen (1995) and van de Geer (1995b) consider sieves. A very general approach, and adaptive estimation, is given in Barron, Birgé and Massart (1995). The fact that linear estimators are suboptimal in certain cases was pointed out by Donoho (1990). Korostelev and Tsybakov (1993b) study the estimation of the support of a density.

10.6. Problems and complements

10.1. (Smooth functions with unbounded support) Let $\mathscr{Z} = [0, \infty)$, and

$$Y_i = g_0(z_i) + W_i, \quad i = 1, \ldots, n,$$

where $W_1, \ldots, W_n, \ldots$ are independent, centered, uniformly sub-Gaussian random variables, and where

$$I^2(g_0) = \int_0^\infty \left(g^{(m)}(z)\right)^2 dz \le M_0^2 < \infty.$$

Consider the penalized least squares estimator

$$\hat{g}_n = \arg\min\left(\|y - g\|_n^2 + \lambda_n^2 I^2(g)\right).$$

Then $\|\hat{g}_n - g_0\|_n = O_{\mathbf{P}}(\lambda_n)$ and $I(\hat{g}_n) = O_{\mathbf{P}}(1)$, under the following conditions:

(a) $\lambda_n^{-1} = O_{\mathbf{P}}\left(n^{\frac{m}{2m+1}}\right)$,

(b) the eigenvalues of $\Sigma_n = \int \psi\psi^T \, dQ_n$ stay away from zero, where $\psi_k(z) = z^{k-1}$, $k = 1, \ldots, m$,

(c) Q_n satisfies the tail-condition of Problem 2.6, uniformly in n, i.e., there is a constant $\tilde{C}$ such that for all $T < \infty$ there are $0 = b_0 < \cdots < b_N = T$, such that $b_j - b_{j-1} \ge 1$, $j = 1, \ldots, N$, and such that

$$\sum_{j=1}^N (b_j - b_{j-1})^{(2m-1)/(2m+1)} Q_n^{1/(2m+1)}(b_{j-1}, b_j] \le \tilde{C}.$$

10.2. Let $\mathscr{Z} = [0,1]^2$ and

$$Y_i = 1_{D_0}(z_i) + W_i, \quad i = 1, \ldots, n,$$

where $W_1, \ldots, W_n, \ldots$ are independent, centered, uniformly sub-Gaussian random variables, and where

$$D_0 = D_{f_0} \in \mathscr{D} = \{D_f = \{(u,v) : 0 \le v \le f(u)\}\},$$
$$I^2(f_0) = \int_0^1 (f_0^{(m)}(u))^2 \, du \le M_0^2 < \infty.$$

Consider the penalized least squares estimator

$$\hat{D}_n = D_{\hat{f}_n} = \arg\min_{D_f \in \mathscr{D}} \left(\|y - 1_{D_f}\|_n^2 + \lambda_n^2 I^2(f)\right).$$

Verify that

$$H(\delta, \{1_{D_f} : D_f \in \mathscr{D}, \ I(f) \le M\}, Q_n) \le A\left(\frac{M}{\delta}\right)^{-2/m}, \quad \delta > 0, \ M \ge 1.$$

This implies that for $m > 1$ and $\lambda_n^{-1} = O_{\mathbf{P}}(n^{m/2(m+1)})$, one has $Q_n(\hat{D}_n \Delta D_0) = O_{\mathbf{P}}(\lambda_n^2)$ and $I(\hat{f}_n) = O_{\mathbf{P}}(1)$.

10.3. (Local rates) Let $\mathscr{Z} = [0,1]$, $z_i = i/n$, $i = 1, \ldots, n$, and

$$Y_i = g_0(z_i) + W_i, \quad i = 1, \ldots, n,$$

where $W_1, \ldots, W_n, \ldots$ are independent, centered, uniformly sub-Gaussian random variables, and where

$$I^2(g_0) = \int_0^1 \left(g_0^{(m)}(z)\right)^2 dz \leq M_0^2 < \infty.$$

Consider the penalized least squares estimator

$$\hat{g}_n = \arg\min \left(\|y - g\|_n^2 + \lambda_n^2 I^2(g)\right),$$

where we take $\lambda_n = n^{-m/(2m+1)}$. Let $z_0 \in [n^{-1/(2m+1)}, 1 - n^{-1/(2m+1)}]$ be fixed, and define

$$\tilde{g}_n = \arg\{\min \left(\|g_0 - g\|_n^2 + \lambda_n^2 I^2(g)\right), \text{ s.t. } g(z) = \hat{g}_n(z), \text{ for } |z - z_0| \geq n^{-1/(2m+1)}\}.$$

Show that

$$\|\hat{g}_n - \tilde{g}_n\|_n^2 + \lambda_n^2 I^2(\hat{g}_n - \tilde{g}_n) = (w, \hat{g}_n - \tilde{g}_n)_n.$$

Thus, we have localized the problem near z_0. Deduce that

$$\|\hat{g}_n - \tilde{g}_n\|_n = O_{\mathbf{P}}\left(n^{-1/2}\right),$$

and

$$I^2(\hat{g}_n - \tilde{g}_n) = O_{\mathbf{P}}\left(n^{-1/(2m+1)}\right).$$

10.4. Let

$$\mathscr{G} = \{g : \mathbf{R} \to [0,1], \ g \text{ increasing}\}.$$

Show that for $n = md$, $m, d \in \{1, 2, \ldots\}$, we have for Kolmogorov's linear d-width

$$K(\mathscr{G}, d, Q_n) \leq d^{-1/2}.$$

(Hint: Assume without loss of generality that $z_1, \ldots, z_n$ are distinct. Take $b_0 < \cdots < b_d$ in such a way that $Q_n(b_{j-1}, b_j] = 1/d$, $j = 1, \ldots, d$, and define for $g \in \mathscr{G}$,

$$\tilde{g}(z) = d \int_{b_{j-1}}^{b_j} g \, dQ_n, \ z \in (b_{j-1}, b_j].$$

Then $\|g - \tilde{g}\|_n^2 \leq 1/d$.)

10.5. (Exponential families as sieves) Let $X_1, \ldots, X_n$ be i.i.d. random variables with distribution P on $[0,1]$, and

$$\mathscr{P}_n^* = \{p : [0,1] \to \mathbf{R},\ p(x) = \gamma(x) - b(\gamma),\ \gamma \in \Gamma_n\},$$

where $b(\gamma) = \log \int e^{\gamma(x)}\, dx$, and Γ_n is the d_n-dimensional linear space

$$\Gamma_n = \left\{ \gamma(x) = \theta^T \psi_n(x) = \sum_{j=1}^{d_n} \theta_j \psi_{j,n}(x) \ :\ \theta \in \mathbf{R}^{d_n} \right\}.$$

We normalize the functions $\psi_{j,n}$ to have mean zero:

$$\int \psi_{j,n}\, dP = 0, \quad j = 1, \ldots, d_n.$$

Let

$$p_n^* = \exp[\gamma_n^* - b(\gamma_n^*)] = \arg\max_{p \in \mathscr{P}_n^*} \int \log p\, dP,$$

and

$$\hat{p}_n = \exp[\hat{\gamma}_n - b(\hat{\gamma}_n)] = \arg\max_{p \in \mathscr{P}_n^*} \int \log p\, dP_n.$$

We assume that for all n sufficiently large

$$|\gamma_n^*|_\infty \leq K,$$

that P has a density p_0 with respect to Lebesgue measure, where

$$p_0(x) \geq \eta_0^2 > 0, \quad x \in [0,1],$$

and that

$$|\gamma|_\infty \leq A_n \|\gamma\|, \quad \gamma \in \Gamma_n,$$

where $A_n^2 d_n / n \to 0$ (compare with Problem 5.5). Then $\|\hat{\gamma}_n - \gamma_n^*\| = O_{\mathbf{P}}\big(\sqrt{d_n/n}\big)$. To see this, first note that by the convexity of $b(\gamma)$, we have for all $0 \leq \alpha \leq 1$,

$$b\big(\alpha \hat{\gamma}_n + (1-\alpha)\gamma_n^*\big) - b(\gamma_n^*) \leq \alpha \int (\hat{\gamma}_n - \gamma_n^*)\, dP_n.$$

Now, take

$$\alpha = \hat{\alpha}_n = \frac{1}{1 + B_n \|\hat{\gamma}_n - \gamma_n^*\|},$$

where $A_n / B_n \to 0$ and $B_n^2 d_n / n \to 0$. Because

$$\frac{|\hat{\gamma}_n - \gamma_n^*|_\infty}{1 + B_n \|\hat{\gamma}_n - \gamma_n^*\|} = o(1),$$

we have for some constant c_0,

$$b\left(\gamma_n^* + \frac{\hat{\gamma}_n - \gamma_n^*}{1 + B_n\|\hat{\gamma}_n - \gamma_n^*\|}\right) - b(\gamma_n^*) \geq \frac{\|\hat{\gamma}_n - \gamma_n^*\|^2}{c_0^2\left(1 + B_n\|\hat{\gamma}_n - \gamma_n^*\|\right)^2}.$$

Here, we used the fact that for $\gamma_n^* = (\theta_n^*)^T\psi_n$,

$$\frac{\partial}{\partial\theta} b(\theta^T\psi_n)|_{\theta=\theta_n^*} = 0,$$

since γ_n^* minimizes $b(\gamma)$. We now find

$$\frac{\|\hat{\gamma}_n - \gamma_n^*\|}{1 + B_n\|\hat{\gamma}_n - \gamma_n^*\|} = O_{\mathbf{P}}\left(\sqrt{\frac{d_n}{n}}\right).$$

The assumption $B_n^2 d_n/n \to 0$ ensures that this implies

$$\|\hat{\gamma}_n - \gamma_n^*\| = O_{\mathbf{P}}\left(\sqrt{\frac{d_n}{n}}\right).$$

See, for example, Stone (1990) and Stone, Hansen, Kooperberg and Truong (1995) for related results.

11

Some Applications to Semiparametric Models

This chapter consists of three parts, with a common structure. First, a global rate of convergence is proved, using the results of the previous chapters. This global rate is then applied to establish asymptotic normality of certain aspects of the estimator. The latter involves the computation of a worst possible one-dimensional subdirection. We shall see that in each case, a global rate of order δ_n is required, with $\delta_n = o(n^{-1/4})$. Recall that this rate occurred earlier, whenever the entropy integral converges.

In Section 11.1, partial linear models are investigated, and the asymptotic normality of a penalized estimator of the parametric component. Section 11.2 provides some methods for proving asymptotic normality of certain linear functions of the maximum likelihood estimator of a mixing distribution. Finally, Section 11.3 presents a one-step estimator of the parametric component in a single-indexed model.

11.1. Partial linear models

Let $(Y_1, Z_1), \ldots, (Y_n, Z_n) \ldots$ be independent copies of (Y, Z), where Y is a real-valued response variable, and $Z \in [0,1]^d$ is a covariate. The conditional expectation of Y given $Z = z$ is denoted by

$$\mu_0(z) = E(Y \mid Z = z).$$

In a partial linear model, the covariate consists of two variables, say $Z = (U, V)$, with $U \in [0,1]^{d_1}$, $V \in [0,1]^{d_2}$, $d_1 + d_2 = d$, and the conditional expectation is modelled as

$$\mu_0(u,v) = F\big(\theta_0^T u + \gamma_0(v)\big),$$

where $F : \mathbf{R} \to \mathbf{R}$ is a given link function, $\theta_0 \in \mathbf{R}^{d_1}$ is an unknown parameter, and γ_0 is an unknown function in a given class of smooth functions. Thus, the model has two parts: a parametric part containing the finite-dimensional parameter θ_0, and a non-parametric part containing the infinite-dimensional parameter γ_0. To ease the notation, we take $d_1 = d_2 = 1$.

Notice that in principle, the range of $\theta_0^T u + \gamma_0(v)$ is not restricted, whereas the range of the response variable Y often is. For instance, think of the situation where Y is a binary variable (for example, yes/no answers to a questionnaire). Then $\mu_0(z)$ can only take values between 0 and 1. The link function F allows one to take this into account.

We shall assume that γ_0 is smooth, in the sense that $I(\gamma_0) < \infty$, where

$$I^2(\gamma) = \int_0^1 \left|\gamma^{(m)}(v)\right|^2 dv,$$

denotes the roughness of the function γ. Using the quasi-likelihood method, with a penalty on the roughness, one obtains estimators of θ_0 and γ_0. We shall first derive a global rate of convergence for these estimators, and then prove asymptotic normality of the estimator of the parametric component θ_0. In both stages, results from empirical process theory are inserted: the modulus of continuity of the empirical process in the first stage, and asymptotic equicontinuity in the second stage.

The general definition of the penalized quasi-likelihood estimator is given in Subsection 11.1.3. We shall study two special cases: partial splines in Subsection 11.1.1, and the partially linear binary choice model in Subsection 11.1.2.

Throughout, we use the following notation. Take $\mathcal{G}$ to be the class of all regression functions g of the form $g(u,v) = \theta u + \gamma(v)$. The distribution of Z is denoted by Q, and Q_n is the empirical distribution of $Z_1, \ldots, Z_n$. Write

$$\|g\|^2 = \int g^2 \, dQ,$$

and

$$\|g\|_n^2 = \int g^2 \, dQ_n.$$

For a function γ depending on v only, we use the same notation, i.e.,

$$\|\gamma\|^2 = \int \gamma(v) \, dQ(u,v),$$

and

$$\|\gamma\|_n^2 = \int \gamma(v)\, dQ_n(u,v).$$

We shall moreover write

$$(11.1) \qquad I^2(g) = I^2(\gamma) = \int_0^1 \left(\gamma^{(m)}(v)\right)^2 dv, \quad g(u,v) = \theta u + \gamma(v).$$

Define

$$\psi_0(u,v) = u,$$
$$\psi_k(u,v) = v^{k-1}, \quad k = 1, \ldots, m,$$
$$\psi(z) = \left(\psi_0(z), \ldots, \psi_m(z)\right)^T,$$

and let

$$\Sigma = \int \psi\psi^T\, dQ.$$

11.1.1. Partial splines

The model in this subsection is

$$Y = g_0(Z) + W,$$

with $E(W \mid Z) = 0$, and with

$$g_0(Z) = \theta_0 U + \gamma_0(V), \quad Z = (U, V).$$

As in Subsection 10.1.1, we define the penalized least squares estimator as

$$(11.2) \qquad \hat{g}_n = \arg\min_{g \in \mathcal{G}} \left(\frac{1}{n} \sum_{i=1}^n (Y_i - g(Z_i))^2 + \lambda_n^2 I^2(g) \right),$$

where $\mathcal{G}$ is now the class of all regression functions g of the form $g(u,v) = \theta u + \gamma(v)$, and where $I^2(g) = I^2(\gamma) = \int_0^1 (\gamma^{(m)}(v))^2\, dv$, $g(u,v) = \theta u + \gamma(v)$.

Now, assume that for all $z \in [0,1]^2$,

$$(11.3) \qquad 2K^2 E\left(e^{|W|/K} - 1 - (|W|/K) \mid Z = z\right) \le \sigma_0^2.$$

Moreover, suppose

$$(11.4) \qquad I(g_0) < \infty.$$

Lemma 11.1 *Suppose that* (11.3) *and* (11.4) *hold and that* Σ *is non-singular. Then for* $\lambda_n^{-1} = O_{\mathbf{P}}\left(n^{m/(2m+1)}\right)$, *we have*

$$(11.5) \qquad \|\hat{g}_n - g_0\| = O_{\mathbf{P}}(\lambda_n),$$

$$\|\hat{g}_n - g_0\|_n = O_{\mathbf{P}}(\lambda_n), \tag{11.6}$$

and

$$I(\hat{g}_n) = O_{\mathbf{P}}(1). \tag{11.7}$$

Proof This follows from the same arguments as the ones used to obtain Corollary 10.4. The weaker assumption (11.3), instead of the assumption of sub-Gaussian errors, suffices because one can use entropy with bracketing (see also Chapter 9, where this issue is treated in detail). In Corollary 10.4, Lemma 8.4 is used. Replace this by Lemma 5.13. Due to Lemma 5.16, the theoretical norm $\|\cdot\|$ and the empirical norm $\|\cdot\|_n$ are of the same order on the relevant subsets of $\mathcal{G}$. □

Now, $\hat{g}_n(u,v) = \hat{\theta}_n u + \hat{\gamma}_n(v)$, and under identifiability conditions, $\|\hat{g}_n - g_0\| = o_{\mathbf{P}}(1)$ implies $|\hat{\theta}_n - \theta_0| = o_{\mathbf{P}}(1)$ and $\|\hat{\gamma}_n - \gamma_0\| = o_{\mathbf{P}}(1)$. We shall show that the rate of convergence for $\hat{\theta}_n$ is in fact $O_{\mathbf{P}}(n^{-1/2})$, i.e. the rate for the parametric part is much faster than the global rate. Moreover, the estimator $\hat{\theta}_n$ is asymptotically normal. To arrive at this result, we suppose that the regression of U on V is sufficiently smooth. Write this regression as

$$h(v) = E(U \mid V = v),$$

and suppose that

$$I^2(h) = \int_0^1 \left(h^{(m)}(v)\right)^2 dv < \infty. \tag{11.8}$$

Let

$$\tilde{h}(u,v) = u - h(v).$$

Furthermore, we let $W_i = Y_i - E(Y_i \mid Z_i)$, $i = 1, \ldots, n$.

Lemma 11.2 *Assume that the conditions of Lemma* 11.1 *are met. Moreover, suppose that* (11.8) *holds, that* $\lambda_n = o_{\mathbf{P}}(n^{-1/4})$ *and that* $\|\tilde{h}\| > 0$. *Then*

$$\sqrt{n}(\hat{\theta}_n - \theta_0) = \frac{\frac{1}{\sqrt{n}}\sum_{i=1}^n W_i \tilde{h}(Z_i)}{\|\tilde{h}\|^2} + o_{\mathbf{P}}(1). \tag{11.9}$$

Proof Since $\|\hat{g}_n - g_0\| = o_{\mathbf{P}}(1)$ and $\|\tilde{h}\| > 0$, it follows that $|\hat{\theta}_n - \theta_0| = o_{\mathbf{P}}(1)$ and $\|\hat{\gamma}_n - \gamma_0\| = o_{\mathbf{P}}(1)$. Now, define for $t \in \mathbf{R}$,

$$\hat{g}_{n,t}(u,v) = \hat{g}_n(u,v) + t\tilde{h}(u,v) = (\hat{\theta}_n + t)u + \left(\hat{\gamma}_n(v) - th(v)\right).$$

Because $\hat{g}_{n,t} \in \mathcal{G}$ for all t, we find that

$$\frac{d}{dt}\left(\frac{1}{n}\sum_{i=1}^{n}(Y_i - \hat{g}_{n,t}(Z_i))^2 + \lambda_n^2 I^2(\hat{g}_{n,t})\right)\Bigg|_{t=0} = 0.$$

But

$$\begin{aligned}\frac{1}{2}\frac{d}{dt}&\left(\frac{1}{n}\sum_{i=1}^{n}(Y_i - \hat{g}_{n,t}(Z_i))^2 + \lambda_n^2 I^2(\hat{g}_{n,t})\right)\Bigg|_{t=0} \\ &= -(w,\tilde{h})_n + (\hat{g}_n - g_0, \tilde{h})_n + \lambda_n^2 I(\hat{\gamma}_n, h) \\ &= -\mathrm{I} + \mathrm{II} + \mathrm{III},\end{aligned}$$

where

$$(w,\tilde{h})_n = \frac{1}{n}\sum_{i=1}^{n} W_i\tilde{h}(Z_i), \qquad (g,\tilde{h})_n = \int g\tilde{h}\,dQ_n,$$

and

$$I(\gamma, h) = \int_0^1 \gamma^{(m)}(v)h^{(m)}(v)\,dv.$$

Because $\lambda_n = o_{\mathbf{P}}(n^{-1/4})$, $I(\hat{\gamma}_n) = O_{\mathbf{P}}(1)$, and $I(h) < \infty$, we find

$$\mathrm{III} = \lambda_n^2 I(\hat{\gamma}_n, h) \le \lambda_n^2 I(\hat{\gamma}_n) I(h_1) = o_{\mathbf{P}}(n^{-1/2}).$$

Moreover,

$$\begin{aligned}\mathrm{II} = (\hat{g}_n - g_0, \tilde{h})_n &= (\hat{\theta}_n - \theta_0)\|\tilde{h}\|_n^2 + (\hat{\theta}_n - \theta_0)(h,\tilde{h})_n + (\hat{\gamma}_n - \gamma_0, \tilde{h})_n \\ &= \mathrm{i} + \mathrm{ii} + \mathrm{iii}.\end{aligned}$$

Clearly, by the law of large numbers,

$$\mathrm{i} = (\hat{\theta}_n + \theta_0)(\|\tilde{h}\|_2 + o(1)).$$

Because $E(h(V)\tilde{h}(Z)) = 0$, the law of large numbers also gives

$$\mathrm{ii} = (\hat{\theta}_n - \theta_0)o(1).$$

Moreover, $E\big(\gamma(V)\tilde{h}(Z)\big) = 0$ also. Because $I(\hat{\gamma}_n) = O_{\mathbf{P}}(1)$, asymptotic equicontinuity (see Section 5.5) yields

$$(\hat{\gamma}_n - \gamma_0, \tilde{h})_n = o_{\mathbf{P}}\big(n^{-\frac{1}{2}}\big),$$

since $\|\hat{\gamma}_n - \gamma_0\| = o_{\mathbf{P}}(1)$. Thus, we may conclude that

$$\mathrm{II} = (\hat{\theta}_n - \theta_0)\big(\|\tilde{h}\|^2 + o(1)\big) + o_{\mathbf{P}}\big(n^{-\frac{1}{2}}\big).$$

Combining the results gives

$$0 = -\text{I} + \text{II} + \text{III} = -(w, \tilde{h})_n + (\hat{\theta}_n - \theta_0)(\|\tilde{h}\|^2 + o(1)) + o_{\mathbf{P}}(n^{-\frac{1}{2}}).$$

Rewriting this yields (11.9). □

It follows that $\hat{\theta}_n$ is asymptotically normal:

$$\sqrt{n}(\hat{\theta}_n - \theta_0) \rightarrow^{\mathscr{L}} \mathcal{N}\left(0, \frac{\|\sigma\tilde{h}\|^2}{\|\tilde{h}\|^4}\right),$$

where $\sigma^2(z) = E(W^2 \mid Z = z)$. If Z and $W = Y - E(Y \mid Z)$ are independent, and W is normally distributed, then $\hat{\theta}_n$ is an asymptotically efficient estimator of θ_0. We shall not prove this here: it involves showing regularity of the estimator.

11.1.2. Partially linear binary choice model

Suppose that $Y \in \{0, 1\}$, with

$$P(Y = 1 \mid Z = z) = 1 - P(Y = 0 \mid Z = z) = F(\theta_0 u + \gamma_0(v)), \; z = (u, v).$$

The link function $F : \mathbf{R} \rightarrow (0, 1)$ is assumed to be known (the case where F is unknown is treated in Section 11.3). We require it to be differentiable, with derivative $f(\xi) = dF(\xi)/d\xi$, $\xi \in \mathbf{R}$. Moreover, we suppose

$$|f|_\infty = \sup_{\xi \in \mathbf{R}} |f(\xi)| < \infty. \tag{11.10}$$

To estimate g_0, we use a penalized maximum likelihood estimator $\hat{g}_n(u, v) = \hat{\theta}_n u + \hat{\gamma}_n(v)$, defined as

$$\hat{g}_n = \arg \max_{g \in \mathscr{G}} \left(\frac{1}{n} \sum_{i=1}^n (Y_i \log F(g(Z_i)) + (1 - Y_i) \log(1 - F(g(Z_i))) - \lambda_n^2 I^2(g) \right).$$

First, we shall establish a rate of convergence for $\hat{g}_n$. We use the same arguments as in Section 10.2. One of the ingredients is the calculation of entropy.

Lemma 11.3 *Under condition* (11.10), *we have*

$$H\left(\delta, \left\{\frac{F(g)}{1 + I(g)} : g \in \mathscr{G}\right\}, Q_n\right) \leq A\delta^{-1/m}, \quad \textit{for all } \delta > 0, \; n \geq 1. \tag{11.11}$$

Proof As in Subsection 9.3.2, write for $g \in \mathscr{G}$,

$$g = \vartheta^T \psi + g_2,$$

with $\vartheta \in \mathbf{R}^{m+1}$ and $|g_2|_\infty \le I(g_2) = I(g)$. Now, let $\tilde{g}$ be a fixed function, and consider the class

$$\{F(\vartheta^T\psi + \tilde{g}) : \vartheta \in \mathbf{R}^{m+1}\}. \tag{11.12}$$

The collection $\{\vartheta^T\psi + \tilde{g} : \vartheta \in \mathbf{R}^{m+1}\}$ is a Vapnik–Chervonenkis subgraph class (see Example 3.7.4d). Since F is of bounded variation, the collection in (11.12) is therefore also a Vapnik–Chervonenkis subgraph class (see the remark following Definition 3.3). By Theorem 3.11, we thus have for all δ sufficiently small

$$H\big(\delta, \{F(\vartheta^T\psi+\tilde{g}) : \vartheta \in \mathbf{R}^{m+1}\}, Q_n\big) \le A_1 \log\left(\frac{1}{\delta}\right). \tag{11.13}$$

Here, we use the fact that F is bounded by 1, so that the class has envelope 1.

Define for $g = \vartheta^T\psi + g_2$,

$$v(g) = \left\lfloor \frac{1}{(1+I(g))\delta} \right\rfloor \delta,$$

where $\lfloor x \rfloor$ denote the integer part of $x \ge 0$. Then

$$\{v(g)g_2\} \subset \{\gamma : |\gamma|_\infty \le I(\gamma) \le 1\},$$

so by Theorem 2.4,

$$H_\infty\big(u, \{v(g)g_2\}\big) \le A_2 u^{-1/m}, \quad \text{for all } u > 0. \tag{11.14}$$

Of course, if we replace here the supremum norm by the $L_2(Q_n)$-norm, the result remains true and holds uniformly in $n \ge 1$.

Together, (11.13) and (11.14) give the required result. To see this, let $g \in \mathcal{G}$, $g = \vartheta^T\psi + g_2$, and let $v_j = v(g)$. Suppose that γ_j is such that

$$\|v_j g_2 - \gamma_j\|_n \le \delta,$$

and that ϑ_j is such that

$$\left\| F\left(\vartheta^T\psi + \frac{\gamma_j}{v_j}\right) - F\left(\vartheta_j^T\psi + \frac{\gamma_j}{v_j}\right) \right\|_n \le \delta.$$

Then

$$\begin{aligned}
&\left\| \frac{F(\vartheta^T\psi + g_2)}{1+I(g)} - F\left(\vartheta_j^T\psi + \frac{\gamma_j}{v_j}\right) v_j \right\|_n \\
&\le v_j \left\| F(\vartheta^T\psi + g_2) - F\left(\vartheta^T\psi + \frac{\gamma_j}{v_j}\right) \right\|_n + \left| \frac{1}{1+I(g)} - v_j \right| \\
&\quad + \left\| F\left(\vartheta^T\psi + \frac{\gamma_j}{v_j}\right) - F\left(\vartheta_j^T\psi + \frac{\gamma_j}{v_j}\right) \right\|_n \\
&\le |f|_\infty \delta + \delta + \delta.
\end{aligned}$$

□

Now, we can proceed as in Theorem 10.6. As there, we need the assumption that the density stays away from zero. In this context, this assumption reads: for some $\eta_0 > 0$,

$$\eta_0^2 \leq F(g_0(z)) \leq 1-\eta_0^2, \quad \text{for all } z \in [0,1]^2. \tag{11.15}$$

Now, recall that if the errors are sub-Gaussian, one may use the empirical entropy without bracketing, instead of entropy with bracketing for the theoretical norm. Indeed, in this case the error $W = Y - E(Y \mid Z)$ is bounded by 1, so is certainly sub-Gaussian.

Lemma 11.4 *Suppose* (11.10) *and* (11.15) *hold. Then for* $\lambda_n^{-1} = O_{\mathbf{P}}\left(n^{m/(2m+1)}\right)$, *we have*

$$\|F(\hat{g}_n) - F(g_0)\|_n = O_{\mathbf{P}}(\lambda_n), \tag{11.16}$$

and

$$I(\hat{g}_n) = O_{\mathbf{P}}(1). \tag{11.17}$$

Proof Write

$$p_g(y \mid z) = F\big(g(z)\big)^y \big(1 - F(g(z))\big)^{1-y},$$

and

$$\bar{p}_g(y \mid z) = \frac{1}{2}\big(p_g(y \mid z) + p_{g_0}(y \mid z)\big) = \bar{F}\big(g(z)\big)^y \big(1 - \bar{F}(g(z))\big)^{1-y},$$

with

$$\bar{F}(g) = \frac{1}{2}\big(F(g) + F(g_0)\big).$$

Let

$$h_n^2(\bar{p}_g, \bar{p}_{g_0}) = \frac{1}{2}\left\|\sqrt{\bar{F}(\hat{g}_n)} - \sqrt{F(g_0)}\right\|_n^2 + \frac{1}{2}\left\|\sqrt{1-\bar{F}(\hat{g}_n)} - \sqrt{1-F(g_0)}\right\|_n^2.$$

Note that $h_n(\bar{p}_g, p_{g_0})$ is the (normalized) Hellinger distance between $\bar{p}_g$ and p_{g_0}, given the covariates $Z_1, \ldots, Z_n$. A conditional variant of the Basic Inequality in Lemma 10.5 (using moreover that $\log x \leq 2(\sqrt{x}-1)$ for $x > 0$) reads

$$\begin{aligned} &h_n^2(\bar{p}_{\hat{g}_n}, p_{g_0}) + \lambda_n^2 I^2(\hat{g}_n) \\ &\leq \frac{8}{n}\sum_{i=1}^n W_i \left(\frac{\sqrt{\bar{F}(\hat{g}_n(Z_i))} - \sqrt{F(g_0(Z_i))}}{\sqrt{F(g_0(Z_i))}}\right) \\ &\quad + \frac{8}{n}\sum_{i=1}^n W_i \left(\frac{\sqrt{1-\bar{F}(\hat{g}_n(Z_i))} - \sqrt{1-F(g_0(Z_i))}}{\sqrt{1-F(g_0(Z_i))}}\right) + \lambda_n^2 I^2(g_0). \end{aligned}$$

Because $F \le 1$, and $F(g_0) \ge \eta_0^2$, we find from Lemma 11.3 that

$$H\left(\delta, \left\{\frac{\sqrt{F(g)} - \sqrt{F(g_0)}}{\sqrt{F(g_0)}(1+I(g))} : g \in \mathcal{G}\right\}, Q_n\right) \le A'\delta^{-1/m}, \quad \text{for all } \delta > 0.$$

The same is true for the class

$$\left\{\frac{\sqrt{1-F(g)} - \sqrt{1-F(g_0)}}{\sqrt{1-F(g_0)}(1+I(g))} : g \in \mathcal{G}\right\}.$$

Thus, from Lemma 8.4,

$$\begin{aligned} &h_n^2(\bar{p}_{\hat{g}_n}, p_{g_0}) + \lambda_n^2 I^2(\hat{g}_n) \\ &\quad \le O_{\mathbf{P}}(n^{-\frac{1}{2}}) h_n^{1-\frac{1}{2m}}(\bar{p}_{\hat{g}_n}, p_{g_0}) \left(1 + I(\hat{g}_n)\right)^{\frac{1}{2m}} + \lambda_n^2 I^2(g_0), \end{aligned}$$

which gives $h_n(\bar{p}_{\hat{g}_n}, p_{g_0}) = O_{\mathbf{P}}(\lambda_n)$ and $I(\hat{g}_n) = O_{\mathbf{P}}(1)$. In turn, this implies $\|F(\hat{g}_n) - F(g_0)\|_n = O_{\mathbf{P}}(\lambda_n)$. □

Now that we have established $\|F(\hat{g}_n) - F(g_0)\|_n = O_{\mathbf{P}}(\lambda_n)$ and $I(\hat{g}_n) = O_{\mathbf{P}}(1)$, we can apply a ULLN (see Theorem 3.7), to conclude that $\|F(\hat{g}_n) - F(g_0)\| = o_{\mathbf{P}}(1)$. So we have convergence in the theoretical norm, but not yet a rate. Lemma 5.6 presents the ratio of empirical and theoretical norm, under conditions on the empirical entropy. However, it is of no help here, because it only says that the empirical norm is small, whenever the theoretical norm is small, and nothing about the reverse. Lemma 5.16 is more to the point, but it uses conditions on the entropy with bracketing. Fortunately, this latter entropy can be handled easily if the parameters are identifiable. Therefore, we impose the identifiability conditions

$$\inf_{\|g-g_0\|>\eta,\ I(g)\le M} \|F(g) - F(g_0)\| > 0, \quad \text{for all } \eta > 0, \text{ and } M < \infty, \tag{11.18}$$

and

$$\begin{aligned} &E\mathrm{var}(U \mid V) > 0, \quad \text{and } Q \text{ has density } q \\ &\text{w.r.t. Lebesgue measure, with } q(z) \ge \eta_0^2. \end{aligned} \tag{11.19}$$

If we moreover assume that for some $\eta_1 > 0$ and for all $z \in [0,1]^2$ we have for $\xi_0 = g_0(z)$,

$$|f(\xi)| \ge \eta_1 > 0, \quad \text{for all } |\xi - \xi_0| \le \eta_1, \tag{11.20}$$

we obtain the same rate $O_{\mathbf{P}}(\lambda_n)$ for $\hat{g}_n$.

Lemma 11.5 *Suppose the conditions of Lemma* 11.4 *are met. Furthermore, assume the identifiability conditions* (11.18) *and* (11.19), *and also assume* (11.20). *Then*

(11.21) $$\|F(\hat{g}_n) - F(g_0)\| = O_{\mathbf{P}}(\lambda_n),$$

(11.22) $$\|\hat{g}_n - g_0\| = O_{\mathbf{P}}(\lambda_n),$$

and

(11.23) $$\|\hat{g}_n - g_0\|_n = O_{\mathbf{P}}(\lambda_n).$$

Proof Since $\|F(\hat{g}_n) - F(g_0)\|_n = o_{\mathbf{P}}(1)$, together with $I(\hat{g}_n) = O_{\mathbf{P}}(1)$, implies $\|F(\hat{g}_n) - F(g_0)\| = o_{\mathbf{P}}(1)$, we know from (11.18) that $\|\hat{g}_n - g_0\| = o_{\mathbf{P}}(1)$. Condition (11.19) now implies that $|\hat{g}_n - g_0|_\infty = o_{\mathbf{P}}(1)$ (use Lemma 10.9). Now, use condition (11.20) to obtain that $\|\hat{g}_n - g_0\|_n = O_{\mathbf{P}}(\lambda_n)$. Exploit the entropy with bracketing for the class $\{F(g) : g \in \mathcal{G}, |g|_\infty \le C, I(g) \le C\}$ and $\{g : |g|_\infty \le C, I(g) \le C\}$, to find from Lemma 5.16 that $\|F(\hat{g}_n) - F(g_0)\| = O_{\mathbf{P}}(\lambda_n)$ and $\|\hat{g}_n - g_0\| = O_{\mathbf{P}}(\lambda_n)$ respectively. □

Remark In view of Lemma 10.9, the conditions of Lemma 11.5 imply that for $k \in \{0, \dots, m\}$,

(11.24) $$\left\|\hat{\gamma}_n^{(k)} - \gamma_0^{(k)}\right\| = O_{\mathbf{P}}\left(\lambda_n^{\frac{m-k}{k}}\right).$$

Moreover,

(11.25) $$|\hat{\gamma}_n - \gamma_0|_\infty = O_{\mathbf{P}}\left(\lambda_n^{1-\frac{1}{2m}}\right).$$

Once we have obtained consistency of the estimator $\hat{g}_n$ in the supremum norm, it is not so hard to deduce asymptotic normality of $\hat{\theta}_n$. The reason is that, locally, the model is approximately linear, provided that the appropriate expansions are valid. To this end, we assume that for all $z \in [0,1]^2$ and for $\xi_0 = g_0(z)$,

(11.26) $$\left|\frac{d}{d\xi} f(\xi)\right| \le \frac{1}{\eta_2}, \quad \text{for all } |\xi - \xi_0| \le \eta_2.$$

We can apply the same arguments as in the previous section. Some notation is needed to state the result. Define

$$l(\xi) = \frac{f(\xi)}{F(\xi)(1 - F(\xi))}, \quad \xi \in \mathbf{R},$$

and $l_0 = l(g_0)$, $f_0 = f(g_0)$. Let

$$h(v) = \frac{E\big(Uf_0(U,V)l_0(U,V) \mid V = v\big)}{E\big(f_0(U,V)l_0(U,V) \mid V = v\big)},$$

and

$$\tilde{h}(u,v) = u - h(v).$$

Lemma 11.6 *Assume that the conditions of Lemma* 11.5 *hold, that* (11.26) *is met, and that*

$$\| (f_0 l_0)^{1/2} \tilde{h} \| > 0. \tag{11.27}$$

Moreover, assume $\lambda_n = o_{\mathbf{P}}\big(n^{-1/4}\big)$ *and* $I(h) < \infty$. *Then*

$$\sqrt{n}(\hat{\theta}_n - \theta_0) = \frac{\frac{1}{\sqrt{n}}\sum_{i=1}^n W_i l_0(Z_i)\tilde{h}(Z_i)}{\| \big(f_0 l_0\big)^{1/2}\tilde{h} \|^2} + o_{\mathbf{P}}(1). \tag{11.28}$$

Proof For $t \in \mathbf{R}$, let

$$\hat{g}_{n,t} = \hat{g}_n + t\tilde{h} \in \mathcal{G}.$$

The derivative of the penalized log-likelihood is zero in this direction. Writing $\hat{l}_n = l(\hat{g}_n)$, this derivative is equal to

$$\begin{aligned} 0 &= \frac{1}{n}\sum_{i=1}^n W_i \hat{l}_n(Z_i)\tilde{h}(Z_i) - \frac{1}{n}\sum_{i=1}^n [F(\hat{g}_n(Z_i)) - F(g_0(Z_i))]\hat{l}_n(Z_i)\tilde{h}(Z_i) \\ &\quad - \lambda_n^2 \int_0^1 \hat{\gamma}_n^{(m)}(v) h^{(m)}(v)\, dv \\ &= \tilde{\mathrm{I}} - \widetilde{\mathrm{II}} - \mathrm{III}. \end{aligned}$$

As in Lemma 11.1, we find

$$\mathrm{III} = o_{\mathbf{P}}(n^{-\frac{1}{2}}).$$

Let us write

$$\begin{aligned} \widetilde{\mathrm{II}} &= \frac{1}{n}\sum_{i=1}^n [(\hat{g}_n(Z_i) - g_0(Z_i))f_0(Z_i)]\, l_0(Z_i)\tilde{h}(Z_i) \\ &\quad + \frac{1}{n}\sum_{i=1}^n \Big([F(\hat{g}_n(Z_i)) - F(g_0(Z_i))]\hat{l}_n(Z_i)\tilde{h}(Z_i) \\ &\qquad - [(\hat{g}_n(Z_i) - g_0(Z_i))f_0(Z_i)]\, l_0(Z_i)\tilde{h}(Z_i)\Big) \\ &= \mathrm{II} + \mathrm{ii}. \end{aligned}$$

Using (11.26), we see that

$$\mathrm{ii} = O(1)\,\|\hat{g}_n - g_0\|_n^2 = o_{\mathbf{P}}\big(n^{-\frac{1}{2}}\big).$$

The term II can be treated in the same way as in Lemma 11.1. One obtains

$$\mathrm{II} = (\hat{\theta}_n - \theta_0)(\|(f_0 l_0)^{1/2}\tilde{h}\|^2 + o(1)) + o_{\mathbf{P}}(n^{-\frac{1}{2}}).$$

Finally,

$$\tilde{\mathrm{I}} = \frac{1}{n}\sum_{i=1}^n W_i l_0(Z_i)\tilde{h}(Z_i) + \frac{1}{n}\sum_{i=1}^n W_i\big(\hat{l}_n(Z_i) - l_0(Z_i)\big)\tilde{h}(Z_i) = \mathrm{I} + \mathrm{i}.$$

Use the asymptotic equicontinuity argument to obtain

$$\mathrm{i} = o_{\mathbf{P}}\big(n^{-\frac{1}{2}}\big). \qquad \square$$

Remark It can be shown that $\hat{\theta}_n$ is an asymptotically efficient estimator of θ_0.

Example Suppose that one uses the canonical parameter

$$\xi = \log\left(\frac{\mu}{1-\mu}\right)$$

to model the linear dependence. The link function is then

$$\mu = F(\xi) = \frac{e^{\xi}}{1+e^{\xi}}.$$

So in that case, $l(\xi) = 1$ for all $\xi \in \mathbf{R}$, and

$$\sqrt{n}(\hat{\theta}_n - \theta_0) \to^{\mathcal{L}} \mathcal{N}\big(0, \|f_0^{1/2}\tilde{h}\|^{-2}\big). \tag{11.29}$$

We note moreover that in this case the log-likelihood $y\log F\big(g(z)\big) + (1-y)\log\big(1 - F(g(z))\big)$ is a concave function of $g(z)$. This means that for establishing the rate of convergence, one may use exactly the same arguments as the ones used in Theorem 10.10, i.e., by renormalizing one may restrict attention to a neighbourhood $\{g \in \mathcal{G} :\ \|g - g_0\| \le 1\}$.

11.1.3. Penalized quasi-likelihood estimation

The estimators studied in Subsections 11.1.1 and 11.1.2 are special cases of so-called penalized quasi-likelihood estimators. Recall that the general model is

$$E(Y \mid Z = z) = \mu_0(z),$$

with

$$\mu_0(z) = F\big(g_0(z)\big),$$

where g_0 is in a given class $\mathcal{G}$. The quasi-(log-)likelihood is now

$$\mathcal{Q}(y;\mu) = \int_y^\mu \frac{(y-s)}{V(s)}\,ds,$$

with $V : F(\mathbf{R}) \to (0,\infty)$ some given function, often referred to as the variance function. The penalized quasi-likelihood estimator is

$$\hat{g}_n = \arg\max_{g\in\mathcal{G}} \left(\frac{1}{n}\sum_{i=1}^n \mathcal{Q}\big(Y_i; F(g(Z_i))\big) - \lambda_n^2 I^2(g) \right). \tag{11.30}$$

For a rate of convergence of the penalized quasi-likelihood estimator, certain conditions are needed on the link function F, the variance function V, and on the distribution of the error W.

In Subsection 11.1.1, we studied the case $F(s) = V(s) = s$, and we assumed sub-Gaussian errors. In Subsection 11.1.2, we assumed some regularity on F. We moreover took $V(s) = s(1-s)$, so that in fact the quasi-likelihood is the true likelihood.

In general, the following situation is easily handled. Suppose that F is a strictly increasing function, with derivative $f > 0$. Let $V(s) = f(F^{-1}(s))$. Then, the quasi-likelihood is a concave function of g. If it is moreover in fact the true likelihood, one can use exactly the same arguments as in Theorem 10.10 to arrive at a rate of convergence. An example is the Poisson regression model, where Y given Z has a Poisson distribution with mean $\exp g_0(Z)$ (i.e., $F(\xi) = \exp \xi, V(s) = s$).

Once consistency of the estimators is obtained, one may linearize, locally at g_0, in order to arrive at asymptotic normality of the estimator $\hat{\theta}_n$ of θ_0. The arguments of the previous two subsections need only minor adjustments.

11.2. Mixture models

11.2.1. Introduction

Let $X_1, \ldots, X_n, \ldots$ be i.i.d. random variables on $(\mathcal{X}, \mathcal{A})$ with distribution P. The distribution P is assumed to be an unknown mixture of a given collection of distributions, dominated by a σ-finite measure μ. Let $\{k(\cdot \mid y) : y \in \mathcal{Y}\}$ be the given collection of densities w.r.t. μ. Here $(\mathcal{Y}, \mathcal{B})$ is a measurable space. We denote by Λ, the class of all probability measures on $(\mathcal{Y}, \mathcal{B})$. Then the density of an observation X_i is

$$p_0 = \frac{dP}{d\mu} = \int k(\cdot \mid y)\, dF_0(y) = p_{F_0}, \quad F_0 \in \Lambda.$$

Thus $p_0 \in \mathscr{P}$, with $\mathscr{P}$ the class of all mixtures

$$\mathscr{P} = \left\{ p_F = \frac{dP_F}{d\mu} = \int k(\cdot \mid y)\, dF(y) :\ F \in \Lambda \right\}. \tag{11.31}$$

The maximum likelihood estimator of F_0 is (not necessarily uniquely) defined by

$$\hat{F}_n = \arg\max_{F \in \Lambda} \int \log p_F \, dP_n. \tag{11.32}$$

In this section, we are interested in a linear function

$$\theta_F = \int a \, dF,$$

with $a : \mathscr{Y} \to \mathbf{R}$ given. For example, if $\mathscr{Y} = \mathbf{R}$ and we want to estimate the mean of F_0, take $a(y) = y$ for all $y \in \mathbf{R}$.

We shall often use the shorthand notation $\theta_0 = \theta_{F_0}$ and $\hat{\theta}_n = \theta_{\hat{F}_n}$. So $\hat{\theta}_n$ is the maximum likelihood estimator of θ_0.

In Example 4.2.4, we studied consistency of the maximum likelihood estimator, and the general theory in Chapter 7 can be applied to obtain a rate of convergence in Hellinger distance. Here, we show how the consistency and rates can be applied to obtain that, for certain differentiable functions θ_F, the maximum likelihood estimator $\hat{\theta}_n$ converges with parametric rate $n^{-1/2}$, and is asymptotically normally distributed.

To begin with, we assume throughout that $h(p_{\hat{F}_n}, p_{F_0}) \to 0$, a.s., and that $|\hat{\theta}_n - \theta_0| \to 0$. Conditions on rates will appear later on.

Let us first see why, even in non-parametric models, there may exists functionals that can be estimated with rate $n^{-1/2}$. A simple example, in the case of real-valued observations, is the mean of the observations

$$EX_1 = \int x \, dP(x) = \int a \, dF_0,$$

where $a(y) = \int x k(x \mid y)\, d\mu(x)$. Clearly, if the variance of the observations is finite, the sample mean $\sum_{i=1}^n X_i/n$ is a $\sqrt{n}$-consistent and asymptotically normal estimator of the theoretical mean. However, the sample mean may not be the maximum likelihood estimator. It does not use the information that we are dealing with a mixture model at all. So one may hope that the maximum likelihood estimator of the theoretical mean is also $\sqrt{n}$-consistent, and is in fact better than the sample mean.

Returning to the general problem, suppose that for some function b,

$$E(b(X) \mid Y = y) = a(y), \quad \text{for all } y. \tag{11.33}$$

Because $Eb(X) = \theta_0$, the estimator $\int b\,dP_n$ is an unbiased estimator of θ_0. If moreover $b \in L_2(P)$, this estimator is $\sqrt{n}$-consistent and asymptotically normal. We say that θ_F is *differentiable* at all F if a solution b of (11.33) exists.

Note that if $\{k(\cdot \mid y) : y \in \mathcal{Y}\}$ is a complete family, then there is at most one solution of (11.33). In general, there may be several solutions, in which case one would like to take the one with the smallest variance. But such a solution possibly depends on F_0. Under the conditions of Proposition 11.7 below, the maximum likelihood estimator automatically chooses the best solution, albeit with $\hat{F}_n$ plugged in for F_0.

A crucial role here is played by the so-called *worst possible subfamily*. To describe this concept, we first need some definitions. For a function $b : \mathcal{X} \to \mathbf{R}$, write

$$A^*b(y) = E(b(X) \mid Y = y) = \int k(x \mid y) b(x)\, d\mu(x). \tag{11.34}$$

Furthermore, for $h : \mathcal{Y} \to \mathbf{R}$, define

$$A_F h(x) = \frac{\int k(x \mid y) h(y)\, dF(y)}{p_F(x)}. \tag{11.35}$$

Thus, $A_{F_0} h(x) = E(h(Y) \mid X = x)$.

If for η sufficiently small and for some function $h_{F_0} : \mathcal{Y} \to \mathbf{R}$,

$$\{F_{0,t} : dF_{0,t} = (1 + t h_{F_0}) dF_0,\ |t| \le \eta\} \in \Lambda, \tag{11.36}$$

then any regular estimator of $t_0 = 0$ will have asymptotic variance (in the sense that the asymptotic distribution has this variance) at least $\int b_{F_0}^2\, dP$, where $b_{F_0}(x) = A_{F_0} h_{F_0}$. To see this, note that $\int b_{F_0}^2\, dP$ is the inverse of the Fisher information in the parametric submodel.

In view of (11.36), we must have $\int h_{F_0}\, dF_0 = 0$. This implies that also $\int b_{F_0}\, dP = 0$, which is just the fact that the derivative of the log-density has mean zero.

Now, we want to estimate θ_0. This means that we have to be able to choose h_{F_0} in such a way that $b_{F_0} = A_{F_0} h_{F_0}$ satisfies

$$A^* b_{F_0} = a(y) - \theta_0, \quad \text{for } F_0\text{-almost all } y. \tag{11.37}$$

In that case, we say that h_{F_0} is the worst possible subdirection (along which one can consider a submodel) for estimating θ_0.

Let us now summarize the concepts. We say that θ_F is *differentiable at F* if for some b, we have

$$A^*b(y) = a(y) - \theta_F, \quad \text{for } F\text{-almost all } y \tag{11.38}$$

The function $a(y) - \theta_F$ is called the *gradient* of θ_F. If for some $h_F \in L_\infty(F)$ with $\int h_F \, dF = 0$,

$$A^*A_F h_F(y) = a(y) - \theta_F, \quad \text{for } F\text{-almost all } y, \tag{11.39}$$

we call h_F the *worst possible subdirection* for estimating θ_F and $b_F = A_F h_F$ the *efficient influence curve*. We shall assume throughout that the efficient influence curve b_{F_0} and the worst possible subdirection h_{F_0} exist at F_0, and that the information for estimating θ_0 is positive and finite, i.e., $0 < \int b_{F_0}^2 \, dP < \infty$.

Remark As an extension, let for each $m \in \{1, 2, \ldots\}$, $h_{F,m} \in L_\infty$ satisfy $\int h_{m,F} \, dF = 0$ and define $b_{F,m} = A_F h_{F,m}$. Suppose that

$$\lim_{m \to \infty} \| b_{F,m} - b_F \| = 0,$$

where b_F satisfies

$$A^* b_F(y) = a(y) - \theta_F, \quad \text{for } F\text{-almost all } y.$$

Then b_F is still called the efficient influence curve at F. However, to avoid digressions, we shall not use this extended definition.

11.2.2. Asymptotic normality

Since we assumed consistency of $\hat{F}_n$, further conditions are only needed in a neighbourhood of F_0. We let Λ_0 be such a neighbourhood, i.e., $F_0 \in \Lambda_0$, and $\hat{F}_n \in \Lambda_0$ a.s., for all n sufficiently large.

Proposition 11.7 below explains the idea behind proving asymptotic normality. Its conditions are virtually never met, and we shall present three methods to cope with this difficulty (see Theorems 11.8, 11.9 and 11.11). However, the fundaments of these methods are still based on the argument developed in Proposition 11.7. It may be helpful to compare the proposition with Lemma 12.7, where its finite-dimensional counterpart is presented.

The examples will facilitate understanding what is going on. Example 11.2.3a illustrates Proposition 11.7. As a motivation for Theorem 11.8, one could proceed by looking at Example 11.2.3b. Example 11.2.3.d illustrates that also Theorem 11.8 does not always work, which is why we stated Theorem 11.9. We do not present any illustrations of Theorem 11.11, thus indicating that it is not suitable here.

To avoid the notational confusion between h_F as a direction and the Hellinger distance h, we write $d(F, F_0) = h(p_F, p_{F_0})$.

Proposition 11.7 *Suppose that*

(i) *the efficient influence curves b_F and worst possible subdirections h_F exist for all $F \in \Lambda_0$, and $\int b_F \, dP_{F_0} = -(\theta_F - \theta_{F_0})$;*

(ii) *$\{b_F : F \in \Lambda_0\}$ forms a P-Donsker class;*

(iii) $\lim_{d(F,F_0)\to 0} \|b_F - b_{F_0}\| = 0$.

Then

$$\sqrt{n}(\hat{\theta}_n - \theta_0) = \sqrt{n} \int b_{F_0} \, dP_n + o_{\mathbf{P}}(1) \left(\to^{\mathscr{L}} \mathscr{N}(0, \|b_{F_0}\|^2)\right). \tag{11.40}$$

Proof Because for η sufficiently small, $\hat{F}_n$ is an interior point of

$$\{\hat{F}_{n,t} : d\hat{F}_{n,t} = (1 + t h_{\hat{F}_n}) d\hat{F}_n, \ |t| \leq \eta\},$$

we have

$$\frac{d}{dt} \int \log p_{\hat{F}_{n,t}} \, dP_n \bigg|_{t=0} = 0.$$

In other words,

$$\int b_{\hat{F}_n} \, dP_n = 0.$$

But

$$\int b_{\hat{F}_n} \, dP_n = \int b_{\hat{F}_n} \, d(P_n - P) + \int b_{\hat{F}_n} \, dP$$
$$= \int b_{\hat{F}_n} \, d(P_n - P) - (\hat{\theta}_n - \theta_0).$$

So we have shown that

$$\hat{\theta}_n - \theta_0 = \int b_{\hat{F}_n} \, d(P_n - P).$$

Asymptotic equicontinuity arguments (see Section 6.2) yield that

$$\int b_{\hat{F}_n} \, d(P_n - P) = \int b_{F_0} \, d(P_n - P) + o_{\mathbf{P}}(n^{-1/2}). \qquad \square$$

The second part of condition (i) of Proposition 11.7 follows from the first part, if F dominates F_0 for all $F \in \Lambda_0$. To see this, note that changing the order of integration yields

$$\int b_F \, dP = \int A^* b_F \, dF_0 = \int \big(a(y) - \theta_F\big) \, dF(y) = -(\theta_F - \theta_{F_0}).$$

However, the assumption that $\hat{F}_n$ dominates F_0 is hardly ever true. For example, F_0 may be continuous, and $\hat{F}_n$ discrete.

A major problem in Proposition 11.7 is that the condition that the efficient influence curve $b_{\hat{F}_n}$ exists, is often not fulfilled. In Theorems 11.8, 11.9 and 11.11 respectively, we shall provide three methods to establish asymptotic normality without this condition. Each method will have its own merits and drawbacks. The first method concentrates on the problem that $\hat{F}_n$ might not dominate F_0, and employs a convex combination of $\hat{F}_n$ and F_0 to handle this. Method number two is based on the idea that, even when the efficient influence curve at $\hat{F}_n$ does not exist, one can often construct a function that behaves very much like one. Finally, the third method relies on approximating the efficient influence curve b_{F_0} by $\bar{b}_F = A_F \bar{h}_F$, where $\bar{h}_F$ is an appropriate subdirection.

Before proceeding, we summarize the rates of convergence that we shall need. (Note that Proposition 11.7 does not need any rates of convergence.) Now, we shall require a fast enough rate of convergence, in order to be able to prove that remainder terms are negligible. The remainder terms are often quadratic, in which case the rate $o_{\mathbf{P}}(n^{-1/4})$ is sufficient. In Chapter 7, we established such a rate for the Hellinger distance. Because our class of densities $\mathscr{P}$ is convex, it is possible to use the results of Section 7.3. Recall the notation

$$J_B\big(\delta, \mathscr{G}^{(\mathrm{conv})}, P\big) = \int_{\delta^2/c}^{\delta} H_B^{1/2}\big(u, \mathscr{G}^{(\mathrm{conv})}, P\big) \, du \vee \delta,$$

where

$$\mathscr{G}^{(\mathrm{conv})} = \left\{ \frac{2p}{p + p_0} \; : \; p \in \mathscr{P} \right\}.$$

Assume that the entropy integral is finite, and take

$$\Psi(\delta) \geq \int_0^{\delta} H_B^{1/2}\big(u, \mathscr{G}^{(\mathrm{conv})}, P\big) \, du \vee \delta,$$

in such a way that $\Psi(\delta)/\delta^2$ is a non-increasing function of δ. Then for $\sqrt{n}\delta_n^2 \geq c\Psi(\delta_n)$, we have by Theorem 7.6,

$$d(\hat{F}_n, F_0) = O_{\mathbf{P}}(\delta_n), \tag{11.41}$$

and by Corollary 7.8,

$$\int \log \frac{2\hat{p}_n}{\hat{p}_n + p_0}\, dP_n = O_{\mathbf{P}}(\delta_n^2). \tag{11.42}$$

Moreover, because the entropy integral converges, (11.41) and (11.42) hold with $\delta_n = o(n^{-1/4})$.

Now we are ready to start our three methods program. The first method is based on the idea that for any $\alpha \in (0, 1]$, the convex combination

$$\bar{F}_\alpha = (1-\alpha)F + \alpha F_0$$

dominates F_0.

Theorem 11.8 *Assume that the following conditions are met:*

(i) *for all $0 < \alpha \le 1$ and $F \in \Lambda_0$, we have that the worst possible subdirections $h_{\bar{F}_\alpha}$ and the efficient influence curves $b_{\bar{F}_\alpha} = A_{\bar{F}_\alpha} h_{\bar{F}_\alpha}$ exist;*

(ii) *$\{b_{\bar{F}_\alpha} : F \in \Lambda_0,\ 0 < \alpha \le 1\}$ forms a P-Donsker class;*

(iii) $\lim_{d(\bar{F}_\alpha, F_0)\to 0} \|b_{\bar{F}_\alpha} - b_{F_0}\| = 0$;

(iv) *(rate of convergence) we have*

$$\int \log \frac{2\hat{p}_n}{\hat{p}_n + p_0}\, dP_n = O_{\mathbf{P}}(\delta_n^2),$$

with $\delta_n = o(n^{-1/4})$;

(v) *(control on the efficient influence curves $b_{\bar{F}_\alpha}$) the efficient influence curves are uniformly bounded:*

$$\sup_{F\in\Lambda_0}\ \sup_{0<\alpha\le 1} |b_{\bar{F}_\alpha}|_\infty < \infty; \tag{11.43}$$

(vi) *(control on the subdirections $h_{\bar{F}_\alpha}$) for some $M < \infty$,*

$$\sup_{F\in\Lambda_0}\ \sup_{0<\alpha\le 1} |\alpha h_{\bar{F}_\alpha}|_\infty \le M. \tag{11.44}$$

Then

$$\hat{\theta}_n - \theta_0 = \int b_{F_0}\, dP_n + o_{\mathbf{P}}(n^{-1/2}). \tag{11.45}$$

Proof Take an increasing function $\gamma : (0,\infty) \to (0,\infty)$, satisfying

$$\lim_{x\to 0} \gamma(x) = 0, \tag{11.46}$$

$$\lim_{x\to 0} \frac{x}{\gamma(x)} = 0, \tag{11.47}$$

$$\frac{x}{\gamma(x)} \leq \frac{1}{2}, \quad \text{for all } x > 0, \tag{11.48}$$

and

$$\delta_n^2 = o\big(n^{-1/2}\big)\gamma\big(n^{-1/2}\big).$$

Since $\delta_n = o\big(n^{-1/4}\big)$, this is indeed possible. By condition (iv), we have

$$\int \log \frac{2p_{\hat{F}_n}}{p_{\hat{F}_n} + p_{F_0}}\, dP_n = O_{\mathbf{P}}\big(\delta_n^2\big) = o_{\mathbf{P}}\big(n^{-1/2}\big)\gamma\big(n^{-1/2}\big). \tag{11.49}$$

Choose

$$\alpha_n = \frac{|\hat{\theta}_n - \theta_0| + n^{-1/2}}{\gamma(|\hat{\theta}_n - \theta_0| + n^{-1/2})}.$$

Then $\alpha_n > 0$ and by (11.48), $\alpha_n \leq 1/2$. Since $|\hat{\theta}_n - \theta_0| = o(1)$, (11.47) implies that $\alpha_n = o(1)$. Furthermore, (11.46) yields

$$\frac{|\hat{\theta}_n - \theta_0|}{\alpha_n} = o(1), \tag{11.50}$$

as well as

$$\frac{n^{-1/2}}{\alpha_n} = o(1). \tag{11.51}$$

Define

$$\tilde{F}_n = (1-\alpha_n)\hat{F}_n + \alpha_n F_0.$$

Without loss of generality, we may assume $\hat{F}_n \in \Lambda_0$. Because $\alpha_n > 0$, the efficient influence curve $b_{\tilde{F}_n} = A_{\tilde{F}_n} h_{\tilde{F}_n}$ exists. Since $\tilde{F}_n$ dominates F_0, we find

$$\int b_{\tilde{F}_n}\, dP = -(\theta_{\tilde{F}_n} - \theta_0).$$

Now,

$$\theta_{\tilde{F}_n} = (1-\alpha_n)\hat{\theta}_n + \alpha_n\theta_0,$$

so that

$$\int b_{\tilde{F}_n}\,dP = -(1-\alpha_n)(\hat{\theta}_n - \theta_0) = -(\hat{\theta}_n - \theta_0)\big(1 + o_{\mathbf{P}}(1)\big).$$

Together with conditions (ii) and (iii), this implies

$$\begin{aligned}(11.52)\quad \int b_{\tilde{F}_n}\,dP_n &= \int b_{\tilde{F}_n}\,d(P_n - P) - (\hat{\theta}_n - \theta_0)\big(1 + o_{\mathbf{P}}(1)\big)\\ &= \int b_{F_0}\,dP_n - (\hat{\theta}_n - \theta_0)\big(1 + o_{\mathbf{P}}(1)\big) + o_{\mathbf{P}}\big(n^{-1/2}\big).\end{aligned}$$

Let

$$t_n = \frac{\int b_{\tilde{F}_n}\,dP_n}{\alpha_n \int b_{F_0}^2\,dP}.$$

Equality (11.52), together with (11.50) and (11.51), implies that

$$(11.53)\qquad t_n = o_{\mathbf{P}}(1).$$

Take M as in (11.44). Then from (11.53), we may, without loss of generality, assume that $|t_n| \le 1/M$. So $\tilde{F}_{n,t_n} \in \Lambda$, where

$$d\tilde{F}_{n,t_n} = (1 + t_n\alpha_n h_{\tilde{F}_n})d\tilde{F}_n.$$

By conditions (ii) and (v),

$$\begin{aligned}(11.54)\quad \int \log\frac{p_{\tilde{F}_{n,t_n}}}{p_{\tilde{F}_n}}\,dP_n &= t_n\alpha_n \int b_{\tilde{F}_n}\,dP_n - \frac{1}{2}t_n^2\alpha_n^2 \int b_{\tilde{F}_n}^2\,dP_n(1 + o_{\mathbf{P}}(1))\\ &= \frac{1}{2}\frac{\left(\int b_{\tilde{F}_n}\,dP_n\right)^2}{\int b_{F_0}^2\,dP}\big(1 + o_{\mathbf{P}}(1)\big).\end{aligned}$$

Since $\hat{F}_n$ maximizes the likelihood,

$$(11.55)\qquad \int \log p_{\hat{F}_n}\,dP_n \ge \int \log p_{\tilde{F}_{n,t_n}}\,dP_n.$$

Furthermore, the concavity of the log-function, and the fact that $\alpha_n \le 1/2$, yield

$$(11.56)\quad \int \log p_{\tilde{F}_n}\,dP_n \ge (1 - 2\alpha_n)\int \log P_{\hat{F}_n}\,dP_n + 2\alpha_n \int \log\frac{p_{\hat{F}_n} + p_{F_0}}{2}\,dP_n.$$

Combine (11.54), (11.55) and (11.56) to find

$$(11.57)\qquad 2\alpha_n \int \log\frac{2p_{\hat{F}_n}}{p_{\hat{F}_n} + p_{F_0}}\,dP_n \ge \frac{1}{2}\frac{\left(\int b_{\tilde{F}_n}\,dP_n\right)^2}{\int b_{F_0}^2\,dP}\big(1 + o_{\mathbf{P}}(1)\big).$$

From (11.49), we know that the left-hand side of this inequality is

$$\alpha_n o_{\mathbf{P}}(n^{-1/2})\gamma(n^{-1/2}) = (|\hat{\theta}_n - \theta_0| + n^{-1/2}) o_{\mathbf{P}}(n^{-1/2}).$$

In view of (11.52), the right-hand side of (11.57) is of the form

$$\frac{\left(O_{\mathbf{P}}(n^{-1/2}) - (\hat{\theta}_n - \theta_0)(1 + o_{\mathbf{P}}(1))\right)^2}{\int b_{F_0}^2 \, dP}(1 + o_{\mathbf{P}}(1)).$$

So we find from (11.57),

$$|\hat{\theta}_n - \theta_0|^2 \leq \max\{O_{\mathbf{P}}(n^{-1}), (|\hat{\theta}_n - \theta_0| + n^{-1/2}) o_{\mathbf{P}}(n^{-1/2})\},$$

which implies $|\hat{\theta}_n - \theta_0| = O_{\mathbf{P}}(n^{-1/2})$. But then, the left-hand side of (11.57) is $o_{\mathbf{P}}(n^{-1})$, so that it reads

(11.58)

$$\left(\int b_{F_0} \, dP_n + o_{\mathbf{P}}(n^{-1/2}) - (\hat{\theta}_n - \theta_0)(1 + o_{\mathbf{P}}(1))\right)^2 (1 + o_{\mathbf{P}}(1)) = o_{\mathbf{P}}(n^{-1}).$$

In other words,

$$\hat{\theta}_n - \theta_0 = \int b_{F_0} \, dP_n + o_{\mathbf{P}}(n^{-1/2}). \qquad \square$$

The advantage of Theorem 11.8 is that, as far as rates are concerned, it only needs the appropriate order of the log-likelihood ratio, and that there is a general theory for this. In our further results, we encounter conditions on the rate of convergence, in some metric which depends on the problem at hand. In applications, this means that one might have to insert ad hoc arguments.

Theorem 11.9 is based on the idea that even if the influence curve at $\hat{F}_n$ does not exist, there may well be something very close to it. We introduce a (pseudo-)metric $\bar{d}$ on Λ, which in some applications is equal to d. If $\hat{F}_n$ and F_0 are close for the metric $\bar{d}$, where $\bar{d}$ is sufficiently strong, then it is to be expected that something not unlike an influence curve exists at $\hat{F}_n$.

Theorem 11.9 *Suppose that the following conditions are met:*

(ia) *for all $F \in \Lambda_0$, there exists a function b_F^*, with $\int b_F^* \, dP_F = 0$, and with $\int b_F^* \, dP_{F_0} = -(\theta_F - \theta_{F_0})$;*

(ib) *for all $F \in \Lambda_0$, there is a direction $\bar{h}_F \in L_\infty$, with $\int \bar{h}_F \, dF = 0$, such that for $\bar{b}_F = A_F \bar{h}_F$, we have*

(11.59) $$\|\bar{b}_F - b_F^*\|_{\bar{P}_F} \leq c_0 \bar{d}(F, F_0),$$

where $\bar{P}_F = (P_F + P_{F_0})/2$;

(ii) $\{\bar{b}_F : F \in \Lambda_0\}$ *forms a P-Donsker class;*
(iii) $\lim_{d(F,F_0)\to 0} \|\bar{b}_F - b_{F_0}\| = 0$;
(iv) $d(\hat{F}_n, F_0) = o_{\mathbf{P}}(n^{-1/4})$, *and* $\bar{d}(\hat{F}_n, F_0) = o_{\mathbf{P}}(n^{-1/4})$.
Then

$$\hat{\theta}_n - \theta_0 = \int b_{F_0}\, dP_n + o_{\mathbf{P}}(n^{-1/2}). \tag{11.60}$$

Proof As in the proof of Proposition 11.7, one can deduce that

$$0 = \int \bar{b}_{\hat{F}_n}\, d(P_n - P) - (\hat{\theta}_n - \theta_0) + \int (\bar{b}_{\hat{F}_n} - b^*_{\hat{F}_n})\, dP.$$

But, since $\int b^*_{\hat{F}_n} p_{\hat{F}_n}\, d\mu = \int \bar{b}_{\hat{F}_n} p_{\hat{F}_n}\, d\mu = 0$, we find

$$\begin{aligned}\left|\int (\bar{b}_{\hat{F}_n} - b^*_{\hat{F}_n})\, dP\right| &= \left|\int (\bar{b}_{\hat{F}_n} - b^*_{\hat{F}_n})(p_{F_0} - p_{\hat{F}_n})\, d\mu\right| \\ &\le 2\sqrt{2}\|\bar{b}_{\hat{F}_n} - b^*_{\hat{F}_n}\|_{P_{\hat{F}_n}} d(\hat{F}_n, F_0) \le 2\sqrt{2}C\bar{d}(\hat{F}_n, F_0) d(\hat{F}_n, F_0) \\ &= o_{\mathbf{P}}(n^{-1/2}).\end{aligned}$$

By conditions (ii) and (iii), this completes the proof. □

In applications (see Example 11.2.3d), one can often proceed as follows. Take b^*_F of the form

$$b^*_F(x) = \frac{\int k(x \mid y)\, dH^*_F(y)}{p_F(x)},$$

where H^*_F is of bounded variation, and $\int dH^*_F(y) = 0$. We require that

$$\int b^*_F\, dP = -(\theta_F - \theta_0),$$

but we do not assume that dH^*_F/dF exists. Instead, we look for an approximation $\bar{H}_F$ of H^*_F, with $\int d\bar{H}_F = 0$, such that $\bar{h}_F = d\bar{H}_F/dF$ does exist.

For the case $(\mathscr{Y}, \mathscr{B}) = (\mathbf{R}, \text{Borel sets})$, the following lemma presents a recipe on how to take $\bar{H}_F$.

Lemma 11.10 *Suppose that for some function* $\xi : \mathbf{R} \to \mathbf{R}$, $d\xi/dF_0$ *exists and satisfies*

$$\left|\frac{d\xi}{dF_0}\right|_\infty = c_1 < \infty. \tag{11.61}$$

Then for all $F \in \Lambda$ there exists a function $\bar{\xi}_F$ such that $d\bar{\xi}_F/dF$ exists, and such that

$$|\bar{\xi}_F(y) - \xi(y)| \leq c_1|F(y) - F_0(y)|, \quad \text{for all } y \in \mathbf{R}. \tag{11.62}$$

Moreover, if F has finite support,

$$\left|\frac{d\bar{\xi}_F}{dF}\right|_\infty < \infty. \tag{11.63}$$

Proof If $F(y) < F(\tau)$ for all $y < \tau$, we define $\bar{\xi}_F(\tau) = \xi(\tau)$. If $F(y) = F(\tau_0)$ on $[\tau_0, \tau_1)$, $F(y) > F(\tau_0)$ for all $y \geq \tau_1$, and $F(y) < F(\tau_0)$ for all $y < \tau_0$, we define for $y \in [\tau_0, \tau_1)$,

$$\bar{\xi}_F(y) = \begin{cases} \text{(i)} \ \ \xi(s), & \text{if there is an } s \in [\tau_0, \tau_1) \text{ with } F_0(s) = F(s), \\ \text{(ii)} \ \ \xi(\tau_0), & \text{if } F_0(y) > F(y) \text{ for all } y \in [\tau_0, \tau_1), \\ \text{(iii)} \ \ \xi(\tau_1-), & \text{if } F_0(y) < F(y) \text{ for all } y \in [\tau_0, \tau_1). \end{cases}$$

In the three cases, we find

(i) $$|\bar{\xi}_F(y) - \xi(y)| = |\xi(s) - \xi(y)| \leq c_1|F_0(s) - F_0(y)| = c_1|F(s) - F_0(y)| \\ = c_1|F(y) - F_0(y)|,$$

(ii) $$|\bar{\xi}_F(y) - \xi(y)| = |\xi(\tau_0) - \xi(y)| \leq c_1\big(F_0(y) - F_0(\tau_0)\big) \leq c_1\big(F_0(y) - F(\tau_0)\big) \\ = c_1(F_0(y) - F(y)),$$

(iii) $$|\bar{\xi}_F(y) - \xi(y)| = |\xi(\tau_1-) - \xi(y)| \leq c_1\big(F_0(\tau_1-) \\ - F_0(y)\big) \leq c_1\big(F(\tau_0) - F_0(y)\big) = c_1\big(F(y) - F_0(y)\big).$$

Clearly, $d\bar{\xi}_F/dF$ exists, and if F has finite support, $d\bar{\xi}_F/dF$ takes only finitely many values. □

As a final approach, we directly approximate the efficient influence curve at F_0. We present it for completeness, but shall not apply it in our examples. It seems that Theorem 11.11 is more appropriate for partially censored models, i.e., models where one has a good deal of direct observations from F_0, so that only a proportion of the observations is a mixture.

Theorem 11.11 *Suppose that the following conditions are met:*

(ia) *for all $F \in \Lambda_0$, we have $\int b_{F_0}\, dP_F = \theta_F - \theta_{F_0}$;*

(ib) *for all $F \in \Lambda_0$, there exists a direction $\bar{h}_F \in L_\infty$ with $\int \bar{h}_F \, dF = 0$, such that for $\bar{b}_F = A_F \bar{h}_F$,*

$$\|\bar{b}_F - b_{F_0}\|_{\bar{P}_F} \leq c_0 \bar{d}(F, F_0), \tag{11.64}$$

where $\bar{P}_F = (P_F + P_{F_0})/2$;

(ii) $\{\bar{b}_F : F \in \Lambda_0\}$ *forms a P-Donsker class;*

(iii) $d(\hat{F}_n, F_0) = o_{\mathbf{P}}(n^{-1/4})$, *and* $\bar{d}(\hat{F}_n, F_0) = o_{\mathbf{P}}(n^{-1/4})$.

Then

$$\hat{\theta}_n - \theta_0 = \int b_{F_0} \, dP_n + o_{\mathbf{P}}(n^{-1/2}). \tag{11.65}$$

Proof Again, write

$$0 = \int \bar{b}_{\hat{F}_n} \, dP_n = \int \bar{b}_{\hat{F}_n} \, d(P_n - P) + \int \bar{b}_{\hat{F}_n} \, dP.$$

But for $F \in \Lambda_0$,

$$\int \bar{b}_F \, dP_{F_0} = -\int b_{F_0} \, dP_F + \int (\bar{b}_F - b_{F_0})(p_F - p_{F_0}) \, d\mu,$$

and

$$\left| \int (\bar{b}_F - b_{F_0})(p_F - p_{F_0}) \, d\mu \right| \leq 2\sqrt{2} \|\bar{b}_F - b_{F_0}\|_{\bar{P}_F} d(F, F_0).$$

From condition (i) combined with (iii), it follows that

$$\hat{\theta}_n - \theta_0 = \int \bar{b}_{\hat{F}_n} \, d(P_n - P) + o_{\mathbf{P}}(n^{-1/2}).$$

By condition (ii), this completes the proof. □

Remark Asymptotic normality of a linear function θ_F can be used to establish asymptotic normality of non-linear functions γ_F. For example, suppose $\gamma_F = g(\theta_F)$ with $g(\theta)$ differentiable in at θ_0 with non-zero derivative $\dot{g}(\theta_0)$. Then asymptotic normality of $\hat{\theta}_n$ immediately implies asymptotic normality of $\hat{\gamma}_n = \gamma_{\hat{F}_n}$. More generally, the result holds if for some function $a_{F_0} : \mathcal{Y} \to \mathbf{R}$, we have for $\theta_F = \int a_{F_0} \, dF$,

$$\frac{|\gamma_{\hat{F}_n} - \gamma_{F_0} - (\theta_{\hat{F}_n} - \theta_{F_0})|}{|\theta_{\hat{F}_n} - \theta_{F_0}|} = o_{\mathbf{P}}(1).$$

11.2.3. Examples

Proposition 11.7 is applied in Example 11.2.3a. In Examples 11.2.3b, 11.2.3c and 11.2.3f, we use Theorem 11.8. Examples 11.2.3d and 11.2.3e illustrate Theorem 11.9. Finally, we end with an open problem in Example 11.2.3g.

Example 11.2.3a. The binary choice model Let $X_1 = (Y_1, Z_1), \ldots$ be i.i.d. copies of $X = (Y, Z)$ with $Y \in \{0, 1\}$ a binary response variable, and $Z \in \mathbf{R}$ a covariate with distribution Q. Consider the model

$$P(Y = 1 \mid Z = z) = F_0(z), \quad z \in \mathbf{R}, \ F_0 \in \Lambda.$$

The density of Y w.r.t. $\mu =$ (counting measure on$\{0,1\}$) $\times\, Q$ is

$$\begin{aligned} p_{F_0}(y, z) &= yF_0(z) + (1 - y)(1 - F_0(z)) \\ &= \int k(y, z \mid u)\, dF_0(u), \end{aligned}$$

with

$$k(y, z \mid u) = y\mathrm{l}\{u \le z\} + (1 - y)\mathrm{l}\{u > z\}.$$

So this is indeed a mixture model. (In fact, any model where the density is in some convex class can be regarded as a mixture model.)

Suppose we want to estimate $EY = \int F_0(z)\, dQ(z)$, and suppose that Q is known. We then take $\theta_F = \int F(z)\, dQ(z) = \int (1 - Q(z-)\, dF(z)$, so that the gradient is $a(z) - \theta_F$, with $a(z) = 1 - Q(z-)$.

We shall apply Proposition 11.7. The equation $b_F = A_F h_F$ in this example is

$$b_F(y, z) = y\frac{H_F(z)}{F(z)} - (1 - y)\frac{H_F(z)}{1 - F(z)}, \qquad h_F = \frac{dH_F}{dF}.$$

Moreover,

$$A^* b_F(u) = \int_{u-}^{\infty} b_F(0, z)\, dQ(z) + \int_{-\infty}^{u-} b_F(1, z)\, dQ(z).$$

We have to solve the equation

$$A^* b_F(u) = a(u) - \theta_F.$$

Taking the derivative w.r.t. Q on both sides gives

$$-b_F(0, u) + b_F(1, u) = -1,$$

or, using $b_F = A_F h_F$,

$$-\frac{H_F(u)}{F(u)} - \frac{H_F(u)}{1 - F(u)} = -1.$$

Hence, $H_F(u) = F(u)(1 - F(u))$. Note that indeed $h_F = dH_F/dF$ exists. Moreover,

$$b_F(y, z) = y(1 - F(z)) - (1 - y)F(z),$$

and since F is a monotone function uniformly bounded by 1, we know that $\{b_F : F \in \Lambda\}$ is a P-Donsker class (condition (ii) of Proposition 11.7). Condition (iii) is met as well.

Since we solved the equation $A^* b_F(u) = a(u) - \theta_F$ for *all* u, condition (i) of Proposition 11.7 also holds true. So we have

$$\hat{\theta}_n - \theta_0 = \int b_{F_0}\, dP_n + o_{\mathbf{P}}(n^{-\frac{1}{2}}).$$

The efficient asymptotic variance is

$$\int b_{F_0}^2\, dP = \int F_0(z)(1 - F_0(z))\, dQ(z). \tag{11.66}$$

The variance of the naive estimator $\bar{Y} = \sum_{i=1}^n Y_i/n$ is

$$\text{var}(\bar{Y}) = \frac{1}{n}\text{var}(Y),$$

with
(11.67)

$$\text{var}(Y) = \int F_0(z)\big(1 - F_0(z)\big)\, dQ(z) + \left(\int F_0^2(z)\, dQ(z) - \left(\int F_0(z)\, dQ(z)\right)^2\right).$$

Indeed, this is larger than (11.66).

Observe that the equality

$$\int b_{\hat{F}_n}\, dP_n = 0,$$

reads

$$\bar{Y} = \frac{1}{n}\sum_{i=1}^n \hat{F}_n(Z_i).$$

In other words, $\hat{\theta}_n = \int \hat{F}_n(z)\, dQ(z)$ is an asymptotically efficient estimator of θ_0, whereas $\bar{Y} = \frac{1}{n}\sum_{i=1}^n \hat{F}_n(z_i)$ is inefficient. This is the result we already announced in Example 1.3.

Example 11.2.3b. Convolution with the uniform Let μ be Lebesgue measure, $k(x \mid y) = k(x - y)$ with $k(x) = \mathrm{l}\{0 \le x \le 1\}$, and $\mathcal{Y} = [0, 1]$ (the case where

$\mathscr{Y} = [0, r]$ with $r > 1$ will be treated in Example 11.2.3e). The densities with respect to Lebesgue measure are

$$p_F(x) = F(x) - F(x-1) = \begin{cases} F(x), & \text{if } 0 \leq x \leq 1 \\ 1 - F(x-1), & \text{if } 1 \leq x \leq 2. \end{cases}$$

We want to estimate $\theta_{F_0} = \int a\,dF_0$. Let us apply Theorem 11.8 here. We assume that $\dot{a}(y) = da(y)/dy$ exists, that $\dot{a} \in L_\infty$ and that

$$\left|\frac{d\dot{a}}{dF_0}\right|_\infty \leq c.$$

Write for $x \in [0, 1]$,

$$b_{F_0}(x) = \frac{H_{F_0}(x)}{F_0(x)}, \tag{11.68}$$

$$b_{F_0}(x+1) = -\frac{H_{F_0}(x)}{1 - F_0(x)}, \tag{11.69}$$

where $H_{F_0}(0-) = H_{F_0}(1) = 0$. Now, we have to solve the equation

$$A^* b_{F_0}(y) = a(y) - \theta_{F_0},$$

or

$$\int_{y-}^{1} b_{F_0}(x)\,dx + \int_0^{y-} b_{F_0}(x+1)\,dx = a(y) - \theta_{F_0}. \tag{11.70}$$

Differentiating (11.70) yields

$$-b_{F_0}(y) + b_{F_0}(y+1) = \dot{a}(y).$$

Now, b_{F_0} should be of the form given in (11.68) and (11.69). Inserting this gives

$$-\frac{H_{F_0}(y)}{F_0(y)} - \frac{H_{F_0}(y)}{1 - F_0(y)} = \dot{a}(y),$$

or

$$H_{F_0}(y) = -F_0(y)\big(1 - F_0(y)\big)\dot{a}(y). \tag{11.71}$$

The derivative $h_{F_0} = dH_{F_0}/dF_0$ exists, since $d\dot{a}/dF_0$ exists. We can now apply Theorem 11.8. Let $\bar{F}_\alpha = (1-\alpha)F + \alpha F_0$, $\alpha \in (0, 1]$, replace in (11.71), F_0 by $\bar{F}_\alpha$, and call the result $H_{\bar{F}_\alpha}$. Then for some constant M

$$\left|\frac{dH_{\bar{F}_\alpha}}{d\bar{F}_\alpha}\right|_\infty \leq \frac{M}{\alpha}.$$

So condition (vi) of Theorem 11.8 is satisfied.

Moreover, for $0 \leq x \leq 1$,

$$b_{\bar{F}_\alpha}(x) = -\big(1 - \bar{F}_\alpha(x)\big)\dot{a}(x),$$
$$b_{\bar{F}_\alpha}(x+1) = \bar{F}_\alpha(x)\dot{a}(x).$$

Condition (v) of Theorem 11.8 holds. Moreover, the collection $\{b_{\bar{F}_\alpha} : F \in \Lambda,\ \alpha \in (0,1]\}$ forms a P-Donsker class. Also condition (iv) is met: we may take $\Psi(\delta) = c\sqrt{\delta}$, so that $\delta_n = O\big(n^{-1/3}\big) = o\big(n^{-1/4}\big)$ (see Example 7.4.5.). It now follows from Theorem 11.8 that

$$\hat{\theta}_n - \theta_0 = \int b_{F_0}\, dP_n + o_{\mathbf{P}}\big(n^{-1/2}\big).$$

The efficient asymptotic variance is

$$\int b_{F_0}^2\, dP = \int F_0(x)\big(1 - F_0(x)\big)\big(\dot{a}(x)\big)^2\, dx.$$

Example 11.2.3c. Interval censoring case I Recall the situation of Example 7.4.3: $X_i = (\Delta_i, Z_i)$, with $\Delta_i = \mathrm{l}\{Y_i \leq Z_i\}$, Y_i independent of Z_i, Y_i has distribution F_0 and Z_i has distribution Q. The densities with respect to (counting measure on $\{0,1\}) \times Q$ are

$$p_F(\Delta, z) = \Delta F(z) + (1 - \Delta)\big(1 - F(z)\big).$$

Note the similarity with the previous two examples. The only difference is that Lebesgue measure on $[0,1]$ is replaced by the measure Q on $\mathbf{R}$. Therefore, we now assume that da/dQ exists and lies in L_∞, and that

$$\left|\frac{d(da/dQ)}{dF_0}\right|_\infty \leq c.$$

Then

$$\hat{\theta}_n - \theta_0 = \int b_{F_0}\, dP_n + o_{\mathbf{P}}\big(n^{-1/2}\big),$$

with

$$b_{F_0}(\Delta, z) = -\Delta\big(1 - F_0(z)\big)\frac{da}{dQ}(z) + (1 - \Delta)F_0(z)\frac{da}{dQ}(z).$$

The efficient asymptotic variance becomes

$$\int b_{F_0}^2\, dP = \int F_0(1 - F_0)\left(\frac{da}{dQ}\right)^2 dQ.$$

Example 11.2.3d. Interval censoring case II As in Example 7.4.4, consider observations $X_i = (\beta_i, \gamma_i, U_i, V_i)$, with $V_i \geq U_i$, $\beta_i = 1\{Y_i \leq U_i\}$, $\gamma_i = 1\{U_i < Y_i \leq V_i\}$, where Y_i and (U_i, V_i) are independent, Y_i has distribution F_0 on $[0,2]$ and (U_i, V_i) has distribution $\tilde{Q}$ on $[0,2]$. We want to estimate $\theta_{F_0} = \int a\,dF_0$. In general, no explicit expression for the influence curve b_{F_0} can be given. The equation $A^* b_{F_0} = a - \theta_{F_0}$ becomes

$$\int\int_{u\geq y} b_{F_0}(1,0,u,v)\,d\tilde{Q}(u,v) + \int_{v\geq y}\int_{u<y} b_{F_0}(0,1,u,v)\,d\tilde{Q}(u,v)$$
$$\int_{v<y}\int b_{F_0}(0,0,u,v)\,d\tilde{Q}(u,v) = a(y) - \theta_{F_0}.$$

Moreover, we must have

$$\begin{aligned} b_{F_0}(\beta,\gamma,u,v) &= A_{F_0} h_{F_0}(\beta,\gamma,u,v) \\ &= \beta \frac{H_{F_0}(u)}{F_0(u)} + \gamma \frac{H_{F_0}(v) - H_{F_0}(u)}{F_0(v) - F_0(u)} - (1-\beta-\gamma)\frac{H_{F_0}(v)}{1 - F_0(v)}, \end{aligned}$$

where $dH_{F_0}/dF_0 = h_{F_0}$. In Geskus and Groeneboom (1996), it is shown that the solution exists, under certain regularity conditions. In fact, it can be shown that Theorem 11.8 is applicable under certain regularity conditions. One of the conditions is that the two examination times U and V can be arbitrarily close, i.e., for all $\epsilon > 0$,

$$\tilde{Q}(V_i - U_i \leq \epsilon) > 0. \tag{11.72}$$

Here, we shall consider a situation where (11.72) does not hold, and where explicit expressions are available, namely, the case $U_i \in [0,1]$ and $V_i = U_i + 1$. We shall apply Theorem 11.9. Let Q be the distribution of U_i. Assume that $da/dQ \in L_\infty$ and that for all $0 \leq y \leq 1$,

$$\left|\frac{d(da/dQ)(y)}{dF_0(y)}\right| \leq c, \quad \left|\frac{d(da/dQ)(y)}{dF_0(y+1)}\right| \leq c,$$

$$\left|\frac{d(da/dQ)(y+1)}{dF_0(y)}\right| \leq c, \quad \left|\frac{d(da/dQ)(y+1)}{F_0(y+1)}\right| \leq c,$$

$$\left|\frac{dF_0(y+1)}{dF_0(y)}\right| \leq c, \quad \left|\frac{dF_0(y)}{dF_0(y+1)}\right| \leq c,$$

and

$$F_0(y+1) - F_0(y) \geq \eta_0^2 > 0.$$

Then,

$$(11.73)\quad b_{F_0}(\beta,\gamma,u,u+1)$$
$$= -\beta\left((1-F_0(u))\frac{da}{dQ}(u)+(1-F_0(u+1))\frac{da}{dQ}(u+1)\right)$$
$$+\gamma\left(F_0(u)\frac{da}{dQ}(u)-(1-F_0(u+1))\frac{da}{dQ}(u+1)\right)$$
$$+(1-\beta-\gamma)\left(F_0(u)\frac{da}{dQ}(u)+F_0(u+1)\frac{da}{dQ}(u+1)\right).$$

Moreover,

$$(11.74)\quad H_{F_0}(y)=$$
$$\begin{cases} -F_0(y)\left((1-F_0(y))\frac{da}{dQ}(y)+(1-F_0(y+1))\frac{da}{dQ}(y+1)\right), & 0\le y\le 1,\\ (1-F_0(y))\left(F_0(y-1)\frac{da}{dQ}(y-1)+F_0(y)\frac{da}{dQ}(y)\right), & 1\le y\le 2.\end{cases}$$

One may replace F_0 in (11.73) and (11.74) by any $F\in\Lambda$. The result will be denoted by b_F^* and H_F^* respectively. If dH_F^*/dF does not exist, b_F^* is not the efficient influence function at θ_F. However, by construction,

$$\int b_F^*\,dP=-(\theta_F-\theta_0),$$

so that condition (ia) of Theorem 11.9 holds. Now, let $\xi_F(y)$ be defined by

$$H_F^*(y)=\begin{cases}F(y)\xi_F(y), & 0\le y\le 1\\ (1-F(y))\xi_F(y), & 1\le y\le 2.\end{cases}$$

Take $\bar\xi_F$ according to the recipe of Lemma 11.10 with $\xi=\xi_{F_0}$. Then we obtain

$$|\xi_F(y)-\bar\xi_F(y)|\le|\xi_F(y)-\xi_{F_0}(y)|+c_1|F_0(y)-F(y)|.$$

Let

$$\bar H_F(y)=\begin{cases}F(y)\bar\xi_F(y), & 0\le y\le 1,\\ (1-F(y))\bar\xi_F(y), & 1\le y\le 2.\end{cases}$$

Then $\bar h_F=d\bar H_F/dF$ exists. If $F(y+1)-F(y)\ge\eta_0^2/2$, we find that for some constant c_1, and for $\bar b_F=A_F\bar h_F$,

$$\|\bar b_F-b_F^*\|_{\bar P_F}\le c_1 d(F,F_0).$$

Moreover, as in Example 11.2.3c, $d(\hat F_n,F_0)=O_{\mathbf P}(n^{-1/3})$. We conclude that the conditions of Theorem 11.9 are fulfilled provided we take

$$\Lambda_0=\{F\in\Lambda:\ F\text{ has finite support, }F(y+1)-F(y)\ge\eta_0^2/2,\text{ for all }y\in[0,1]\}.$$

We have thus shown that

$$\hat{\theta}_n - \theta_0 = \int b_{F_0}\, dP_n + o_{\mathbf{P}}(n^{-\frac{1}{2}}).$$

Example 11.2.3e. Convolution with the uniform revisited Consider again the model

$$X = Y + Z,$$

with Y and Z independent and Z uniformly distributed on $[0, 1]$. The distribution F_0 of Y is unknown. We suppose it has support contained in a bounded interval:

$$\text{supp}(F_0) \subset [0, r].$$

Without loss of generality, we assume that $r \geq 1$ is integer. We want to estimate $\theta_{F_0} = \int a\, dF_0$, where we assume that $\dot{a}(x) = da(x)/dx$ exists, and $\dot{a} \in L_\infty$. Moreover, we assume

$$\left| \frac{d\dot{a}(y+j)}{dF_0(y+l)} \right| \leq c, \quad \text{for all } j, l = 0, \ldots, r-1, \ y \in [0,1],$$
$$\left| \frac{dF_0(y+j)}{dF_0(y+l)} \right| \leq c, \quad \text{for all } j, l = 0, \ldots, r-1, \ y \in [0,1],$$

and

$$F_0(y+1) - F_0(y) \geq \eta_0^2 > 0, \quad \text{for all } y \in [0, r-1].$$

The efficient influence curve is now

$$b_{F_0}(x) = -\sum_{l=0}^{r-1} (1 - F_0(x+l))\dot{a}(x+l), \quad x \in [0,1],$$
$$b_{F_0}(x+j) = \sum_{l=0}^{j-1} \dot{a}(x+l) + b_{F_0}(x), \quad x \in [0,1], \ j = 1, \ldots, r.$$

The function H_{F_0} is defined by

$$H_{F_0}(y) = F_0(y) b_{F_0}(y), \quad y \in [0,1],$$
$$H_{F_0}(y+j) = H_{F_0}(y+j-1) + (F_0(y+j) - F_0(y+j-1)) b_{F_0}(y+j),$$
$$y \in [0,1], \ j = 1, \ldots, r-1.$$

Note that for $r = 2$, the problem is of the same form as in the previous example. (In fact, the model is similar to the interval censoring case r, with r equally spaced examination times (regular checkups).) Therefore, we obtain

$$\hat{\theta}_n - \theta_0 = \int b_{F_0}\, dP_n + o_{\mathbf{P}}(n^{-1/2}).$$

Example 11.2.3f. Convolution with the exponential distribution Let μ be Lebesgue measure on $(0,\infty)$ and

$$k(x \mid y) = e^{-(x-y)} 1\{x > y\}.$$

Assume $\dot{a}(y) = da(y)/dy$ exists, $\dot{a} \in L_\infty$ and that for all $y \in \text{support}(F_0) \subset (0,\infty)$,

$$\left| \frac{d(e^{-y}\dot{a}(y))}{dF_0(y)} \right| \leq c.$$

Moreover, assume $\int_0^\infty e^y \, dF_0(y) < \infty$. The equation $A^* b_{F_0}(y) = a(y) - \theta_{F_0}$ becomes

$$\int_{y-}^\infty e^{-(x-y)} b_{F_0}(x) \, dx = a(y) - \theta_{F_0},$$

or

$$b_{F_0}(y) = a(y) - \dot{a}(y) - \theta_{F_0}.$$

Notice that there is only one solution. We may use the simple estimator

$$\tilde{\theta}_n = \int (a - \dot{a}) \, dP_n.$$

Solving the equation $A_{F_0} h_{F_0} = b_{F_0}$ gives $h_{F_0} = dH_{F_0}/dF_0$ with

$$H_{F_0}(y) = \int_0^y \left(a(u) - \theta_{F_0} - e^{-(y-u)} \dot{a}(y) \right) dF_0(u).$$

So,

$$h_{F_0}(y) = a(y) - \dot{a}(y) - \frac{d(e^{-y}\dot{a}(y))}{dF_0(y)} \int_0^y e^u \, dF_0(u) - \theta_{F_0}.$$

One may verify that the conditions of Theorem 11.8 are met. It follows that the maximum likelihood estimator $\hat{\theta}_n$ is asymptotically equivalent to the simple estimator $\tilde{\theta}_n$, and that they are both asymptotically efficient.

Example 11.2.3g. Mixture with an exponential family Let $(\mathcal{Y}, \mathcal{B}) = (\mathbf{R}, \text{Borel sets})$, and

$$k(x \mid y) = \exp\left[\psi(y) T(x) - c(\psi(y)) \right],$$

where

$$c(\psi) = \log \int \exp\left[\psi T(x) \right] d\mu(x).$$

Write

$$\dot{c}(\psi) = \frac{dc(\psi)}{d\psi}.$$

We assume that $\{\psi(y) : y \in \mathbf{R}\}$ has a non-empty interior, and that this interior is a subset of the support of F_0. Then $\{k(\cdot \mid y) : y \in \text{supp}(F_0)\}$ is a complete exponential family, so that there is at most one solution $b(x)$ to the equation

$$A^* b(y) = a(y) - \theta_0, \quad F_0\text{-a.s..}$$

Now, take $a(y) = \dot{c}(\psi(y))$, so that $\theta_{F_0} = \int \dot{c}(\psi)\, dF_0$. Then clearly, $b_{F_0}(x) = T(x) - \theta_{F_0}$ is the unique solution:

$$A^*(T - \theta_{F_0}) = \dot{c}(\psi(y)) - \theta_{F_0}.$$

Therefore, if one can solve the equation $A_{F_0} h_{F_0}(x) = T(x) - \theta_{F_0}$, the simple estimator

$$\tilde{\theta}_n = \int T \, dP_n$$

is asymptotically efficient. The following lemma investigates this further.

Lemma 11.12 *Suppose that $dF_0/d\psi$ exists and is a continuous function of y, of bounded variation. Then for*

$$H_{F_0}(y) = \int_{-\infty}^{y} \dot{c}(\psi)\, dF_0 - \frac{dF_0}{d\psi}(y) - \theta_{F_0} F_0(y),$$

we have

$$\frac{\int k(x \mid y)\, dH_{F_0}(y)}{p_{F_0}(y)} = T(x) - \theta_{F_0}. \tag{11.75}$$

Moreover, if each $F \in \Lambda$ is identifiable, H_{F_0} is also the unique solution of (11.75).

Proof Note first of all that H_{F_0} is of bounded variation, so that the integrals are well defined. Partial integration gives

$$\begin{aligned}
\int k(x \mid y)\, dH_{F_0}(y) &= \int k(x \mid y) \dot{c}(\psi(y))\, dF_0(y) \\
&\quad - \int k(x \mid y)\, d\left(\frac{dF}{d\psi}\right)(y) - p_{F_0}(x)\theta_{F_0} \\
&= \int k(x \mid y) \dot{c}(\psi(y))\, dF_0(y) \\
&\quad + \int \frac{dF_0}{d\psi}(y) k(x \mid dy) - p_{F_0}(x)\theta_{F_0} \\
&= \int k(x \mid y) \dot{c}(\psi(y))\, dF_0(y) \\
&\quad + \int \frac{dF_0}{d\psi}(y) k(x \mid y)(T(x) - \dot{c}(\psi(y)))\, d\psi(y) - p_{F_0}(x)\theta_{F_0} \\
&= p_{F_0}(x)(T(x) - \theta_{F_0}).
\end{aligned}$$

If H_1 and H_2 are two solutions of (11.75), both of bounded variation, then it follows that

$$\int k(x \mid y)\, d\big(H_1(y) - H_2(y)\big) = 0.$$

But since $H_1 - H_2$ is also of bounded variation, each $F \in \Lambda$ identifiable together with (11.75) implies $H_1 = H_2$. □

Corollary 11.13 *Suppose that for some constant* c_2

$$\left| \frac{d(dF_0/d\psi)}{dF_0} \right|_\infty \leq c_2.$$

Then $T - \theta_{F_0}$ *is the efficient influence curve for estimating* $\theta_{F_0} = \int \dot{c}(\psi)\, dF_0$, *and*

$$h_{F_0} = \dot{c}(\psi) - \frac{d(dF_0/d\psi)}{dF_0} - \theta_{F_0} \tag{11.76}$$

is a worst possible subdirection.

We conclude that under the condition of the above corollary, $\tilde{\theta}_n = \int T\, dP_n$ is asymptotically efficient. The question arises whether the same is true for the maximum likelihood estimator $\hat{\theta}_n = \int \dot{c}(\psi)\, d\hat{F}_n$. We leave this as an open problem.

11.3. A single-indexed model with binary explanatory variable

Let $(Y_1, Z_1), \ldots, (Y_n, Z_n), \ldots$ be i.i.d. copies of (Y, Z), where $Y \in \{0, 1\}$ is a binary response variable, and $Z = (U, V)$ is a covariate consisting of a binary variable $U \in \{0, 1\}$ and a continuous variable $V \in (0, 1)$. For example, Y indicates whether a person has a job, U indicates whether he/she has children, and V is education (measured on a continuous scale).

We transform the random variable V to an $\mathbf{R}$-valued random variable, using the transformation $V \mapsto \Phi^{-1}(V)$, where Φ is the distribution function of the standard normal distribution, and consider the single-indexed model

$$P(Y = 1 \mid U = u,\ V = v) = uF_0(\Phi^{-1}(v)) + (1 - u)F_0(\Phi^{-1}(v) + \theta_0),$$

where F_0 is an unknown function on $\mathbf{R}$, and $\theta_0 \in \mathbf{R}$ is an unknown shift. We shall investigate asymptotic normality (and efficiency) of a one-step estimator of θ_0. This estimator is based on an estimator of the efficient influence function. The efficient influence function depends on the unknown parameters θ_0 and F_0, as well as on the density of (U, V). Therefore, all these quantities need to be estimated well enough, with a fast enough rate. We shall employ penalized maximum likelihood as the estimation method, and then plug in the estimators in the expression for the efficient influence curve. Other plug-in estimators are also allowed, as long as they are sufficiently good, in the sense that they converge with a fast enough rate.

11.3.1. Rates of convergence

Let Q be the distribution of $Z = (U, V)$. Suppose that the density (with respect to Lebesgue measure) q_u of V given $U = u$ exists. Define $\pi_1 = P(U = 1)$ and $\pi_0 = P(U = 0) = 1 - \pi_1$, $0 < \pi_1 < 1$. The density $p_0 = p_{\theta_0, F_0}$, with respect to $\mu =$ (counting measure on $\{0, 1\}) \times Q$ of Y, is in the class

$$\begin{aligned}\mathscr{P} = \big\{p_{\theta,F}(y, u, v) &= yuF\big(\Phi^{-1}(v)\big) + (1 - y)u\big(1 - F(\Phi^{-1}(v))\big) \\ &+ y(1 - u)F(\Phi^{-1}(v) + \theta) \\ &\quad + (1 - y)(1 - u)\big(1 - F(\Phi^{-1}(v) + \theta)\big) \,:\, \theta \in \Theta, F \in \Lambda\big\}.\end{aligned}$$

Here, the parameter space (Θ, Λ) has to be sufficiently small for our purposes. To avoid digressions, we suppose that Θ is a bounded open set. Now, it turns out that for the asymptotic normality of an estimator of θ_0, one needs to be able to estimate the derivative $f_0(\xi) = dF_0(\xi)/d\xi$ of F_0. Therefore, we assume that each F in Λ is twice differentiable. Because the functions F have unbounded support, we use the transformation $\xi \mapsto \Phi(\xi)$. Let

$$\Gamma = \left\{\gamma : (0, 1) \to (0, 1) : \int_0^1 \left(\gamma^{(2)}(v)\right)^2 dv < \infty\right\}, \tag{11.77}$$

and

$$\Lambda = \{F(\xi) = \gamma(\Phi(\xi)) \,:\, \gamma \in \Gamma\}.$$

Thus, we require that

$$F_0 \in \Lambda. \tag{11.78}$$

For $F = \gamma(\Phi)$, we define

$$I^2(F) = \int_0^1 (\gamma^{(2)}(v))^2 \, dv.$$

Let $\lambda_n > 0$ be a smoothing parameter, and consider the penalized maximum likelihood estimator

$$(\hat{\theta}_n, \hat{F}_n) = \arg \max_{\theta \in \Theta,\, F \in \Lambda} \left(\int \log p_{\theta,F} - \lambda_n^2 I^2(F)\right). \tag{11.79}$$

Write $F_0 = \gamma_0(\Phi)$, $\hat{F}_n = \hat{\gamma}_n(\Phi)$, and $\hat{f}_n(\xi) = d\hat{F}_n(\xi)/d\xi = \hat{\gamma}_n^{(1)}(\Phi(\xi))\phi(\xi)$, with ϕ the density of the standard normal distribution.

We shall assume that for some constants $\epsilon_0 > 0$, $0 < \eta_0 < 1$, $K > 0$,

$$q_u(v) \geq \epsilon_0^2, \quad v \in (0,1),\ u \in \{0,1\}, \tag{11.80}$$

$$\eta_0^2 \leq F_0(\xi) \leq 1 - \eta_0^2, \quad \xi \in \mathbf{R}, \tag{11.81}$$

$$f_0(\xi) \geq \epsilon_0^2, \quad \text{for all } |\xi| \leq K. \tag{11.82}$$

Lemma 11.14 *Take* $\lambda_n \asymp n^{-2/5}$. *Then under conditions* (11.78), (11.80), (11.81) *and* (11.82),

$$h(\hat{p}_n, p_0) = O_{\mathbf{P}}(n^{-2/5}), \tag{11.83}$$

$$I(\hat{F}_n) = O_{\mathbf{P}}(1), \tag{11.84}$$

$$|\hat{F}_n - F_0|_\infty = O_{\mathbf{P}}(n^{-3/10}), \tag{11.85}$$

$$\int \left(\hat{f}_n(\xi) - f_0(\xi)\right)^2 d\xi = O_{\mathbf{P}}(n^{-2/5}), \tag{11.86}$$

$$|\hat{f}_n - f_0|_\infty = O_{\mathbf{P}}(n^{-1/10}), \tag{11.87}$$

and

$$|\hat{\theta}_n - \theta_0| = O_{\mathbf{P}}(n^{-2/5}), \tag{11.88}$$

Proof By Lemma 10.5, we have for $g_p = \frac{1}{2}\log(p + p_0)/2p_0$,

$$h^2(\hat{p}_n, p_0) + 4\lambda_n^2 I^2(\hat{F}_n) \leq 16 \int g_{\hat{p}_n}\, d(P_n - P) + 4\lambda_n^2 I(F_0).$$

Define for $M \geq 1$, $\mathscr{P}_M = \{p_{F,\theta} \in \mathscr{P} : I(F) \leq M\}$ and

$$\bar{\mathscr{P}}_M^{1/2} = \left\{ \sqrt{\frac{p_{F,\theta} + p_0}{2}} : p_{F,\theta} \in \mathscr{P}_M \right\}.$$

Since $p_0 \geq \eta_0^2$, it is easy to see that Theorem 2.4 implies

$$H_B(\delta, \bar{\mathscr{P}}_M^{1/2}, \mu) \leq A\left(\frac{M}{\delta}\right)^{1/2}, \quad M \geq 1,\ \delta > 0.$$

Proceed as in Theorem 10.6, to find (11.83) and (11.84).

Now, $h(\hat{p}_n, p_0) = O_{\mathbf{P}}(n^{-2/5})$ gives

$$\int \left(\hat{F}_n(\Phi^{-1}(v)) - F_0(\Phi^{-1}(v))\right)^2 q_1(v)\, dv = O_{\mathbf{P}}(n^{-4/5}),$$

which in turn implies

$$\int (\hat{\gamma}_n(v)-\gamma_0(v))^2\,dv \le \frac{1}{\epsilon_0^2}\int \left(\hat{F}_n(\Phi^{-1}(v)) - F_0(\Phi^{-1}(v))\right)^2 q_1(v)\,dv = O_{\mathbf{P}}(n^{-4/5}).$$

Since $I^2(\hat{F}_n) = \int (\hat{\gamma}_n^{(2)}(v))^2\,dv = O_{\mathbf{P}}(1)$, we conclude from Lemma 10.9 that

$$|\hat{\gamma}_n - \gamma_0|_\infty = O_{\mathbf{P}}(n^{-3/10}),$$

$$\int \left(\hat{\gamma}_n^{(1)}(v) - \gamma_0^{(1)}(v)\right)^2 dv = O_{\mathbf{P}}(n^{-2/5}),$$

and hence

$$|\hat{\gamma}_n^{(1)} - \gamma_0^{(1)}|_\infty = O_{\mathbf{P}}(n^{-1/10}).$$

But then, also

$$|\hat{F}_n - F_0|_\infty = |\hat{\gamma}_n(\Phi) - \gamma_0(\Phi)|_\infty = O_{\mathbf{P}}(n^{-3/10}),$$

$$\int \left(\hat{f}_n(\xi) - f_0(\xi)\right)^2 d\xi = O_{\mathbf{P}}(n^{-2/5}),$$

and

$$|\hat{f}_n - f_0|_\infty = \left|\left(\hat{\gamma}_n^{(1)}(\Phi) - \gamma_0^{(1)}(\Phi)\right)\phi\right|_\infty = O_{\mathbf{P}}(n^{-1/10}).$$

It remains to verify (11.88). Since $h(\hat{p}_n, p_0) = O_{\mathbf{P}}(n^{-2/5})$, we know that

$$\begin{aligned}
&\int \left(\hat{F}_n(\Phi^{-1}(v) + \hat{\theta}_n) - F_0(\Phi^{-1}(v) + \theta_0)\right)^2 dv \\
&\quad \le \frac{1}{\epsilon_0^2}\int \left(\hat{F}_n(\Phi^{-1}(v) + \hat{\theta}_n) - F_0(\Phi^{-1}(v) + \theta_0)\right)^2 q_0(v)\,dv = O_{\mathbf{P}}(n^{-4/5}).
\end{aligned}$$

Hence,

$$\begin{aligned}
&\int \left|F_0\left(\Phi^{-1}(v) + \hat{\theta}_n\right) - F_0\left(\Phi^{-1}(v) + \theta_0\right)\right| dv \\
&\quad \le \int \left|F_0\left(\Phi^{-1}(v) + \hat{\theta}_n\right) - \hat{F}_n\left(\Phi^{-1}(v) + \hat{\theta}_n\right)\right| dv + O_{\mathbf{P}}(n^{-2/5}) \\
&\quad = \int \left|F_0(\xi) - \hat{F}_n(\xi)\right| \phi(\xi - \hat{\theta}_n)\,d\xi + O_{\mathbf{P}}(n^{-2/5}) \\
&\quad \le \left(\int (\gamma_0(v) - \hat{\gamma}_n(v))^2\,dv\right)^{1/2} \left(\int \frac{\phi^2(\xi - \hat{\theta}_n)}{\phi(\xi)}\,d\xi\right)^{1/2} + O_{\mathbf{P}}(n^{-2/5}) \\
&\quad = O_{\mathbf{P}}(n^{-2/5}).
\end{aligned}$$

Condition (11.82) now yields that $|\hat{\theta}_n - \theta_0| = O_{\mathbf{P}}(n^{-2/5})$. □

Note that we did not require $\hat{F}_n$ to be a monotone function. However, it follows from the consistency of $\hat{f}_n$, that $\hat{F}_n$ will be monotone with large probability.

We also need sufficiently good estimators of π_u and q_u, $u \in \{0,1\}$. Let $\hat{\pi}_{1,n} = \frac{1}{n}\{\text{number of } U_i = 1,\ 1 \le i \le n\}$, $\hat{\pi}_{0,n} = 1 - \hat{\pi}_{1,n}$, $N_u = n\hat{\pi}_{u,n}$, $u \in \{0,1\}$, and

$$Q_{u,n} = \frac{1}{N_u}\sum_{i=1}^{n} \delta_{\{U_i=u,V_i\}}, \quad u \in \{0,1\}.$$

The empirical distribution of $V_1, \dots, V_n$ is then

$$Q_n = \hat{\pi}_{1,n}Q_{1,n} + \hat{\pi}_{0,n}Q_{0,n}.$$

We require that

$$(11.89) \qquad q_u \in \Gamma, \quad u \in \{0,1\},$$

and let $\hat{q}_{u,n}$ be the penalized maximum likelihood estimator
(11.90)

$$\hat{q}_{u,n} = \arg \max_{\tilde{q}\in\Gamma,\ \int \tilde{q}(v)\,dv=1} \left(\int \log \tilde{q}\, dQ_{u,n} - \lambda_n^2 \int_0^1 (\tilde{q}^{(2)}(v))^2\, dv \right), \quad u \in \{0,1\}.$$

Lemma 11.15 *Assume* (11.80) *and* (11.89). *Then for* $\lambda_n \asymp n^{-2/5}$, *we have for* $u \in \{0,1\}$,

$$(11.91) \qquad |\hat{q}_{u,n} - q_u|_\infty = O_{\mathbf{P}}(n^{-3/10}),$$

and

$$(11.92) \qquad \int_0^1 (\hat{q}_{u,n}^{(2)}(v))^2\, dv = O_{\mathbf{P}}(1).$$

Proof It follows from Theorem 10.6, combined with Lemma 10.9, that, given $U_1, \dots, U_n, \dots$,

$$|\hat{q}_{u,n} - q_u|_\infty = O_{\mathbf{P}}(N_u^{-3/10}),$$

and

$$\int_0^1 \hat{q}_{u,n}^{(2)}(v))^2\, dv = O_{\mathbf{P}}(1).$$

Clearly, since $N_u = n\pi_u(1 + O_{\mathbf{P}}(n^{-1/2}))$, this implies (11.91) and (11.92). □

The density of $\Phi^{-1}(V)$ given $U = u$ is

$$g_u(\xi) = q_u(\Phi(\xi))\phi(\xi), \quad \xi \in \mathbf{R}.$$

It can be estimated by

$$\hat{g}_{u,n}(\xi) = \hat{q}_{u,n}(\Phi(\xi))\phi(\xi), \quad \xi \in \mathbf{R}.$$

It follows that

$$|\hat{g}_{u,n} - g_u|_\infty = O_\mathbf{P}(n^{-3/10}), \quad u \in \{0,1\}.$$

11.3.2. Asymptotic normality

We shall now introduce the efficient influence function. Because we have already proved convergence of our estimators, this function need only be considered for a subset of the possible parameter values. Therefore, we define

$$\Lambda_M = \left\{ F(\xi) = \gamma(\Phi(\xi)) \; : \; \frac{\eta_0^2}{2} \le \gamma \le 1 - \frac{\eta_0^2}{2}, \; \int_0^1 \left(\gamma^{(2)}(v)\right)^2 dv \le M^2 \right\},$$

and

$$\mathscr{G}_M = \left\{ \tilde{g}(\xi) = \tilde{q}(\Phi(\xi))\phi(\xi) : \frac{\epsilon_0^2}{2} \le \tilde{q} \le 1 - \frac{\epsilon_0^2}{2}, \int \tilde{q}(v)\, dv = 1, \right.$$
$$\left. \int_0^1 (\tilde{q}^{(2)}(v))^2 \, dv \le M^2 \right\}.$$

Then we know that under the conditions of Lemmas 11.14 and 11.15, for M large, n large, $\hat{F}_n$ and $\hat{g}_{u,n}$ are with large probability in Λ_M and $\mathscr{G}_M$, respectively.

For $\theta \in \Theta$, $\tilde{g}_u \in \mathscr{G}_M$, $F \in \Lambda_M$, $\tilde{\pi}_1 \in (0,1)$, $\tilde{\pi}_0 = 1 - \tilde{\pi}_1$, we let

$$z(\xi; \tilde{g}_1, \tilde{g}_2, \tilde{\pi}_1, \theta) = -\frac{\tilde{\pi}_0 \tilde{g}_0(\xi - \theta)}{\tilde{\pi}_1 \tilde{g}_1(\xi) + \tilde{\pi}_0 \tilde{g}_0(\xi - \theta)}.$$

Consider the functions

$$\begin{aligned}
&b(y, u, v; \tilde{g}_1, \tilde{g}_0, \tilde{\pi}_1, \theta, F) \\
&\quad = yu \frac{z(\xi; \tilde{g}_1, \tilde{g}_2, \tilde{\pi}_1, \theta) f(\xi)}{F(\xi)} - (1-y)u \frac{z(\xi; \tilde{g}_1, \tilde{g}_2, \tilde{\pi}_1, \theta) f(\xi)}{1 - F(\xi)} \\
&\qquad + y(1-u) \frac{(1 + z(\xi + \theta; \tilde{g}_1, \tilde{g}_2, \tilde{\pi}_1, \theta) f(\xi + \theta)}{F(\xi + \theta)} \\
&\qquad - (1-y)(1-u) \frac{(1 + z(\xi + \theta; \tilde{g}_1, \tilde{g}_2, \tilde{\pi}_1, \theta) f(\xi + \theta)}{1 - F(\xi + \theta)}, \qquad \xi = \Phi^{-1}(v).
\end{aligned}$$

We use the shorthand notation

$$\begin{aligned}\hat{z}_n(\xi) &= z(\xi;\hat{g}_{1,n},\hat{g}_{0,n},\hat{\pi}_{1,n},\hat{\theta}_n),\\ z_0(\xi) &= z(\xi,g_1,g_0,\pi_1,\theta_0),\\ \hat{b}_n(y,u,v) &= b(y,u,v;\hat{g}_{1,n},\hat{g}_{0,n},\hat{\pi}_{1,n},\hat{\theta}_n,\hat{F}_n),\end{aligned}$$

and

$$b_0(y,u,v) = b(y,u,v,;g_1,g_0,\pi_1,\theta_0,F_0).$$

It can be shown that b_0 is the efficient influence function at θ_0. Finally define

$$\hat{\tau}_n^2 = \int\left(\frac{\hat{f}_n^2(\xi)}{\hat{F}_n(\xi)(1-\hat{F}_n(\xi))}\right)\left(\frac{\hat{\pi}_{0,n}\hat{\pi}_{1,n}\hat{g}_{0,n}(\xi-\hat{\theta}_n)\hat{g}_{1,n}(\xi)}{\hat{\pi}_{1,n}\hat{g}_{1,n}(\xi)+\hat{\pi}_{0,n}\hat{g}_{0,n}(\xi-\hat{\theta}_n)}\right)d\xi,$$

and

$$\tau_0^2 = \int\left(\frac{f_0^2(\xi)}{F_0(\xi)(1-F_0(\xi))}\right)\left(\frac{\pi_0\pi_1 g_0(\xi-\theta_0)g_1(\xi)}{\pi_1 g_1(\xi)+\pi_0 g_0(\xi-\theta_0)}\right)d\xi.$$

One can think of the efficient influence function b_0 as the derivative of the log-likelihood in the worst possible subdirection. It is therefore easily verified that $\int b_0\,dP = 0$. A Taylor expansion now gives

Lemma 11.16 *Assume the conditions of Lemma* 11.14 *and Lemma* 11.15 *are met. Then*

$$\int \hat{b}_n\,dP = -(\hat{\theta}_n-\theta_0)\hat{\tau}_n^2 + o_P(n^{-1/2}). \tag{11.93}$$

Proof We have

$$\begin{aligned}\int \hat{b}_n\,dP &= \int \frac{\hat{z}_n(\xi)\hat{f}_n(\xi)}{\hat{F}_n(\xi)}F_0(\xi)\pi_1 g_1(\xi)\,d\xi\\ &\quad - \int \frac{\hat{z}_n(\xi)\hat{f}_n(\xi)}{1-\hat{F}_n(\xi)}(1-F_0(\xi))\pi_1 g_1(\xi)\,d\xi\\ &\quad + \int \frac{(1+\hat{z}_n(\xi+\hat{\theta}_n))\hat{f}_n(\xi+\hat{\theta}_n)}{\hat{F}_n(\xi+\theta_n)}F_0(\xi+\theta_0)\pi_0 g_0(\xi)\,d\xi\\ &\quad - \int \frac{(1+\hat{z}_n(\xi+\hat{\theta}_n))\hat{f}_n(\xi+\hat{\theta}_n)}{1-\hat{F}_n(\xi+\hat{\theta}_n)}(1-F_0(\xi+\theta_0))\pi_0 g_0(\xi)\,d\xi\\ &= \mathrm{I}-\mathrm{II}+\mathrm{III}-\mathrm{IV}.\end{aligned}$$

Since $I(F_0) < \infty$,

$$F_0(\xi+\theta_0) = F_0(\xi+\hat{\theta}_n) - (\hat{\theta}_n-\theta_0)f_0(\xi+\hat{\theta}_n) + O\big(|\hat{\theta}_n-\theta_0|^{3/2}\big).$$

This implies that

$$\begin{aligned}
\text{III} &= \int \frac{(1+\hat{z}_n(\xi+\hat{\theta}_n))\hat{f}_n(\xi+\hat{\theta}_n)}{\hat{F}_n(\xi+\hat{\theta}_n)} F_0(\xi+\hat{\theta}_n)\pi_0 g_0(\xi)\,d\xi \\
&\quad - (\hat{\theta}_n - \theta_0)\int \frac{(1+\hat{z}_n(\xi+\hat{\theta}_n))\hat{f}_n(\xi+\hat{\theta}_n)}{\hat{F}_n(\xi+\hat{\theta}_n)} f_0(\xi+\hat{\theta}_n)\pi_0 g_0(\xi)\,d\xi \\
&\quad + O_{\mathbf{P}}\big(|\hat{\theta}_n-\theta_0|^{3/2}\big) \\
&= \int \frac{(1+\hat{z}_n(\xi))\hat{f}_n(\xi)}{\hat{F}_n(\xi)} F_0(\xi)\pi_0 g_0(\xi-\hat{\theta}_n)\,d\xi \\
&\quad - (\hat{\theta}_n - \theta_0)\int \frac{(1+\hat{z}_n(\xi))\hat{f}_n(\xi)}{\hat{F}_n(\xi)} f_0(\xi)\pi_0 g_0(\xi-\hat{\theta}_n))\,d\xi + O_{\mathbf{P}}\big(|\hat{\theta}_n-\theta_0|^{3/2}\big).
\end{aligned}$$

Similarly,

$$\begin{aligned}
\text{IV} &= \int \frac{(1+\hat{z}_n(\xi))\hat{f}_n(\xi)}{1-\hat{F}_n(\xi)} (1-F_0(\xi))\pi_0 g_0(\xi-\hat{\theta}_n)\,d\xi \\
&\quad + (\hat{\theta}_n - \theta_0)\int \frac{(1+\hat{z}_n(\xi))\hat{f}_n(\xi)}{1-\hat{F}_n(\xi)} f_0(\xi)\pi_0 g_0(\xi-\hat{\theta}_n)\,d\xi + O_{\mathbf{P}}\big(|\hat{\theta}_n-\theta_0|^{3/2}\big).
\end{aligned}$$

It follows that

$$\begin{aligned}
&(11.94)\qquad \int \hat{b}_n\,dP = \\
&\int \Big(\hat{z}_n(\xi)\pi_1 g_1(\xi) + \big(1+\hat{z}_n(\xi)\big)\pi_0 g_0\big(\xi-\hat{\theta}_n\big)\Big)\left(\frac{F_0(\xi)}{\hat{F}_n(\xi)} - \frac{1-F_0(\xi)}{1-\hat{F}_n(\xi)}\right)\hat{f}_n(\xi)\,d\xi \\
&\quad - (\hat{\theta}_n - \theta_0)\int \frac{(1+\hat{z}_n(\xi))\hat{f}_n(\xi)}{\hat{F}_n(\xi)(1-\hat{F}_n(\xi))} f_0(\xi)\pi_0 g_0(\xi-\hat{\theta}_n)\,d\xi + O_{\mathbf{P}}(|\hat{\theta}_n-\theta_0|^{3/2}) \\
&= \int \left(\frac{\hat{\pi}_{1,n}\hat{g}_{1,n}(\xi)\pi_0 g_0(\xi-\hat{\theta}_n) - \pi_1 g_1(\xi)\hat{\pi}_{0,n}\hat{g}_{0,n}(\xi-\hat{\theta}_n)}{\hat{\pi}_{1,n}\hat{g}_{1,n}(\xi) + \hat{\pi}_{0,n}\hat{g}_{0,n}(\xi-\hat{\theta}_n)}\right) \\
&\qquad \times \left(\frac{F_0(\xi)}{\hat{F}_n(\xi)} - \frac{1-F_0(\xi)}{1-\hat{F}_n(\xi)}\right)\hat{f}_n(\xi)\,d\xi - (\hat{\theta}_n-\theta_0)\tilde{\tau}_n^2 + O_{\mathbf{P}}\big(|\hat{\theta}_n-\theta_0|^{3/2}\big),
\end{aligned}$$

where

$$\tilde{\tau}_n^2 = \int \left(\frac{f_0(\xi)}{\hat{F}_n(\xi)(1-\hat{F}_n(\xi))}\right)\left(\frac{\hat{\pi}_{1,n}\pi_0\hat{g}_{1,n}(\xi)g_0(\xi-\hat{\theta}_n)}{\hat{\pi}_{1,n}\hat{g}_{1,n}(\xi) + \hat{\pi}_{0,n}\hat{g}_{0,n}(\xi-\hat{\theta}_n)}\right)\hat{f}_n(\xi)\,d\xi.$$

Since $\int(\hat{f}_n(\xi)-f_0(\xi))^2\,d\xi = O_{\mathbf{P}}(n^{-2/5})$ and $|\hat{g}_{u,n}-g_u|_\infty = O_{\mathbf{P}}(n^{-3/10})$, we have

$$|\tilde{\tau}_n^2 - \hat{\tau}_n^2| = O_{\mathbf{P}}(n^{-1/5}).$$

Moreover,

$$|\hat{F}_n - F_0|_\infty = O_{\mathbf{P}}(n^{-3/10}),$$

as well as

$$|\hat{g}_{u,n} - g_u|_\infty = O_{\mathbf{P}}(n^{-3/10}).$$

This implies that the first term on the left-hand side of (11.94) is $O_{\mathbf{P}}(n^{-3/5})$. Moreover, $|\hat{\theta}_n - \theta_0| = O_{\mathbf{P}}(n^{-2/5})$, so that $|\hat{\theta}_n - \theta_0|^{3/2} = O_{\mathbf{P}}(n^{-3/5})$, as well as $(\hat{\theta}_n - \theta_0)(\hat{\tau}_n^2 - \tilde{\tau}_n^2) = O_{\mathbf{P}}(n^{-3/5})$. Insert these rates in (11.94) and note that $n^{-3/5} = o(n^{-1/2})$, to complete the proof. □

As in the previous subsection, it is important that the set of efficient influence functions is P-Donsker. This allows one to conclude that

Lemma 11.17 *Assume that the conditions of Lemma* 11.14 *and Lemma* 11.15 *are met. Then*

$$\int \hat{b}_n d(P_n - P) = \int b_0 \, d(P_n - P) + o_{\mathbf{P}}(n^{-1/2}). \tag{11.95}$$

Proof For any $M < \infty$, the collection

$$\{b(\cdot\,; \tilde{g}_1, \tilde{g}_0, \tilde{\pi}_1, \theta, F) : \ \tilde{g}_u \in \mathcal{G}_M, \ u \in \{0,1\}, \tilde{\pi}_1 \in (0,1), \ \theta \in \Theta, \ F \in \Lambda_M\}$$

is a P-Donsker class. Here, we use the fact that

$$H_\infty(\delta, \mathcal{G}_M) \le A_{1,M}\delta^{-1/2},$$

and

$$H_\infty(\delta, \Lambda_M) \le A_{2,M}\delta^{-1/2},$$

and for

$$\mathcal{F}_M = \{f(\xi) = dF(\xi)/d\xi \ : \ F \in \Lambda_M\},$$

we have

$$H_\infty(\delta, \mathcal{F}_M) \le A_{3,M}\delta^{-1}.$$

Since $\hat{g}_{u,n} \in \mathcal{G}_M$, $u \in \{0,1\}$ and $\hat{F}_n \in \Lambda_M$ with arbitrary large probability for all n sufficiently large, whenever M is sufficiently large, and

$$\|\hat{b}_n - b_0\| = o_{\mathbf{P}}(1),$$

the lemma follows from asymptotic equicontinuity arguments. □

Now, it is not clear whether $\int \hat{b}_n \, dP_n = o_{\mathbf{P}}(n^{-1/2})$ in general. We avoid checking this by considering the one-step estimator

$$\tilde{\theta}_n = \hat{\theta}_n + \hat{\tau}_n^{-2} \int \hat{b}_n \, dP_n. \tag{11.96}$$

Theorem 11.18 *Suppose that the conditions of Lemmas* 11.14 *and* 11.15 *are met. Then*

$$\sqrt{n}(\tilde{\theta}_n - \theta_0) \to^{\mathscr{L}} \mathscr{N}(0, \tau_0^{-2}). \tag{11.97}$$

Proof We have

$$\int \hat{b}_n \, dP_n = \int \hat{b}_n \, d(P_n - P) + \int \hat{b}_n \, dP$$
$$= \int b_0 \, d(P_n - P) - (\hat{\theta}_n - \theta_0)\hat{\tau}_n^2 + o_{\mathbf{P}}(n^{-1/2}).$$

It follows that

$$\tilde{\theta}_n - \theta_0 = \hat{\tau}_n^{-2} \int b_0 \, dP_n + o_{\mathbf{P}}(n^{-1/2}) = \tau_0^{-2} \int b_0 \, dP_n + o_{\mathbf{P}}(n^{-1/2}).$$

One easily verifies that $\|b_0\|^2 = \tau_0^2$, so the result now follows from the classical central limit theorem. □

Note that $\hat{\tau}_n^{-2}$ is an estimator of the variance of the asymptotic distribution of $\sqrt{n}(\tilde{\theta}_n - \theta_0)$. So, for the studentized version, we have convergence to the standard normal distribution:

$$\sqrt{n}\hat{\tau}_n(\tilde{\theta}_n - \theta_0) \to^{\mathscr{L}} \mathscr{N}(0, 1).$$

This allows one to construct asymptotic confidence intervals for θ_0. In summary, the one-step estimator has the disadvantage that it requires estimators of all unknown parameters, including the density of (U, V), but on the other hand these estimators also supply one with asymptotic confidence limits.

11.4. Notes

For the general theory on semiparametric models, we refer to the book of Bickel, Klaassen, Ritov and Wellner (1993). We did not go into the theory on *regular* estimators (see Problem 11.3 for a definition), but our methods certainly allow one to prove regularity. This will imply asymptotic efficiency of the (penalized) maximum likelihood estimators for which we derived asymptotic normality.

Quasi-likelihood models are studied in, for example, Severini and Staniswalis (1994), and Fan, Heckman and Wand (1995). Results for the penalized quasi-likelihood estimator are obtained in Mammen and van de Geer (1997b).

The results for mixture models, as given in Section 11.2, are extensions of van de Geer (1995c). Interval censoring case I is treated in Groeneboom and Wellner (1992), while Geskus and Groeneboom (1996) study interval censoring case II.

Estimation in single-indexed models is closely related to projection pursuit techniques. The model plays a prominent role in the econometrics literature. In Härdle, Hall and Ichimura (1993), one can find optimal kernel smoothing techniques.

11.5. Problems and complements

11.1. Suppose that the log-likelihood of Y given Z, w.r.t. a σ-finite measure ν, is $Q(y;\mu_0(Z))$, where $\underline{\mu} < \mu_0(Z) < \overline{\mu}$, and where

$$Q(y;\mu) = \int_y^\mu \frac{y-s}{V(s)}\,ds, \quad \underline{\mu} < \mu < \overline{\mu}.$$

Show that (11.3) is met for sufficiently large K and σ_0^2.

11.2. Suppose that Y given Z is exponentially distributed, and

$$E(Y \mid Z) = \mu_0(Z) = e^{g_0(Z)}, \quad Z = (U,V) \in [0,1]^2,$$

where

$$g_0 \in \mathscr{G} = \{g(z) = \theta u + \gamma(v),\ |\theta| < K_1,\ |\gamma|_\infty < K_2,\ I(\gamma) < \infty\},$$

with

$$I^2(\gamma) = \int_0^1 (\gamma^{(m)}(v))^2\,dv.$$

Let

$$\hat{g}_n = \arg \max_{g(z)=\theta u+\gamma(v)\in\mathscr{G}} \left(\frac{1}{n}\sum_{i=1}^n Y_i e^{g(Z_i)} - \frac{1}{n}\sum_{i=1}^n g(Z_i) - \lambda_n^2 I^2(\gamma) \right).$$

Assume that

(a) $\lambda_n = o_{\mathbf{P}}(n^{-1/4})$, $\lambda_n^{-1} = O_{\mathbf{P}}(n^{\frac{m}{2m+1}})$,

(b) $I(h) < \infty$, where $h(v) = E(U \mid V = v)$,

(c) the distribution Q of Z has a density q w.r.t. Lebesgue measure, satisfying $q(z) \geq \eta_0 > 0$,

(d) $\|\tilde{h}\| > 0$, where $\tilde{h}(u,v) = u - h(v)$.

Show that $\|\hat{g}_n - g_0\| = O_{\mathbf{P}}(\lambda_n)$, $I(\hat{\gamma}_n) = O_{\mathbf{P}}(1)$ and that

$$\sqrt{n}(\hat{\theta}_n - \theta_0) \to^{\mathscr{L}} \mathscr{N}(0, \|\tilde{h}\|^{-2}).$$

11.3. Here, we briefly describe the concept of *asymptotic efficiency* of estimators. Let $\mathscr{M}$ be a collection of probability measures on $(\mathscr{X}, \mathscr{A})$. Fix $P \in \mathscr{M}$ and let $\{P_t\} \subset \mathscr{M}$ be a one-dimensional submodel, indexed by $t \in [-\eta, \eta]$. We call this one-dimensional submodel *Hellinger differentiable* at P if for some $b \in L_2(P)$,

$$\int \left(\frac{(dP_t)^{1/2} - (dP)^{1/2}}{t} - \frac{1}{2} b (dP)^{1/2} \right)^2 \to 0, \quad \text{as } t \to 0.$$

Then b is called a score function, and the linear span of all score functions is called the *tangent space* B at P.

Consider now a parameter $\theta : \mathscr{M} \to \mathbf{R}$. Suppose that for all Hellinger differentiable paths $t \mapsto P_{t,b}$, with score b,

$$\frac{\theta(P_{t,b}) - \theta(P)}{t} \to \theta'(b),$$

where θ' is a continuous, linear, real-valued function on B. Then θ is called *differentiable*. Let $\bar{B}$ be the closure of $B \subset L^2(P)$ and let $\bar{\theta}'$ be a continuous linear function on $\bar{B}$, namely the smallest extension of θ' to $\bar{B}$, Then we can write

$$\bar{\theta}'(b) = (b_0, b) = \int b_0 b \, dP, \; b \in \bar{B},$$

for some $b_0 \in \bar{B}$, and b_0 is called the *efficient influence function.*

A sequence of estimators $T_n = T_n(X_1, \ldots, X_n)$ is called *regular* if for all Hellinger differentiable submodels $\{P_{n^{-1/2},b}\}$, and for $X_1, \ldots, X_n$ i.i.d. with distribution $P_{n^{-1/2},b}$, we have that $\sqrt{n}(T_n - \theta(P_{n^{-1/2},b}))$ converges weakly to a fixed random variable.

If the set of all scores is a linear space, then the limiting distribution of any regular estimator T_n of a differentiable parameter θ has variance at least $\|b_0\|^2$, and is called *asymptotically efficient* if the variance of the limiting distribution is equal to $\|b_0\|^2$. (See for example van der Vaart and Wellner (1996, Theorem 3.11.2).)

11.4. Let $\mathscr{P} = \{p_F = \int k(\cdot \mid y) \, dF(y) : \; F \in \Lambda\}$ be a mixture model, and $\theta(P_F) = \int a \, dF$, $P_F \in \mathscr{M} = \{P_F : dP_F/d\mu = p_F \in \mathscr{P}\}$. Here, Λ is the collection of all probability measures on $(\mathscr{Y}, \mathscr{B})$. Show that the tangent space at $P = P_{F_0}$ is $\{b = A_{F_0}h : h \in L_2(F_0), \int h\,dF_0 = 0\}$. (The tangent space is defined in Problem 11.3.)

11.5. Suppose that in Example 11.2.3c, Q and F_0 have support $[0,1]$, and densities q and f_0 respectively, with respect to Lebesgue measure. Assume moreover that for some $c_0 > 0$,

$$\frac{1}{c_0} \leq q(z) \leq c_0, \quad z \in [0,1],$$

and

$$f_0(y) \geq \frac{1}{c_0}, \quad y \in [0,1].$$

Write

$$m_0 = \int y\,dF_0(y).$$

Consider the maximum likelihood estimator

$$\hat{\theta}_n = \int y^2\,d\hat{F}_n(y) - \left(\int y\,d\hat{F}_n(y)\right)^2$$

of the variance

$$\theta_0 = \int y^2\,dF_0(y) - m_0^2.$$

Show that $\sqrt{n}(\hat{\theta}_n - \theta_0)$ converges in distribution to a centered normal random variable, with variance

$$4\int F_0(z)\big(1 - F_0(z)\big)\frac{(z - m_0)^2}{q(z)}\,dz.$$

11.6. Let $X = Y + Z$, with Y and Z independent, Y has an unknown distrubution F_0 on $[0,1]$, and the density k, w.r.t. Lebesgue measure, of Z is known to be

$$k(z) = \frac{e^z}{e-1}, \quad z \in [0,1].$$

Write $\theta_0 = \int y\,dF_0(y)$ and let $\hat{\theta}_n = \int y\,d\hat{F}_n(y)$ be the maximum likelihood estimator of θ_0, based on n independent copies of X. Suppose that for some $c_0 > 0$,

$$\frac{1}{c_0} \leq f_0(y) \leq c_0, \quad y \in [0,1].$$

With the help of Theorem 11.8, prove that $\sqrt{n}(\hat{\theta}_n - \theta_0)$ is asymptotically normal. (See van de Geer (1995c).)

11.7. (Mixture with the Poisson distribution) Let μ be counting measure on $\{0, 1, \ldots\}$, and

$$k(x \mid y) = e^{-y}\frac{y^x}{x!}, \quad x \in \{0, 1, \ldots\},\ y > 0.$$

Consider the model

$$p_0 = p_{F_0} \in \mathscr{P} = \left\{ p_F = \int k(x \mid y)\, dF(y) \ :\ F \in \Lambda \right\},$$

with Λ the collection of all probability measures on $(0, \infty)$. Consider the parameter $\theta_F = \int y\, dF(y)$. Suppose that F_0 has density f_0 w.r.t. Lebesgue measure, where f_0 has derivative $\dot{f}_0$, satisfying

$$\int \frac{\dot{f}_0^2(y)}{f_0(y)}\, dy < \infty.$$

Moreover, assume $y f_0(y) \to 0$ as $y \to 0$. Show that $b_{F_0}(x) = x - \theta_{F_0}$ is the efficient influence function, and that the worst possible subdirection is

$$h(y) = y - 1 - y\frac{\dot{f}_0(y)}{f_0(y)} - \theta_{F_0}.$$

11.8. In Lemma 11.14, one may also use an alternative penalty, for example, if

$$F_0 = \frac{e^{\gamma_0(\Phi)}}{1 + e^{\gamma_0(\Phi)}},$$

with

$$\int \left(\gamma_0^{(2)}(v)\right)^2 dv < \infty,$$

one may take the penalty

$$\tilde{I}^2(F) = \int_0^1 (\gamma^{(2)}(v))^2\, dv,$$

where

$$\gamma(v) = \log \frac{F(\Phi^{-1}(v))}{1 - F(\Phi^{-1}(v))}.$$

12

M-Estimators

We have seen that there are two ingredients that make up a theory on rates for maximum likelihood and least squares estimators: a Basic Inequality, and the behaviour of the increments of empirical processes. In the first section of this chapter, we make this observation in the general framework of M-estimation. For the Basic Inequality, one has various choices, depending on the loss function and on the parameter space.

The matter is investigated in more detail for the regression problem, using a general loss function. This yields, for example, rates of convergence for robust estimators, such as least absolute deviations estimators.

Finally, we return to the more or less classical situation, where the parameter to be estimated is finite-dimensional. Here, we also formulate conditions for asymptotic normality.

12.1. Introduction

Let $X_1, \ldots, X_n$ be independent observations, with values in the measurable space $(\mathscr{X}, \mathscr{A})$. The distribution of X_i is denoted by $P^{(i)}$, $i = 1, \ldots, n$. Moreover, we define $\bar{P} = \sum_{i=1}^n P^{(i)}/n$, and let $P_n = \sum_{i=1}^n \delta_{X_i}/n$ be the empirical distribution. Now, consider a subset Θ of a metric space (Λ, d), and loss functions $\gamma_\theta : \mathscr{X} \to \mathbf{R}$, $\theta \in \Theta$. The M-estimator (M stands for minimum) of

$$\theta_0 = \arg \min_{\theta \in \Theta} \int \gamma_\theta \, d\bar{P}, \tag{12.1}$$

is given by

$$\hat{\theta}_n = \arg\min_{\theta\in\Theta} \int \gamma_\theta \, dP_n. \tag{12.2}$$

To arrive at a rate of convergence for this estimator, we need two ingredients: a Basic Inequality, and the increments of the empirical process appearing in this inequality. In order to formulate a Basic Inequality, we have to investigate the behaviour of $\int \gamma_\theta \, d\bar{P}$. Since θ_0 minimizes this quantity, we have

$$\int (\gamma_\theta - \gamma_{\theta_0}) \, d\bar{P} \geq 0, \quad \text{for all } \theta \in \Theta.$$

In many situations, this difference in fact behaves like the squared distance $d^2(\theta, \theta_0)$ between θ and θ_0. (Other powers are also possible, but this does not change the general line of reasoning, although it *will* have its impact on the rates.) Let us assume this to be true, perhaps after a slight modification. Let $\tau : \Theta \to \Theta$ be some function (depending on θ_0), such that

$$\gamma_\theta - \gamma_{\tau(\theta)} \geq \tilde{\gamma}_\theta, \quad \theta \in \Theta, \tag{12.3}$$

and suppose that for some $\eta > 0$,

$$\int \tilde{\gamma}_\theta \, d\bar{P} \geq \eta d^2(\theta, \theta_0), \quad \theta \in \Theta. \tag{12.4}$$

Lemma 12.1. (Basic Inequality) *Assume that* (12.3) *and* (12.4) *are satisfied. Then,*

$$\eta d^2(\hat{\theta}_n, \theta_0) \leq -\int \tilde{\gamma}_{\hat{\theta}_n} \, d(P_n - \bar{P}). \tag{12.5}$$

Proof Because $\hat{\theta}_n$ minimizes $\int \gamma_\theta \, dP_n$ over all $\theta \in \Theta$, and $\tau(\hat{\theta}_n) \in \Theta$, we have

$$\int \gamma_{\hat{\theta}_n} \, dP_n \leq \int \gamma_{\tau(\hat{\theta}_n)} \, dP_n.$$

So we find

$$\begin{aligned} 0 &\geq \int (\gamma_{\hat{\theta}_n} - \gamma_{\tau(\hat{\theta}_n)}) \, dP_n \geq \int \tilde{\gamma}_{\hat{\theta}_n} \, dP_n = \int \tilde{\gamma}_{\hat{\theta}_n} \, d(P_n - \bar{P}) + \int \tilde{\gamma}_{\hat{\theta}_n} \, d\bar{P} \\ &\geq \int \tilde{\gamma}_{\hat{\theta}_n} \, d(P_n - \bar{P}) + \eta d^2(\hat{\theta}_n, \theta_0). \end{aligned}$$

□

The rest of the program is now clear. We may invoke the theory of Chapter 5 or Section 8.1 to study the behavior of the empirical process

$\{\nu_n(\tilde{\gamma}_\theta) = \sqrt{n}\int \tilde{\gamma}_\theta\, d(P_n - \bar{P}) : \theta \in \Theta\}$. The increments of the empirical process are stated in terms of the $L_2(\bar{P})$ distance, or in terms of the Bernstein difference $\rho_K(\cdot)$. Thus, one needs to translate the result in terms of the distance d.

So far, we have followed this program for the maximum likelihood estimator and the least squares estimator. Let us summarize the Basic Inequalities used there.

12.1.1. *Maximum likelihood estimation*

Suppose that $X_1, \ldots, X_n$ are i.i.d. with density p_0 in a given class of densities $\mathcal{P}$ with respect to some σ-finite measure μ. Let $\bar{P} = P$, and let

$$\hat{p}_n = \arg\max_{p \in \mathcal{P}} \int \log p \, dP_n \tag{12.6}$$

be the maximum likelihood estimator of p_0. Take $\theta = p$, $d(p_1, p_2) = h(p_1, p_2)$, and

$$\gamma_p = -\log p.$$

We have used three Basic Inequalities.

(a) Choose $\tau(p) = p_0$, and observe that

$$\gamma_p - \gamma_{p_0} = -\log\frac{p}{p_0} \geq 2\left(1 - \sqrt{\frac{p}{p_0}}\right) = \tilde{\gamma}_p,$$

and

$$\int \tilde{\gamma}_p \, dP = 2h^2(p, p_0).$$

(b) Choose $\tau(p) = p_0$, and observe that

$$\gamma_p - \gamma_{p_0} = -\log\frac{p}{p_0} \geq -\frac{1}{2}\log\frac{p + p_0}{2p_0} = \tilde{\gamma}_p,$$

and

$$\int \tilde{\gamma}_p \, dP \geq h^2\left(\frac{p + p_0}{2}, p_0\right) \geq \frac{1}{16}h^2(p, p_0).$$

(c) Suppose that $\mathcal{P}$ is convex. Choose $\tau(p) = (p + p_0)/2$, and observe that

$$\gamma_p - \gamma_{\tau(p)} = -\log\frac{2p}{p + p_0} \geq 1 - \frac{2p}{p + p_0} = \tilde{\gamma}_p,$$

and

$$\int \tilde{\gamma}_p \, dP \geq h^2(p, p_0).$$

The case of independent, but not identically distributed random variables, can also be put in this framework. Let $Y_1, \ldots, Y_n$ be independent, and suppose that Y_i has density $p_{i,0} \in \mathcal{P}_i$, where $\mathcal{P}_i$ is a given class of densities with respect to a σ-finite measure μ_i, $i = 1, \ldots, n$. Take $\theta = (p_1, \ldots, p_n)$, and

$$d^2(\theta_1, \theta_2) = \frac{1}{2n} \sum_{i=1}^{n} \int \left(p_{i,1}^{1/2} - p_{i,2}^{1/2}\right)^2 d\mu_i.$$

Moreover, let $X_i = (Y_i, z_i)$, with $z_i = i$, and

$$\gamma_\theta(y, z) = -\log p_z(y).$$

We can formulate similar Basic Inequalities as in (a)–(c).

12.1.2. Least squares estimation

Let

$$Y_i = g_0(z_i) + W_i, \quad i = 1, \ldots, n,$$

where $W_1, \ldots, W_n$ are independent errors with zero expectation, and where g_0 is in a given class of regression functions $\mathcal{G}$. The least squares estimator is

$$\hat{g}_n = \arg\min_{g \in \mathcal{G}} \sum_{i=1}^{n} (Y_i - g(z_i))^2. \tag{12.7}$$

Choose $\theta = g$, $d^2(g_1, g_2) = \|g_1 - g_2\|_n$, and $\tau(g) = g_0$. Then for $\gamma_g(y, z) = (y - g(z))^2$, we have

$$\gamma_g(y, z) - \gamma_{g_0}(y, z) = -2(y - g_0(z))\,(g(z) - g_0(z)) + (g(z) - g_0(z))^2 = \tilde{\gamma}_g,$$

and

$$\int \tilde{\gamma}_g \, d\bar{P} = \|g - g_0\|_n^2.$$

Conditions (12.3) and (12.4) are thus fulfilled in the context of maximum likelihood estimation and least squares estimation, when we make the appropriate choices. Condition (12.4) can cause problems in other contexts: $\int \tilde{\gamma}_\theta \, d\bar{P}$ might have irregular behaviour when θ is too far from θ_0. However, if (Λ, d) is a normed vector space, and if γ_θ is a convex function of θ, then a local version of (12.4) suffices. As an illustration, we shall consider estimation of a regression function using a general loss function.

12.2. Estimating a regression function using a general loss function

Let $Y_1, \ldots, Y_n$ be independent, real-valued random variables, and $z_1, \ldots, z_n$ be covariates in some space $\mathcal{Z}$. Consider a convex function

$$\gamma : \mathbf{R} \to \mathbf{R}$$

and the estimator

$$\hat{g}_n = \arg\min_{g\in\mathcal{G}} \sum_{i=1}^{n} \gamma(Y_i - g(z_i)). \tag{12.8}$$

Let

$$g_0 = \arg\min_{g\in\mathcal{G}} \sum_{i=1}^{n} \mathbf{E}\gamma(Y_i - g(z_i)). \tag{12.9}$$

Define

$$W_i = Y_i - g_0(z_i), \quad i = 1,\dots,n.$$

Then by (12.9), the minimum of $\sum_{i=1}^{n} \mathbf{E}\gamma(W_i - f(z_i))$, over all $f \in \{g - g_0 : g \in \mathcal{G}\}$ is attained at $f \equiv 0$. We shall require a little more, namely that for each $i \in \{1,\dots,n\}$,

$$\mathbf{E}\gamma(W_i - b) \text{ has a unique minimum in } b = 0. \tag{12.10}$$

To see that $\hat{g}_n$ is an M-estimator as defined in (12.2), let us write $\gamma_g(y,z) = \gamma(y - g(z))$ and let $P^{(i)}$ be the distribution of (Y_i, z_i), $i = 1,\dots,n$, and take P_n as the empirical distribution of $(Y_1, z_1),\dots,(Y_n, z_n)$. Then,

$$\hat{g}_n = \arg\min_{g\in\mathcal{G}} \int \gamma_g \, dP_n, \tag{12.11}$$

and

$$g_0 = \arg\min_{g\in\mathcal{G}} \int \gamma_g \, d\bar{P}.$$

The proof of Lemma 12.2 below is based on a local version of (12.4), i.e., we show that the mean loss function locally behaves like a quadratic form. This is enough for our purposes because the loss function is convex. As before, let $Q_n = \sum_{i=1}^{n} \delta_{z_i}/n$ be the empirical distribution of the covariates, and $\|\cdot\|_n$ the $L_2(Q_n)$-norm. Define for some $L_n \geq 0$,

$$\mathcal{F}_n = \left\{ \frac{g - g_0}{1 + L_n\|g - g_0\|_n} : g \in \mathcal{G} \right\}. \tag{12.12}$$

Note that in general, we have no control over $\|g - g_0\|_n$, but that $\|f\|_n \leq 1/L_n$ for all $f \in \mathcal{F}_n$. We use the shorthand notation

$$\hat{f}_n = \frac{\hat{g}_n - g_0}{1 + L_n\|\hat{g}_n - g_0\|_n}. \tag{12.13}$$

Lemma 12.2. (Basic Inequality) *Let γ be a convex loss function satisfying (12.10). Suppose that there exists a constant $K_2 < \infty$, such that*

$$\sup_{f \in \mathcal{F}_n} |f|_\infty \leq K_2. \tag{12.14}$$

Moreover, assume that for some $\epsilon > 0$ and for all $|a| \leq \epsilon$,

$$\min_{1 \leq i \leq n} \mathbf{E}\big(\gamma(W_i + a) - \gamma(W_i)\big) \geq \epsilon a^2. \tag{12.15}$$

Then for some constant $\eta > 0$ depending on K_2 and ϵ,

$$\eta \|\hat{f}_n\|_n^2 \leq -\int \big(\gamma_{g_0+\hat{f}_n} - \gamma_{g_0}\big)\, d(P_n - \bar{P}). \tag{12.16}$$

Proof Because γ is convex, we have for $0 \leq \alpha \leq 1$,

$$\gamma(y-g(z))-\gamma(y-g_0(z)) \geq \frac{1}{\alpha}\big[\gamma((1-\alpha)(y-g_0(z))+\alpha(y-g(z)))-\gamma(y-g_0(z))\big].$$

So taking $\alpha = \alpha_g = 1/(1 + L_n\|g - g_0\|_n)$, we obtain

$$\gamma_g - \gamma_{g_0} \geq \frac{1}{\alpha_g}\gamma_{g_0+f} - \gamma_{g_0},$$

where $f = \alpha_g(g - g_0) \in \mathcal{F}_n$. Thus,

$$\int \big(\gamma_{g_0+\hat{f}_n} - \gamma_{g_0}\big)\, dP_n \leq 0.$$

On the other hand, assuming without loss of generality that $\epsilon/K_2 \leq 1$, by assumption (12.10)

$$\int (\gamma_{g_0+f} - \gamma_{g_0})\, d\bar{P} \geq \frac{1}{n}\sum_{i=1}^n \mathbf{E}\left[\gamma\left(W_i - \frac{\epsilon}{K_2}f(z_i)\right) - \gamma(W_i)\right] \geq \frac{\epsilon^3}{K_2^2}\|f\|_n^2.$$

Thus the lemma follows. □

We are now in the position to prove a rate of convergence for $\hat{g}_n$. Take $\mathcal{F}_n(\delta) = \{f \in \mathcal{F}_n : \|f\|_n \leq \delta\}$, and, for an appropriate constant c_1,

$$J(\delta, \mathcal{F}_n(\delta), Q_n) = \int_{\delta^2/c_1}^{\delta} H^{1/2}\big(u, \mathcal{F}_n(\delta), Q_n\big)\, du \vee \delta. \tag{12.17}$$

We shall apply Lemma 8.5, and therefore need to assume that

$$|\gamma(W_i - f(z_i)) - \gamma(W_i - \tilde{f}(z_i))| \le |V_i| \, |f(z_i) - \tilde{f}(z_i)|, \quad \text{for all } f, \tilde{f} \in \mathscr{F}_n, \tag{12.18}$$

where $V_1, \ldots, V_n$ are uniformly sub-Gaussian:

$$\max_{i=1,\ldots,n} K_1^2 \mathbf{E}\left(e^{|V_i|^2/K_1^2} - 1\right) \le \sigma_0^2. \tag{12.19}$$

Theorem 12.3 *Suppose that the conditions of Lemma* 12.2 *are met, and that* (12.18) *and* (12.19) *hold true. Take* $\Psi(\delta) \ge J(\delta, \mathscr{F}_n(\delta), Q_n)$ *in such a way that* $\Psi(\delta)/\delta^2$ *is a non-decreasing function of* δ, $0 < \delta < c_1$. *Then for a constant* c_2, *depending on* η, ϵ, K_1, K_2 *and* σ_0^2, *and for* $\delta_n \le 1/(2L_n)$ *and*

$$\sqrt{n}\delta_n^2 \ge c_2 \Psi(\delta_n), \tag{12.20}$$

we have for all $1/(2L_n) \ge \delta \ge \delta_n$,

$$\mathbf{P}(\|\hat{g}_n - g_0\|_n > 2\delta) \le c_2 \exp\left[-\frac{n\delta^2}{c_2^2}\right]. \tag{12.21}$$

Proof This follows from the same arguments as the ones used in the proofs of Theorems 9.1 and 9.2. □

Note that Theorem 12.3 imposes conditions on the loss function and on the class of regression functions. The latter are as in Chapter 9, so the same examples as there can be given. We shall therefore only illustrate the theorem by checking the conditions for a particular loss function.

Example: least absolute deviations Let $\gamma(x) = |x|$, $x \in \mathbf{R}$. Clearly, this γ is convex. Assumption (12.10) is satisfied if each W_i has a unique median at 0. Condition (12.15) holds if we assume that for some $\epsilon > 0$, and for all $0 \le a \le \epsilon$, and all $i = 1, \ldots, n$,

$$\mathbf{P}(0 \le W_i \le a) \ge \epsilon a, \tag{12.22}$$

and

$$\mathbf{P}(-a \le W_i \le 0) \ge \epsilon a. \tag{12.23}$$

Conditions (12.18) and (12.19) are trivially fulfilled, with $|V_i| = 1$, $i = 1, \ldots, n$.

12.3. Classes of functions indexed by a finite-dimensional parameter

Let $X_1, \ldots, X_n$ be independent random variables with values in the measurable space $(\mathscr{X}, \mathscr{A})$. The distribution of X_i is denoted by $P^{(i)}$, $i = 1, \ldots, n$, and

$\bar{P} = \sum_{i=1}^{n} P^{(i)}/n$ denotes the average distribution. Consider the empirical process

$$\left\{ \nu_n(\gamma) = \sqrt{n} \int \gamma \, d(P_n - \bar{P}) : \gamma \in \Gamma \right\}, \tag{12.24}$$

with P_n the empirical distribution based on $X_1, \ldots, X_n$, and Γ a class of real-valued functions on $\mathcal{X}$. In this section, we consider the situation where Γ is indexed by a finite-dimensional parameter, say

$$\Gamma = \{\gamma_\theta : \theta \in \Theta\}, \ \Theta \subset \mathbf{R}^d. \tag{12.25}$$

The dimension d is assumed to be fixed, i.e., independent of n. From Theorem 8.13, we know that the behavior of the increments of the empirical process can be deduced from the generalized entropy with bracketing of Γ, provided that certain exponential moments exist. On the other hand, we have seen that in the linear case, one does not need (generalized) entropy with bracketing, and only certain second order moments. The behavior of the increments then follows from a simple application of the Cauchy–Schwarz inequality (see Example 9.3.1). Let us recall here how this works. Let

$$\gamma_\theta(x) = \sum_{k=1}^{d} \theta_k \psi_k(x), \quad \theta \in \mathbf{R}^d.$$

Then

$$\left| \nu_n(\gamma_\theta - \gamma_{\theta_0}) \right| \leq |\theta - \theta_0| \left(\sum_{k=1}^{d} \nu_n^2(\psi_k) \right)^{1/2}.$$

So for

$$\text{var}(\psi_k(X_i)) = \sigma_{ki}^2,$$

and

$$\bar{\sigma}^2 = \frac{1}{dn} \sum_{k=1}^{d} \sum_{i=1}^{n} \sigma_{ki}^2,$$

we find

$$\mathbf{E} \left(\sup_{\theta, \theta_0} \frac{\left| \nu_n(\gamma_\theta - \gamma_{\theta_0}) \right|}{|\theta - \theta_0|} \right)^2 \leq d\bar{\sigma}^2.$$

Consider now the situation where γ_θ is possibly non-linear in $\theta \in \Theta \subset \mathbf{R}^d$. The purpose of this section is to check whether $|\nu_n(\gamma_\theta - \gamma_{\theta_0})|$ still behaves like $|\theta - \theta_0|$. We start with showing this is indeed true under moment conditions of order $p > d$. The consequences ($n^{-1/2}$-rates of convergence) for least squares and maximum likelihood problems are discussed in Subsections 12.3.1 and 12.3.2. Finally, in Subsection 12.3.3, we simply assume that the class $\{(\gamma_\theta - \gamma_{\theta_0})/|\theta - \theta_0| : \theta \neq \theta_0\}$ is a P-Donsker class. Then it trivially follows that $|\nu_n(\gamma_\theta - \gamma_{\theta_0})|$ behaves like $|\theta - \theta_0|$. As we shall see, for M-estimators based on the loss function γ_θ, the consequence is not only a $n^{-1/2}$-rate of convergence, but also asymptotic normality.

Define for $\theta_0 \in \Theta$ fixed,

$$\Theta(R) = \{\theta \in \Theta : |\theta - \theta_0| \leq R\}.$$

Then $\Theta(R)$ can be covered by

$$\left(\frac{4R+\delta}{\delta}\right)^d$$

balls with radius δ (see Lemma 2.5). As a consequence, under certain conditions on the map $\theta \mapsto \gamma_\theta$, the entropy of $\{\gamma_\theta : \theta \in \Theta(R)\}$ is so small that one does not need exponential moments. In the linear case, only second order moments are needed. For non-linear models, we shall require moments up to order p, with $p > d$.

Lemma 12.4 *Suppose that $\theta \mapsto \gamma_\theta(x)$ is continuous in $\theta \in \Theta$ for all x, and that for some $p > d$, and $M < \infty$,*

$$\mathbf{E}\left|\nu_n(\gamma_\theta - \gamma_{\tilde\theta})\right|^p \leq |\theta - \tilde\theta|^p M^p. \tag{12.26}$$

Then for a constant c depending on M, d and p, we have for all $T > 0$

$$\mathbf{P}\left(\sup_{\theta\in\Theta(R)} \left|\nu_n(\gamma_\theta - \gamma_{\theta_0})\right| \geq TR\right) \leq cT^{-p}. \tag{12.27}$$

Proof Let $\tau_0(\theta) = \theta_0$, and for $s = 1, 2, \ldots$, let $\tau_s(\theta) \in \Theta$ satisfy

$$|\tau_s(\theta) - \theta| \leq 2^{-s}R.$$

It is easy to see that we may choose $\Theta_s = \{\tau_s(\theta) : \theta \in \Theta(R)\}$ in such a way that $\Theta_{s-1} \subset \Theta_s$, $s = 1, \ldots$, and such that the cardinality of Θ_s is at most $c_1 2^{sd}$. Insert the chaining argument (see Section 3.2). Since $\theta \mapsto g_\theta(x)$ is assumed to be continuous in θ for all x, we have

$$\gamma_\theta(x) - \gamma_{\theta_0}(x) = \sum_{s=1}^{\infty} \left(\gamma_{\tau_s(\theta)}(x) - \gamma_{\tau_{s-1}(\theta)}(x)\right), \quad x \in \mathscr{X}.$$

Let

$$\eta_s = s2^{-s(p-d)/p}/c_2,$$

where $c_2 = \sum_{s=1}^{\infty} s2^{-s(p-d)/p}$. It follows that

$$\begin{aligned}
\mathbf{P}\left(\sup_{\theta\in\Theta(R)} \left|\nu_n(\gamma_\theta - \gamma_{\theta_0})\right| \geq TR\right) \\
&\leq \sum_{s=1}^{\infty} \mathbf{P}\left(\sup_{\theta\in\Theta(R)} \left|\nu_n(\gamma_{\tau_s(\theta)} - \gamma_{\tau_{s-1}(\theta)})\right| \geq \eta_s TR\right) \\
&\leq \sum_{s=1}^{\infty} c_1 2^{sd} \frac{(M2^{-s+1}R)^p}{(\eta_s TR)^p} = \sum_{s=1}^{\infty} c_1 \left(\frac{2Mc_2}{sT}\right)^p \\
&\leq cT^{-p}.
\end{aligned}$$

□

To check (12.26), we may use a result of Whittle (1960), which says the following. Let $W_1, \ldots, W_n$ be independent, real-valued random variables with expectation zero, and let $b_1, \ldots, b_n$ be constants. Then for $p \geq 2$,

$$\mathbf{E}\left|\frac{1}{\sqrt{n}}\sum_{i=1}^{n} b_i W_i\right|^p \leq C_p^p \left(\frac{1}{n}\sum_{i=1}^{n} b_i^2 (\mathbf{E}|W_i|^p)^{2/p}\right)^{p/2},$$

where C_p is some constant depending on p. So, for example, if

$$\mathbf{E}\,|\gamma_\theta(X_i) - \gamma_{\tilde\theta}(X_i)|^p \leq |\theta - \tilde\theta|^p M_i^p, \tag{12.28}$$

we find that (12.26) holds with $M = C_p(\sum_{i=1}^{n} M_i^2/n)^{1/2}$.

12.3.1. Least squares

Lemma 12.4 can be applied to the non-linear regression model

$$Y_i = g_{\theta_0}(z_i) + W_i, \quad i = 1, \ldots, n,$$

with $\theta_0 \in \Theta \subset \mathbf{R}^d$ an unknown parameter, and $W_1, \ldots, W_n$ independent errors with zero expectation. Let

$$\hat\theta_n = \arg\min_{\theta\in\Theta} \sum_{i=1}^{n} \left(Y_i - g_\theta(z_i)\right)^2 \tag{12.29}$$

be the least squares estimator. Suppose that for some constant c_0,

$$\|g_\theta - g_{\theta_0}\|_n \geq \frac{1}{c_0}|\theta - \theta_0|, \quad \theta \in \Theta, \tag{12.30}$$

and

$$\|g_\theta - g_{\tilde{\theta}}\|_n \leq c_0|\theta - \tilde{\theta}|, \quad \theta, \tilde{\theta} \in \Theta. \tag{12.31}$$

(Clearly, these conditions can be reduced to local versions if we already know that $\hat{\theta}_n$ is consistent, perhaps from convexity arguments.) Theorem 9.1 tells us that under the condition

$$\max_{i=1,\ldots,n} K^2\left(\mathbf{E}e^{|W_i|^2/K^2} - 1\right) \leq \sigma_0^2, \tag{12.32}$$

one has the rate $|\hat{\theta}_n - \theta_0| = O_{\mathbf{P}}(n^{-1/2})$. This can be relaxed to

$$\max_{i=1,\ldots,n} \mathbf{E}|W_i|^p \leq M_0^p, \tag{12.33}$$

for some $p > d$, $p \geq 2$. To see this, apply Lemma 12.4 with $\gamma_\theta(Y_i, z_i) = g_\theta(z_i)(Y_i - g_{\theta_0}(z_i))$, and use the peeling device. In summary, for finite-dimensional non-linear models, we employ moment conditions of order larger than the number of parameters. One could therefore say that for infinite-dimensional non-linear models, the exponential moment condition on the errors comes up quite naturally.

12.3.2. Maximum likelihood

Consider now the maximum likelihood problem. For simplicity, let us restrict ourselves to the i.i.d. case. Let $X_1, \ldots, X_n$ be i.i.d. random variables with distribution P and with density $p_{\theta_0} = dP/d\mu$, with respect to a σ-finite measure μ. Here $\theta_0 \in \Theta \subset \mathbf{R}^d$ is again a finite-dimensional unknown parameter. The maximum likelihood estimator is

$$\hat{\theta}_n = \arg\max_{\theta\in\Theta} \int \log p_\theta \, dP_n. \tag{12.34}$$

Apply Lemma 12.4 to

$$\gamma_\theta = \left(\sqrt{\frac{p_\theta + p_{\theta_0}}{2p_{\theta_0}}} - 1\right) 1\{p_{\theta_0} > 0\},$$

and note that

$$\int (\gamma_\theta - \gamma_{\tilde{\theta}})^2 \, dP = 2h_0^2\left(\frac{p_\theta + p_{\theta_0}}{2}, \frac{p_{\tilde{\theta}} + p_{\theta_0}}{2}\right)$$

(recall that $h_0^2(p_1, p_2) = \frac{1}{2}\int_{p_{\theta_0}>0} (p_1^{1/2} - p_2^{1/2})^2 \, d\mu$). Thus, second order moments exist automatically, but moments of higher order may not be available. Now, assume that for some constant c_0,

$$h(p_\theta, p_{\theta_0}) \geq \frac{1}{c_0}|\theta - \theta_0|, \quad \theta \in \Theta, \tag{12.35}$$

and

$$h_0\left(\frac{p_\theta + p_{\theta_0}}{2}, \frac{p_{\tilde\theta} + p_{\theta_0}}{2}\right) \le c_0|\theta - \tilde\theta|, \quad \theta, \tilde\theta \in \Theta. \tag{12.36}$$

(Of course, it suffices to have local versions of (12.35) and (12.36) if it was already shown that $\hat\theta_n$ is consistent, perhaps by convexity arguments. Notice that (12.35) is only possible if $\Theta \subset \{\theta : |\theta - \theta_0| \le c_0\}$.) Theorem 7.4 cannot be applied without further conditions, because (12.35) and (12.36) tell us nothing about the entropy with bracketing. However, if $\theta \mapsto p_\theta(x)$ is continuous in $\theta \in \Theta$ for all x, and $d = 1$, one finds from Lemma 12.4 the rate $|\hat\theta_n - \theta_0| = O_{\mathbf{P}}(n^{-1/2})$.

12.3.3. Asymptotic normality

We conclude that in non-linear models, moments of order $p > 2$ can play a role whenever the dimension of the parameter space is bigger than one. The question arises: what can one do if such higher order moments are not available. It seems to be unavoidable that one then needs more regularity on the map $\theta \mapsto \gamma_\theta$. To avoid digressions, we only study the i.i.d. case in the remainder of this section. Let $X_1, \ldots, X_n$ be i.i.d. with distribution P.

Define

$$f_\theta = \begin{cases} (\gamma_\theta - \gamma_{\theta_0})/(|\theta - \theta_0|), & \text{if } \theta \ne \theta_0, \\ 0, & \text{if } \theta = \theta_0, \end{cases} \tag{12.37}$$

and let

$$\mathscr{F} = \{f_\theta : \theta \in \Theta\}. \tag{12.38}$$

If $\mathscr{F}$ is P-Donsker, the study of the modulus of continuity of $|\nu_n(\gamma_\theta) - \nu_n(\gamma_{\theta_0})|$, in terms of $|\theta - \theta_0|$, is trivial. In Lemma 12.5 below, we summarize the sufficient conditions of Chapter 6, for a class to be P-Donsker.

Lemma 12.5 *Suppose that*

$$\sup_{\theta \in \Theta} |f_\theta| \le F, \tag{12.39}$$

where $F \in L_2(P)$, and that either

$$\sup_{\delta > 0} \frac{H(\delta, \mathscr{F}, P_n)}{H(\delta)} = O_{\mathbf{P}}(1), \tag{12.40}$$

with $H(\delta)$ a non-increasing function satisfying $\int_0^1 H^{1/2}(u)\,du < \infty$, or that

$$\int_0^1 H_B^{1/2}(u, \mathscr{F}, P)\,du < \infty. \tag{12.41}$$

Then

$$\sup_{\theta\in\Theta} \frac{|\nu_n(\gamma_\theta - \gamma_{\theta_0})|}{|\theta-\theta_0|} = O_{\mathbf{P}}(1). \tag{12.42}$$

Proof This follows immediately from Theorems 6.2 and 6.3 respectively. □

Clearly, this lemma can also be used in regression problems and maximum likelihood problems to obtain the $n^{-1/2}$-rate. But the conditions of the lemma are almost sufficient to prove asymptotic normality of such estimators. We formulate this below for M-estimators in general.

Let

$$\hat{\theta}_n = \arg\min_{\theta\in\Theta} \sum_{i=1}^n \gamma_\theta(X_i), \tag{12.43}$$

be an M-estimator. The parameter space Θ is assumed to be a subset of $\mathbf{R}^d$. We require that $\int \gamma_\theta\, dP$ is minimized at θ_0, where θ_0 is an interior point of Θ.

Condition A Suppose that γ_θ is differentiable in quadratic mean, at θ_0, i.e., for some function $l_0 : \mathscr{X} \to \mathbf{R}^d$, with components in $L_2(P)$, we have

$$\lim_{\theta\to\theta_0} \frac{\|\gamma_\theta - \gamma_{\theta_0} - (\theta-\theta_0)^T l_0\|}{|\theta-\theta_0|} = 0.$$

Condition B The difference $\int(\gamma_\theta - \gamma_{\theta_0})\, dP$ behaves like a quadratic form in a neighbourhood of θ_0, i.e., for some positive definite matrix V_0,

$$\int (\gamma_\theta - \gamma_{\theta_0})\, dP = \frac{1}{2}(\theta-\theta_0)^T V_0 (\theta-\theta_0) + o(|\theta-\theta_0|^2).$$

Condition C Let f_θ be defined as in (12.37). Suppose that for some $\epsilon > 0$, the class $\{f_\theta : |\theta-\theta_0| \le \epsilon\}$ is a P-Donsker class with envelope in $L_2(P)$.

Lemma 12.6 *Suppose that conditions* **A**, **B** *and* **C** *hold. Moreover, assume that the estimator* $\hat{\theta}_n$ *defined in* (12.43) *is a consistent estimator of* θ_0. *Then*

$$\sqrt{n}(\hat{\theta}_n - \theta_0) \to^{\mathscr{L}} \mathscr{N}(0, V_0^{-1} I_0 V_0^{-1}), \tag{12.44}$$

where

$$I_0 = \int l_0 l_0^T\, dP. \tag{12.45}$$

Proof Since $\{f_\theta : |\theta - \theta_0| \le \epsilon\}$ is a P-Donsker class, and $\hat\theta_n$ is consistent, we may write

$$0 \ge \int (\gamma_{\hat\theta_n} - \gamma_{\theta_0})\, dP_n = \int (\gamma_{\hat\theta_n} - \gamma_{\theta_0})\, d(P_n - P) + \int (\gamma_{\hat\theta_n} - \gamma_{\theta_0})\, dP$$
$$= (\hat\theta_n - \theta_0)^T \int l_0\, d(P_n - P) + o_{\mathbf{P}}(n^{-1/2})|\hat\theta_n - \theta_0|$$
$$+ \frac{1}{2}(\hat\theta_n - \theta_0)^T V_0 (\hat\theta_n - \theta_0) + o(|\hat\theta_n - \theta_0|^2).$$

This implies $|\hat\theta_n - \theta_0| = O_{\mathbf{P}}(n^{-1/2})$. But then

$$\left| V_0^{1/2}(\hat\theta_n - \theta_0) + V_0^{-1/2} \int l_0\, d(P_n - P) + o_{\mathbf{P}}(n^{-1/2}) \right|^2 \le o_{\mathbf{P}}\left(\frac{1}{n}\right).$$

Therefore,

$$\hat\theta_n - \theta_0 = -V_0^{-1} \int l_0\, d(P_n - P) + o_{\mathbf{P}}(n^{-1/2}).$$

The result now follows from the classical central limit theorem. Because $\int l_0\, dP = 0$, the asymptotic covariance matrix is $V_0^{-1} I_0 V_0^{-1}$. □

If $\gamma_\theta(x)$ is differentiable in θ for all x, we may also derive asymptotic normality from the score equations. Of course, differentiability for each x is a much stronger requirement than differentiability in quadratic mean. However, the asymptotic normality proof becomes very transparent and only uses the property that the estimator is a solution of the score equations.

Condition AA There exists an $\epsilon > 0$, such that $\theta \mapsto \gamma_\theta$ is differentiable for all x, with derivative $l_\theta = \partial\gamma_\theta/\partial\theta$. Moreover, for $l_0 = l_{\theta_0}$,

$$\lim_{\theta \to \theta_0} \|l_\theta - l_0\| = 0.$$

Condition BB For some positive definite matrix V_0,

$$\int l_\theta\, dP = V_0(\theta - \theta_0) + o|\theta - \theta_0|.$$

Condition CC The class $\{l_\theta : \theta - \theta_0| \le \epsilon\}$ is a P-Donsker class with envelope in $L_2(P)$.

Lemma 12.7 *Suppose that conditions* **AA, BB** *and* **CC** *hold. Assume moreover that* $\hat\theta_n$ *defined in* (12.43) *is a consistent estimator of* θ_0. *Then*

$$\sqrt{n}(\hat\theta_n - \theta_0) \to^{\mathcal{L}} \mathcal{N}(0, V_0^{-1} I_0 V_0^{-1}), \tag{12.46}$$

where I_0 *is defined in* (12.45).

Proof Recall that θ_0 is an interior point of Θ, and minimizes $\int \gamma_\theta \, dP$, so that $\int l_0 \, dP = 0$. Because $\hat{\theta}_n$ is consistent, it is eventually a solution of the score equations

$$\int l_{\hat{\theta}_n} \, dP_n = 0.$$

Rewrite the score equations as

$$\begin{aligned} 0 = \int l_{\hat{\theta}_n} \, dP_n &= \int (l_{\hat{\theta}_n} - l_0) \, dP_n + \int l_0 \, dP_n \\ &= \int (l_{\hat{\theta}_n} - l_0) \, d(P_n - P) + \int l_{\hat{\theta}_n} \, dP + \int l_0 \, dP_n. \end{aligned}$$

Now, use condition **BB** and the asymptotic equicontinuity of $\{l_\theta : |\theta - \theta_0| \leq \epsilon\}$ (see Chapter 6), to obtain

$$0 = o_{\mathbf{P}}(n^{-1/2}) + \int l_0 \, dP_n + V_0(\hat{\theta}_n - \theta_0) + o(|\hat{\theta}_n - \theta_0|).$$

This yields

$$(\hat{\theta}_n - \theta_0) = -V_0^{-1} \int l_0 \, dP_n + o_{\mathbf{P}}(n^{-1/2}). \qquad \square$$

Remark Lemma 12.7 has also infinite-dimensional counterparts. See, for example, Proposition 11.7.

Example Let $X_1, \ldots, X_n$ be independent copies of a real-valued population random variable X. Let F be the distribution function of X. First, we consider estimation of the α-quantile $\theta_0 = F^{-1}(\alpha)$, where $0 < \alpha < 1$. We assume that F is continuously differentiable in a neighbourhood of θ_0, with derivative f, satisfying $f(\theta_0) > 0$. Take

$$\gamma_\theta(x) = -\alpha\theta - (x - \theta)\mathrm{l}\{x \leq \theta\}, \quad \theta \in \mathbf{R}.$$

Use Lemma 12.6 to conclude that

$$\sqrt{n}(\hat{\theta}_n - \theta_0) \rightarrow^{\mathscr{L}} \mathscr{N}\left(0, \frac{\alpha(1-\alpha)}{f^2(\theta_0)}\right).$$

Second, let us consider a robust estimator of location. Let

$$\gamma_\theta(x) = (x - \theta)^2 \mathrm{l}\{|x - \theta| \leq a\} + (2a|x - \theta| - a^2)\mathrm{l}\{|x - \theta| > a\}, \quad \theta \in \mathbf{R},$$

with $a > 0$ fixed. Note that this γ_θ is differentiable in θ for all x. Assume that $\int \gamma_\theta \, dP$ has a unique minimum in $\theta = \theta_0$. Use Lemma 12.7 to conclude that

$$\sqrt{n}(\hat{\theta}_n - \theta_0) \rightarrow^{\mathscr{L}} \mathscr{N}(0, \sigma^2),$$

with

$$\sigma^2 = \frac{a^2F(-a) + \int_{-a}^{a} x^2\,dF(x) + a^2(1-F(a))}{\left(F(a)-F(-a)\right)^2}.$$

12.4. Notes

Rates of convergence for least absolute deviations estimators were obtained in van de Geer (1990). The theory developed in Section 12.2 can be easily adjusted to incorporate estimators with penalty (invoke the same ideas as in Section 10.1 and 10.2).

For the finite-dimensional case, the assumption of moments of order larger than the dimension of parameter space is also in Ibragimov and Has'minskii (1981a). Maximum likelihood estimation of a one-dimensional parameter is in Le Cam (1973). Lemma 12.6 is along the lines of Pollard (1984, Theorem VII.1.5) (see also Pollard (1985)). An infinite-dimensional counterpart of Lemma 12.7 is given by van der Vaart (1995). The loss functions in the example at the end of Section 12.3, can be employed in Section 12.2, to establish, for example, rates for non-parametric quantile regression (see Problem 12.2).

12.5. Problems and complements

12.1. Let $X_1, \ldots, X_n$ be i.i.d. with distribution P on $(\mathcal{X}, \mathcal{A})$, and with density $p_0 = dP/d\mu \in \mathcal{P}$. In Birgé and Massart (1993), the estimator

$$\hat{p}_n = \arg\min_{p \in \mathcal{P}} \left(-2\int p\,dP_n + \|p\|_\mu^2\right)$$

is considered. This is an M-estimator, as given in (12.2), with $\Theta = \mathcal{P}$ and $\gamma_p = -2p + \|p\|_\mu^2$. Deduce the Basic Inequality

$$\|\hat{p}_n - p_0\|_\mu^2 \le 2\int (\hat{p}_n - p_0)\,d(P_n - P).$$

12.2. (Quantile regression) Let $Y_1, \ldots, Y_n$ be independent response variables, and let $z_1, \ldots, z_n$ be covariates. Take $0 < \alpha < 1$ fixed, and suppose that Y_i has a unique α-quantile $F_i^{-1}(\alpha)$, where F_i is the distribution function of Y_i, $i = 1, \ldots, n$. We require that for all i, F_i has a density f_i w.r.t. Lebesgue measure, that satisfies for some $\epsilon > 0$,

$$f_i(y) \ge \epsilon, \quad \text{for all } |y - F_i^{-1}(\alpha)| \le \epsilon.$$

We moreover assume $F_i^{-1}(\alpha) = g_0(z_i)$, where g_0 is known to lie in the class $\mathcal{G}$. Let

$$\hat{g}_n = \arg\min_{g\in\mathcal{G}} \sum_{i=1}^{n} \gamma\big(Y_i - g(z_i)\big),$$

where

$$\gamma(x) = \alpha x - x\mathrm{l}\{x \le 0\}.$$

Clearly, γ is convex. Verify (12.10), (12.15), (12.18) and (12.19). Thus, under condition (12.14) on $\mathcal{G}$, a rate of convergence for $\|\hat{g}_n - g_0\|_n$ follows from Theorem 12.3.

12.3. Let

$$Y_i = g_0(z_i) + W_i, \quad i = 1, \ldots, n,$$

where $W_1, \ldots, W_n$ are independent, have median 0, and satisfy (12.22) and (12.23), and where

$$g_0 \in \mathcal{G} = \left\{ g(x) = \sum_{j=1}^{d} \theta_j \psi_j(z) \;:\; \theta \in \mathbf{R}^d \right\}.$$

Here, the dimension d is fixed (i.e., independent of n). Consider the least absolute deviations estimator

$$\hat{g}_n = \arg\min_{g\in\mathcal{G}} \sum_{i=1}^{n} \big|Y_i - g(z_i)\big|.$$

Define $\Sigma_n = \int \psi\psi^T \, dQ_n$, and let λ_n be the smallest eigenvalue of Σ_n. Write

$$\max_{1\le i\le n} \max_{1\le j\le d} \big|\psi_j(z_i)\big| = c_n,$$

and assume that $n^{-1/2} c_n/\lambda_n \to 0$. Show that by Theorem 12.3, $\|\hat{g}_n - g_0\|_n = O_{\mathbf{P}}(n^{-1/2})$.

12.4. (Penalized least absolute deviations) Let $\mathcal{Z} = [0,1]$ and

$$Y_i = g_0(z_i) + W_i, \quad i = 1, \ldots, n,$$

where $W_1, \ldots, W_n$ are independent, have median 0, and satisfy (12.22) and (12.23), and where g_0 is a fixed function satisfying

$$g_0 \in \mathcal{G} = \{g : [0,1] \to (-K, K),\ I(g) < \infty\},$$

with

$$I^2(g) = \int_0^1 \left(g^{(m)}(z)\right)^2 dz.$$

Consider the penalized least absolute deviations estimator

$$\hat{g}_n = \arg\min_{g \in \mathscr{G}} \left(\frac{1}{n} \sum_{i=1}^{n} |Y_i - g(z_i)| + \lambda_n^2 I^2(g) \right),$$

with $\lambda_n^{-1} = O_{\mathbf{P}}(n^{m/(2m+1)})$. Show that $\|\hat{g}_n - g_0\|_n = O_{\mathbf{P}}(\lambda_n)$ and $I(\hat{g}_n) = O_{\mathbf{P}}(1)$.

12.5. (Linear quantile regression) Let $Z \in [0,1]^d$, and

$$Y = \theta_0^T Z + W,$$

where W has the unique α-quantile $F^{-1}(\alpha) = 0$, F being the distribution function of W. Assume moreover that F has density f with respect to Lebesgue measure, satisfying for some $\epsilon > 0$,

$$f(w) \geq \epsilon, \quad \text{for all } |w| \leq \epsilon.$$

Finally, assume that $\Sigma = \int zz^T \, dQ(z)$ (Q being the distribution of Z) is non-singular. Let $(Y_1, Z_1), \ldots, (Y_n, Z_n)$ be independent copies of (Y, Z), and let $\hat{\theta}_n$ be the quantile regression estimator

$$\hat{\theta}_n = \arg\min_{\theta \in \mathbf{R}^d} \sum_{i=1}^{n} \gamma(Y_i - \theta^T Z_i),$$

where $\gamma(x) = \alpha x - x \mathrm{l}\{x \leq 0\}$. With the help of Lemma 12.6, show that

$$\sqrt{n}(\hat{\theta}_n - \theta_0) \to^{\mathscr{L}} \mathscr{N}\left(0, \frac{\alpha(1-\alpha)}{f^2(0)} \Sigma^{-1}\right).$$

Appendix

Proof of Theorem 8.13 Set

$$S = \min\left\{ s \geq 0 : 2^{-s} \leq \frac{a}{2^4\sqrt{n}R} \right\}.$$

By (8.44),

$$\frac{a}{2^4\sqrt{n}R} \leq \frac{1}{2},$$

so $S \geq 1$, and $R > 2^{-(S+1)}R > a/(2^6\sqrt{n})$. This gives

$$\begin{aligned}
\int_{a/2^6\sqrt{n}}^{R} \mathscr{H}_{B,K}^{1/2}(u, R, \mathbf{F})\,du &\geq \int_{2^{-(S+1)}R}^{R} \mathscr{H}_{B,K}^{1/2}(u, R, \mathbf{F})\,du \\
&= \sum_{s=1}^{S+1} \int_{2^{-s}R}^{2^{-s+1}R} \mathscr{H}_{B,K}^{1/2}(u, R, \mathbf{F})\,du \\
&\geq \sum_{s=1}^{S+1} 2^{-s}R\mathscr{H}_{B,K}^{1/2}(2^{-s+1}R, R, \mathbf{F}) \\
&= \frac{1}{2}\sum_{s=0}^{S} 2^{-s}RH_s^{1/2},
\end{aligned}$$

where $H_s = \mathscr{H}_{B,K}(2^{-s}R, R, \mathbf{F})$, $s = 0, \ldots, S$. In view of (8.45), this means that

$$\sum_{s=0}^{S} 2^{-s}RH_s^{1/2} \leq 2a/C_0, \tag{A.1}$$

so that

$$\text{(A.2)}\qquad \sum_{s=1}^{S} 2^{-s} R \left(\sum_{k=0}^{s} H_k \right)^{1/2} \leq 4a/C_0.$$

For $s = 0, \ldots, S$, let $\{ [\tilde{Z}_{i,j}^{s,L}, \tilde{Z}_{i,j}^{s,U}] \}_{j=1}^{N_s}$ be a $(2^{-s}R)$-bracketing set defined in Definition 8.1, with $N_s = \mathcal{N}_{B,K}(2^{-s}R, R, \mathbf{F})$. So $[\tilde{Z}_{i,j}^{s,L}, \tilde{Z}_{i,j}^{s,U}]$ is $\mathcal{F}_i$-measurable, $i = 1, \ldots, n$, and for each $\theta \in \Theta$, there exists a $j_s(\theta) \in \{1, \ldots, N_s\}$ such that

(i) $\bar{\rho}_K(\tilde{\mathbf{Z}}_{j_s(\theta)}^{s,U} - \tilde{\mathbf{Z}}_{j_s(\theta)}^{s,L}) \leq 2^{-s}R$ on $\{ \bar{\rho}_K(\mathbf{Z}_\theta) \leq R \} \cap \mathbf{F}$,

(ii) $\tilde{Z}_{i,j_s(\theta)}^{s,L} \leq Z_{i,\theta} \leq \tilde{Z}_{i,j_s(\theta)}^{s,U}$, $i = 1, \ldots, n$, on $\{ \bar{\rho}_K(\mathbf{Z}_\theta) \leq R \} \cap \mathbf{F}$.

Without loss of generality, we may assume that $\tilde{Z}_{i,j}^{s,L} \leq \tilde{Z}_{i,j}^{s,U}$ on Ω, for all $i = 1, \ldots, n$, $j = 1, \ldots, N_s$ and $s = 0, \ldots, S$.

Define for $s = 0, \ldots, S$, $\theta \in \Theta$,

$$Z_{i,\theta}^{s,U} = \min_{k \leq s} \tilde{Z}_{i,j_k(\theta)}^{k,U}, \quad Z_{i,\theta}^{s,L} = \max_{k \leq s} \tilde{Z}_{i,j_k(\theta)}^{k,L}, \quad \Delta_{i,\theta}^{s} = Z_{i,\theta}^{s,U} - Z_{i,\theta}^{s,L},$$

and for $s = 1, \ldots, S$,

$$\text{(A.3)}\qquad \eta_s = \max \left\{ 2^{-3} 2^{-s} \left(\sum_{k=0}^{s} H_k \right)^{1/2} C_0 R/a,\ 2^{-(s+3)} \sqrt{s} \right\}.$$

Then by (A.2),

$$\sum_{s=1}^{S} \eta_s \leq 1.$$

For $s = 0, \ldots, S-1$, let

$$\text{(A.4)}\qquad K_s = \frac{2^4 2^{-2s} \sqrt{n} R^2}{\eta_{s+1} a}.$$

For $\theta \in \Theta$, let

$$v_{i,\theta} = \begin{cases} \min\{0 \leq s \leq S-1 : \Delta_{i,\theta}^{s} \geq K_s\}, & \text{if } \Delta_{i,\theta}^{s} \geq K_s \text{ for some } 0 \leq s \leq S-1, \\ S, & \text{else.} \end{cases}$$

Because $v_{i,\theta}$ is a function of $\mathcal{F}_i$-measurable random variables, it is itself $\mathcal{F}_i$ measurable.

Observe the following facts: for all $\theta \in \Theta$, $i = 1, \ldots, n$,

(A.5) $$0 \le \Delta_{i,\theta}^{s} \le \tilde{Z}_{i,j_s(\theta)}^{s,U} - \tilde{Z}_{i,j_s(\theta)}^{s,L}, \quad s = 0, \ldots, S,$$

(A.6) $$\Delta_{i,\theta}^{s} \le \Delta_{i,\theta}^{s-1}, \quad s = 1, \ldots, S,$$

(A.7) $$\Delta_{i,\theta}^{s} 1\{v_{i,\theta} = s\} \ge K_s, \quad s = 0, \ldots, S-1,$$

(A.8) $$\Delta_{i,\theta}^{s} 1\{v_{i,\theta} = s\} \le \Delta_{i,\theta}^{s-1} 1\{v_{i,\theta} = s\} \le K_{s-1}, \quad s = 1, \ldots, S,$$

(A.9) $$\Delta_{i,\theta}^{s-1} 1\{v_{i,\theta} \ge s\} \le K_{s-1}, \quad s = 1, \ldots, S.$$

Moreover, on the set $\{\bar{\rho}_K(\mathbf{Z}_\theta) \le R\} \cap \mathbf{F}$, we have for $i = 1, \ldots, n$,

(A.10) $$0 \le Z_{i,\theta} - Z_{i,\theta}^{s,L} \le \Delta_{i,\theta}^{s}, \quad s = 0, \ldots, S,$$

and

(A.11) $$Z_{i,\theta}^{s,L} - Z_{i,\theta}^{s-1,L} \le \Delta_{i,\theta}^{s-1}, \quad s = 1, \ldots, S.$$

Now, by (A.7), for $s = 0, \ldots, S-1$, and all $\theta \in \Theta$,

$$\frac{1}{n}\sum_{i=1}^{n} \mathbf{E}\big(\Delta_{i,\theta}^{s} 1\{v_{i,\theta} = s\} \mid \mathscr{F}_{i-1}\big) \le \frac{1}{K_s}\frac{1}{n}\sum_{i=1}^{n} \mathbf{E}\big(|\Delta_{i,\theta}^{s}|^2 \mid \mathscr{F}_{i-1}\big) \le \frac{1}{K_s} 2^{-2s} R^2$$
$$= \frac{\eta_{s+1} a}{2^4\sqrt{n}},$$

on $\{\bar{\rho}_K(\mathbf{Z}_\theta) \le R\} \cap \mathbf{F}$. By the definition of S, for all $\theta \in \Theta$,

$$\frac{1}{n}\sum_{i=1}^{n} \mathbf{E}\big(\Delta_{i,\theta}^{S} 1\{v_{i,\theta} = S\} \mid \mathscr{F}_{i-1}\big) \le \left(\frac{1}{n}\sum_{i=1}^{n} \mathbf{E}\big(|\Delta_{i,\theta}^{S}|^2 \mid \mathscr{F}_{i-1}\big)\right)^{1/2} \le 2^{-S} R$$
$$\le \frac{a}{2^4\sqrt{n}},$$

on $\{\bar{\rho}_K(\mathbf{Z}_\theta) \le R\} \cap \mathbf{F}$. Thus, for all $\theta \in \Theta$, on $\{\bar{\rho}_K(\mathbf{Z}_\theta) \le R\} \cap \mathbf{F}$,

$$\sum_{s=0}^{S} \frac{1}{n}\sum_{i=1}^{n} \mathbf{E}\big(\Delta_{i,\theta}^{s} 1\{v_{i,\theta} = s\} \mid \mathscr{F}_{i-1}\big) \le \frac{a}{2^3\sqrt{n}}.$$

Define for an $\mathscr{F}_i$-measurable random variable Z_i, $[Z_i]^c = Z_i - \mathbf{E}(Z_i \mid \mathscr{F}_{i-1})$ (whenever the conditional expectation exists). Then on $\{\bar{\rho}_K(\mathbf{Z}_\theta) \leq R\} \cap \mathbf{F}$,

(A.12)
$$\left|\sum_{s=0}^{S}\frac{1}{n}\sum_{i=1}^{n}\left[\left(Z_{i,\theta} - Z_{i,\theta}^{s,L}\right) 1\{v_{i,\theta} = s\}\right]^c\right| \leq \left|\sum_{s=0}^{S}\frac{1}{n}\sum_{i=1}^{n}\left[\Delta_{i,\theta}^s 1\{v_{i,\theta} = s\}\right]^c\right| + \frac{a}{2^3\sqrt{n}}.$$

We may write for $\theta \in \Theta$

$$Z_{i,\theta} = Z_{i,\theta}^{0,L} + \sum_{s=0}^{S}\left(Z_{i,\theta} - Z_{i,\theta}^{s,L}\right) 1\{v_{i,\theta} = s\} + \sum_{s=1}^{S}\left(Z_{i,\theta}^{s,L} - Z_{i,\theta}^{s-1,L}\right) 1\{v_{i,\theta} \geq s\}.$$

Therefore

$$\begin{aligned}
&\mathbf{P}\left((\sqrt{n}|\bar{Z}_\theta - \bar{A}_\theta| \geq a \wedge \bar{\rho}_K(\mathbf{Z}_\theta) \leq R \text{ for some } \theta) \cap \mathbf{F}\right)\\
&\leq \mathbf{P}\left(\left(\sqrt{n}\left|\frac{1}{n}\sum_{i=1}^{n}\left[Z_{i,\theta}^{0,L}\right]^c\right| \geq \frac{a}{4} \wedge \bar{\rho}_K(\mathbf{Z}_\theta) \leq R \text{ for some } \theta\right) \cap \mathbf{F}\right)\\
&\quad + \mathbf{P}\left(\left(\sqrt{n}\left|\sum_{s=0}^{S}\frac{1}{n}\sum_{i=1}^{n}\left[(Z_{i,\theta} - Z_{i,\theta}^{s,L})1\{v_{i,\theta} = s\}\right]^c\right|\right.\right.\\
&\qquad\qquad \left.\left.\geq \frac{a}{2} \wedge \bar{\rho}_K(\mathbf{Z}_\theta) \leq R \text{ for some } \theta\right) \cap \mathbf{F}\right)\\
&\quad + \mathbf{P}\left(\left(\sqrt{n}\left|\sum_{s=1}^{S}\frac{1}{n}\sum_{i=1}^{n}\left[(Z_{i,\theta}^{s,L} - Z_{i,\theta}^{s-1,L})1\{v_{i,\theta} \geq s\}\right]^c\right|\right.\right.\\
&\qquad\qquad \left.\left.\geq \frac{a}{4} \wedge \bar{\rho}_K(\mathbf{Z}_\theta) \leq R \text{ for some } \theta\right) \cap \mathbf{F}\right)\\
&= \mathbf{P}_{\mathrm{I}} + \mathbf{P}_{\mathrm{II}} + \mathbf{P}_{\mathrm{III}}.
\end{aligned}$$

Now, on $\{\bar{\rho}_K(\mathbf{Z}_\theta) \leq R\} \cap \mathbf{F}$,

(A.13)
$$\bar{\rho}_{4K}(\mathbf{Z}_\theta^{0,L}) \leq 8R^2.$$

Moreover, in view of (8.45),

(A.14)
$$\begin{aligned}
a &\geq C_0 \int_{a/2^6\sqrt{n}}^{R} \mathscr{H}_{B,K}(u, R, \mathbf{F})\, du \geq C_0 H_0^{1/2}(R - a/(2^6\sqrt{n}))\\
&\geq C_0 H_0^{1/2} R/2.
\end{aligned}$$

So the application of Corollary 8.10 and (8.43) gives

(A.15)
$$\begin{aligned}
\mathbf{P}_{\mathrm{I}} &\leq 2\exp\left[H_0 - \frac{a^2}{2^5(aKn^{-1/2} + 8R^2)}\right] \leq 2\exp\left[\frac{4a^2}{C_0^2R^2} - \frac{a^2}{2^8(C_1+1)R^2}\right]\\
&\leq 2\exp\left[-\frac{a^2}{2^9(C_1+1)R^2}\right],
\end{aligned}$$

if we take $C_0^2 \geq C^2(C_1 + 1)$ and C sufficiently large.

Next, we observe that (A.12) yields

$$\mathbf{P}_{\mathrm{II}} \le \mathbf{P}\Bigg(\Bigg(\sqrt{n}\left|\sum_{s=0}^{S}\frac{1}{n}\sum_{i=1}^{n}\left[\Delta_{i,\theta}^{s,L}1\{v_{i,\theta}=s\}\right]^{c}\right| \ge \frac{a}{2}-\frac{a}{2^3}\wedge\bar{\rho}_{\theta}(\mathbf{Z}_{\theta})$$

$$\le R \text{ for some } \theta\Bigg)\cap\mathbf{F}\Bigg)$$

$$\le \mathbf{P}\Bigg(\Bigg(\sqrt{n}\left|\frac{1}{n}\sum_{i=1}^{n}\left[\Delta_{i,\theta}^{0}1\{v_{i,\theta}=0\}\right]^{c}\right| \ge \frac{a}{2^3}\wedge\bar{\rho}_{K}(\mathbf{Z}_{\theta})$$

$$\le R \text{ for some } \theta\Bigg)\cap\mathbf{F}\Bigg)$$

$$+\mathbf{P}\Bigg(\Bigg(\sqrt{n}\left|\sum_{s=1}^{S}\frac{1}{n}\sum_{i=1}^{n}\left[\Delta_{i,\theta}^{s}1\{v_{i,\theta}=s\}\right]^{c}\right| \ge \frac{a}{4}\wedge\bar{\rho}_{K}(\mathbf{Z}_{\theta})$$

$$\le R \text{ for some } \theta\Bigg)\cap\mathbf{F}\Bigg)$$

$$= \mathbf{P}_{\mathrm{II}}^{(\mathrm{i})}+\mathbf{P}_{\mathrm{II}}^{(\mathrm{ii})}.$$

Due to (A.5),

$$\text{(A.16)} \qquad \frac{1}{n}\sum_{i=1}^{n}\rho_K(\Delta_{i,\theta}^{0}) \le R^2$$

on $\{\bar{\rho}_K(\mathbf{Z}_\theta)\le R\}\cap\mathbf{F}$. Corollary 8.10, together with (A.14) and (8.43) yields

$$\text{(A.17)} \quad \mathbf{P}_{\mathrm{II}}^{(\mathrm{i})} \le 2\exp\left[H_0-\frac{a^2}{2^7(aKn^{-1/2}+R^2)}\right] \le 2\exp\left[\frac{4a^2}{C_0^2R^2}-\frac{a^2}{2^7(C_1+1)R^2}\right]$$

$$\le 2\exp\left[-\frac{a^2}{2^8(C_1+1)R^2}\right],$$

if we take $C_0^2 \ge C^2(C_1+1)$ and C sufficiently large.

On $\{\bar{\rho}_K(\mathbf{Z}_\theta)\le R\}\cap\mathbf{F}$,

$$\frac{1}{n}\sum_{i=1}^{n}\mathbf{E}\left(\left|\Delta_{i,\theta}^{s}\right|^2 \mid \mathscr{F}_{i-1}\right) \le 2^{-2s}R^2, \quad s=1,\ldots,S.$$

Moreover, for $i=1,\ldots,n$,

$$\text{(A.18)} \qquad \left|\Delta_{i,\theta}^{s}1\{v_{i,\theta}=s\}\right| \le K_{s-1}, \quad s=1,\ldots,S.$$

So the application of Lemma 8.9 gives

$$\mathbf{P}_{\text{II}}^{(\text{ii})} \le \sum_{s=1}^{S} \mathbf{P}\left(\left(\sqrt{n}\left|\frac{1}{n}\sum_{i=1}^{n}[\Delta_{i,\theta}^{s}1\{v_{i,\theta}=s\}]^{c}\right| \ge \frac{a\eta_s}{4} \wedge \bar{\rho}_K(\mathbf{Z}_\theta)\right.\right.$$
$$\left.\left. \le R \text{ for some } \theta\right) \cap \mathbf{F}\right)$$
$$\le \sum_{s=1}^{S} 2\exp\left[\sum_{k\le s} H_k - \frac{a^2\eta_s^2}{2^5\left(\frac{1}{4}a\eta_s K_{s-1}n^{-1/2} + 2^{-2s}R^2\right)}\right]$$

But

$$\frac{1}{4}a\eta_s\frac{K_{s-1}}{\sqrt{n}} = 2^4 2^{-2s}R^2,$$

and from (A.3)

$$\sum_{k\le s} H_k \le \frac{2^6 a^2\eta_s^2}{2^{-2s}C_0^2 R^2}. \tag{A.19}$$

Thus
(A.20)

$$\mathbf{P}_{\text{II}}^{(\text{ii})} \le 2\sum_{s=1}^{S}\exp\left[\frac{2^6 a^2\eta_s^2}{2^{-2s}C_0^2R^2} - \frac{a^2\eta_s^2}{2^{10}2^{-2s}R^2}\right] \le 2\sum_{s=1}^{S}\exp\left[-\frac{a^2\eta_s^2}{2^{11}2^{-2s}R^2}\right],$$

where in the last inequality, we used (8.46) once more, with C sufficiently large. Now, from by (A.3), $\eta_s \ge 2^{-(s+3)}\sqrt{s}$, $s = 1, \ldots, S$. Using this we obtain

$$\mathbf{P}_{\text{II}}^{(\text{ii})} \le 2\sum_{s=1}^{S}\exp\left[-\frac{a^2 s}{2^{17}R^2}\right] \le 2\left(\frac{2^{17}R^2}{a^2}+1\right)\exp\left[-\frac{a^2}{2^{17}R^2}\right]$$
$$\le 2\left(\frac{2^{19}}{C_0^2}+1\right)\exp\left[-\frac{a^2}{2^{17}R^2}\right],$$

since (see (A.14)) $a \ge C_0 H_0^{1/2}R/2 \ge C_0R/2$. Take C_0 sufficiently large, to find

$$\mathbf{P}_{\text{II}}^{(\text{ii})} \le 4\exp\left[-\frac{a^2}{2^{17}R^2}\right].$$

To handle $\mathbf{P}_{\text{III}}$, we use the fact that on the set $\{\bar{\rho}_K(\mathbf{Z}_\theta) \le R\} \cap \mathbf{F}$

$$\frac{1}{n}\sum_{i=1}^{n}\mathbf{E}\left(\left|Z_{i,\theta}^{s,L} - Z_{i,\theta}^{s-1,L}\right|^2 \mid \mathscr{F}_{i-1}\right) \le 2^{-2(s+1)}R^2, \quad s = 1, \ldots, S,$$

and for $i = 1, \ldots, n$,

(A.21) $$\left| \left(Z_{i,\theta}^{s,L} - Z_{i,\theta}^{s-1,L} \right) 1\{v_{i,\theta} \geq s\} \right| \leq K_{s-1}, \quad s = 1, \ldots, S.$$

Then Lemma 8.9 leads to

$$\mathbf{P}_{\mathrm{III}} \leq \sum_{s=1}^{S} \mathbf{P}\left(\left(\sqrt{n} \left| \frac{1}{n} \sum_{i=1}^{n} \left[\left(Z_{i,\theta}^{s,L} - Z_{i,\theta}^{s-1,L} \right) 1\{v_{i,\theta} \geq s\} \right]^{c} \right| \geq \frac{a\eta_s}{4} \wedge \bar{\rho}_K(\mathbf{Z}_\theta) \leq R \text{ for some } \theta \right) \cap \mathbf{F} \right)$$

$$\leq 2 \sum_{s=1}^{S} \exp\left[\sum_{k \leq s} H_k - \frac{a^2 \eta_s^2}{2^5 \left(\frac{1}{4} a\eta_s K_{s-1} n^{-1/2} + 4 2^{-2s} R^2 \right)} \right]$$

$$\leq 2 \sum_{s=1}^{S} \exp\left[\frac{2^6 a^2 \eta_s^2}{2^{-2s} C_0^2 R^2} - \frac{a^2 \eta_s^2}{2^{10} 2^{-2s} R^2} \right].$$

Comparing this with (A.20), we see that also

$$\mathbf{P}_{\mathrm{III}} \leq 4 \exp\left[-\frac{a^2}{2^{17} R^2} \right]. \quad \square$$

References

AGMON, S. (1965). *Lectures on Elliptic Boundary Value Problems.* Van Nostrand, Princeton, NJ.

ALEXANDER, K. S. (1985). Rates of growth for weighted empirical processes. In *Proceedings of the Berkeley Conference in Honor of Jerzy Neyman and Jack Kiefer* **2** (L. Le Cam and R.A. Olshen eds.) 475–493. University of California Press, Berkeley.

ANDERSEN, N. T., GINÉ, E., OSSIANDER, M. and ZINN, J. (1988). The central limit theorem and the law of the iterated logarithm for empirical processes under local conditions. *Probability Theory and Related Fields* **77** 271–305.

BALL, K. and PAJOR, A. (1990). The entropy of convex bodies with 'few' extreme points. Proceedings of the 1989 Conference in Banach Spaces at Strobl. Austria (P. F. X. Müller and W. Schachermayer eds.). *London Mathematical Society Lecture Note Series* **158** 25–32.

BARRON, A. R., BIRGÉ, L. and MASSART, P. (1995). Risk bounds for model selection via penalization. Technical Report 95.54, Université de Paris-Sud.

BARRON, A. R. and SHEU, C.-H. (1991). Approximation of density functions by sequences of exponential families. *Annals of Statistics* **19** 1347–1369.

BASS, R. F. (1985). Law of the iterated logarithm for set-indexed partial sum processes with finite variance. *Zeitschrift für Wahrscheinlichkeit und Verwandte Gebiete* **65** 181–237.

BENNET, G. (1962). Probability inequalities for sums of independent random variables. *Journal of the American Statistical Association* **57** 33–45.

BERNSTEIN, S. (1924). Sur une modification de l'inégalité de Tchebichef. *Annals Science Institute Sav. Ukraine* Sect. Math. I (Russian, French summary).

BICKEL, P. J., KLAASSEN, C. A. J., RITOV, Y. and WELLNER, J. A. (1993). *Efficient and Adaptive Estimation for Semiparametric Models.* Johns Hopkins University Press, Baltimore.

BILLINGSLEY, P. (1968). *Convergence of Probability Measures.* John Wiley, New York.

BIRGÉ, L. (1983). Approximation dans les espaces métriques et théorie de l'estimation. *Zeitschrift für Wahrscheinlichkeitstheorie und Verwandte Gebiete* **65** 181–237.

BIRGÉ, L. and MASSART, P. (1993). Rates of convergence for minimum contrast estimators. *Probability Theory and Related Fields* **97** 113–150.

BIRGÉ, L. and MASSART, P. (1996). An adaptive compression algorithm in Besov Spaces. Technical Report, Université de Paris-Sud.

BIRMAN, M. Š. and SOLOMJAK, M. Z. (1967). Piece-wise polynomial approximations of functions in the classes W_p^α. *Mathematics of the USSR Sbornik* **73** 295–317.

BLUM, J. R. (1955). On the convergence of empiric distribution functions. *Annals of Mathematical Statistics* **26** 527–729.

BREIMAN, L., FRIEDMAN, J. H., OLSHEN, R. A. and STONE, C. J. (1984). *Tree-Structured Methods for Classification and Regression.* Wadsworth, Belmont, CA.

CLEMENTS, G. F. (1963). Entropies of sets of functions of bounded variation. *Canadian Journal of Mathematics* **15** 422–432.

DEHARDT, J. (1971). Generalizations of the Glivenko–Cantelli theorem. *Annals of Mathematical Statistics* **42** 2050–2055.

DONOHO, D. (1990). Gel'fand n-widths and the method of least squares. Technical Report, University of California, Berkeley.

DUDLEY, R. M. (1966). Weak convergence of measures on nonseparable metric spaces and empirical measures on Euclidean spaces. *Illinois Journal of Mathematics* **10** 109–126.

DUDLEY, R. M. (1967). The sizes of compact subsets of Hilbert Spaces and continuity of Gaussian processes. *Journal of Functional Analysis* **1** 290–330.

DUDLEY, R. M. (1978). Central limit theorems for empirical measures. *Annals of Probability* **6** 899–929 (Correction: *Ann. Probab.* **7** 909–911).

DUDLEY, R. M. (1984). A Course on Empirical Processes. *Lecture Notes in Mathematics* **1097** 2–141, Springer Verlag, New York.

DZHAPARIDZE, K. and VALKEILA, E. (1990). On the Hellinger type distances for filtered experiments. *Probability Theory and Related Fields* **85**, 105–117.

FAN, J., HECKMAN, N. E. and WAND, P. M. (1995). Local polynomial regression for generalized linear models and quasi-likelihood functions. *Journal of the American Statistical Association* **90** 141–150.

GESKUS, R. B. and GROENEBOOM, P. (1996). Asymptotically optimal estimation of smooth functionals for interval censoring, part 1. *Statistica Neerlandica* **50** 69–88.

GINÉ, E. (1996). Empirical processes and applications: an overview. *Bernoulli* **2** 1–28.

GINÉ, E. and ZINN, J. (1984). On the central limit theorem for empirical processes. *Annals of Probability* **12** 929–989.

GROENEBOOM, P. (1985). Estimating a monotone density. In *Proceedings of the Berkeley Conference in Honor of Jerzy Neyman and Jack Kiefer*, edited by L. Le Cam and R. A. Olshen ,**2** 539–555. University of California Press, Berkeley.

GROENEBOOM, P. and WELLNER, J. A. (1992). *Information bounds on non-parametric maximum likelihood estimation.* Birkh äuser, Basel.

HÄRDLE, W., HALL, P. and ICHIMURA, H. (1993). Optimal smoothing in single-index models. *Annals of Statistics* **21** 157–179.

HOEFFDING, W. (1963). Probability inequalities for sums of bounded random variables. *Journal of the American Statistical Association* **58** 13–30.

HOFFMANN-JØRGENSEN, J. (1991). Stochastic processes on Polish spaces. *Various Publications Series* **39**, Aarhus University.

HUANG, J. (1996). Projection estimation in multiple regression with application to functional ANOVA models. Technical Report 451. Department of Statistics, University of California, Berkeley.

HUBER, P. J. (1967). The behaviour of maximum likelihood estimates under nonstandard conditions. *Proceedings of the Fifth Berkeley Symposium on Mathematical Statistics and Probability* **1** 221–233. University of California Press, Berkeley, CA.

IBRAGIMOV, I. A. and HAS'MINSKII, R. Z. (1981a). *Statistical Estimation: Asymptotic Theory.* Springer Verlag, New York.

IBRAGIMOV, I. A. and HAS'MINSKII, R. Z. (1981b). Further results on nonparametric density estimation. In *Investigations in the Mathematical Statistics* **4** (*Zap. Nauch. Sem. Leningrad Otdel. Mat. Inst. Steklov* **108** 73–89).

IBRAGIMOV, I. A. and HAS'MINSKII, R. Z. (1984). On nonparametric estimation of the value of a linear functional in Gaussian white noise. *Theory of Probability and Its Applications* **29** 18–32.

KIEFER, J. and WOLFOWITZ, J. (1956). Consistency of the maximum likelihood estimator in the presence of infinitely many incidental parameters. *Annals of Mathematical Statistics* **27** 887–906.

KOLMOGOROV, A. N. and TIKHOMIROV, V. M. (1959). ϵ-entropy and ϵ-capacity of sets in function spaces. *Uspekhi Mat. Nauk* **14** 3–86 (English translation in *American Mathematical Society Translations* **(2) 17** (1961) 277–364).

KOROSTELEV, A. P. and TSYBAKOV, A. B. (1993a). Minimax Theory of Image Reconstruction. *Lecture Notes in Statistics* **82**, Springer Verlag, New York.

KOROSTELEV, A. P. and TSYBAKOV, A. B. (1993b). Estimation of support of a probability density and estimation of support functionals. *Problems of Information Transmission* **29** 3–18.

KUELBS, J. (1978). Some exponential moments of sums of independent random variables. *Transactions of the American Mathematical Society* **240** 145–162.

LE CAM, L. (1970). On the assumptions used to prove asymptotic normality of maximum likelihood estimates. *Annals of Mathematical Statistics* **41** 802–828.

LE CAM, L. (1973). Convergence of estimates under dimensionality restrictions. *Annals of Statistics* **1** 38–53.

LEDOUX, M. and TALAGRAND, M. (1991). *Probability in Banach Spaces: Isoperimetry and Processes.* Springer Verlag, New York.

MAMMEN, E. (1991). Nonparametric regression under qualitative smoothness assumptions. *Annals of Statistics* **19** 741–759.

MAMMEN, E. and VAN DE GEER, S. A. (1997a). Locally adaptive regression splines. *Annals of Statistics* **25** 387–413.

MAMMEN, E. and VAN DE GEER, S. A. (1997b). Penalized quasi-likelihood estimation in partial linear models. *Annals of Statistics* **25** 1014–1035.

OSSIANDER, M. (1987). A central limit theorem under metric entropy with L_2 bracketing. *Annals of Probability* **15** 897–919.

PFANZAGL, J. (1988). Consistency of the maximum likelihood estimator for certain nonparametric families, in particular: mixtures. *Journal of Statistical Planning and Inference* **19** 137–158.

PINKUS, A. (1985). *n-widths in Approximation Theory.* Springer Verlag, New York.

POLLARD, D. (1982). A central limit theorem for empirical processes. *Journal of the Australian Mathematical Society (Series A)* **33** 235–248.

POLLARD, D. (1984). *Convergence of Stochastic Processes.* Springer Verlag, New York.

POLLARD, D. (1985). New ways to prove central limit theorems. *Economic Theory* **1** 295–314.

POLLARD, D. (1990). Empirical Processes: Theory and Applications. *NSF-CBMS Regional Conference Series in Probability and Statistics* **2**. Institute of Mathematical Statistics and American Statistical Association.

ROCKAFELLAR, R. T. (1970). *Convex Analysis.* Princeton University Press, Princeton, NJ.

SEVERINI, T. A. and STANISWALIS, J. G. (1994). Quasi-likelihood estimation in semiparametric models. *Journal of the American Statistical Association* **89** 501–511.

SHORACK, G. R. and WELLNER, J. A. (1986). *Empirical Processes with Applications to Statistics.* Wiley, New York.

SILVERMAN, B. W. (1982). On the estimation of a probability density function by the maximum penalized likelihood method. *Annals of Statistics* **10** 795–810.

SILVERMAN, B. W. (1985). Some aspects of the spline smoothing approach to nonparametric regression curve fitting (with discussion). *Journal of the Royal Statistical Society Series B* **47** 1–52.

STONE, C. J. (1982). Optimal rates of convergence for nonparametric regression. *Annals of Statistics* **10** 1040–1053.

STONE, C. J. (1990). Large-sample inference for log-spline models. *Annals of Statistics* **18** 717–741.

STONE, C. J., HANSEN, M., KOOPERBERG, C. and TRUONG, Y. (1995). Polynomial splines and their tensor products in extended linear modelling.

Technical Report 437. Department of Statistics, University of California, Berkeley.

TALAGRAND, M. (1987). The Glivenko–Cantelli problem. *Annals of Probability* **15** 837–870.

VAN DE GEER, S. A. (1987). A new approach to least-squares estimation, with applications. *Annals of Statistics* **15** 587–602.

VAN DE GEER, S. A. (1988). Regression Analysis and Empirical Processes. *CWI Tract* 45, Centre for Mathematics and Computer Science, Amsterdam.

VAN DE GEER, S. A. (1990). Estimating a regression function. *Annals of Statistics* **18** 907–924.

VAN DE GEER, S. A. (1991). The entropy bound for monotone functions. Technical Report 91–10, University of Leiden.

VAN DE GEER, S. A. (1993). Hellinger-consistency of certain nonparametric maximum likelihood estimators. *Annals of Statistics* **21** 14–44.

VAN DE GEER, S. A. (1995a). Exponential inequalities for martingales, with application to maximum likelihood estimation for counting processes. *Annals of Statistics* **23** 1779–1801.

VAN DE GEER, S. A. (1995b). The method of sieves and minimum contrast estimators. *Mathematical Methods of Statistics* **4** 20–38.

VAN DE GEER, S. A. (1995c). Asymptotic normality in mixture models. *ESAIM, Probability and Statistics* **2** 17–33 (http://www.emath.fr/ps/).

VAN DE GEER, S. A. (1996). Rates of convergence for the maximum likelihood estimator in mixture models. *Nonparametric Statistics* **6** 293–310.

VAN DE GEER, S. A. and WEGKAMP, M. (1997). Consistency for the least squares estimator in nonparametric regression. *Annals of Statistics* **24** 2513–2523.

VAN DER VAART, A. W. (1995). Efficiency of infinite-dimensional M-estimators. *Statistica Neerlandica* **49** 9–30.

VAN DER VAART, A. W. and WELLNER, J. A. (1996). *Weak Convergence and Empirical Processes, with Applications to Statistics.* Springer Verlag, New York.

VAPNIK, V. N. and CHERVONENKIS, A. Ya. (1971). On the uniform convergence of relative frequencies of events to their probabilities. *Theory of Probability and its Applications* **16** 264–280.

VAPNIK, V. N. and CHERVONENKIS, A. Ya. (1981). Necessary and sufficient conditions for the uniform convergence of means to their expectations. *Theory of Probability and its Applications* **26** 532–553.

WHITTLE, P. (1960). Bounds for the moments of linear and quadratic forms in independent variables. *Theory of Probability and its Applications* **5**, 302–305.

WONG, W. H. and SHEN, X. (1995). Probability inequalities for likelihood ratios and convergence rates for sieve MLE's. *Annals of Statistics* **23** 339–362.

YANG, Y. and BARRON, A. R. (1996). Information-theoretic determination of minimax rates of convergence. Technical Report, Iowa State University and Yale University.

Symbol Index

Author Index

Subject Index